Statistical Size Distributions in Economics and Actuarial Sciences

WILEY SERIES IN PROBABILITY AND STATISTICS

Established by WALTER A. SHEWHART and SAMUEL S. WILKS

A complete list of the titles in this series appears at the end of this volume.

Statistical Size Distributions in Economics and Actuarial Sciences

CHRISTIAN KLEIBER

Universität Dortmund, Germany

SAMUEL KOTZ

The George Washington University

A JOHN WILEY AND SONS, INC., PUBLICATION

Published by John Wiley & Sons, Inc., Hoboken, New Jersey.
Published simultaneously in Canada.

Library of Congress Cataloging-in-Publication Data:

Kleiber, Christian, 1966-
Statistical size distributions in economics and actuarial sciences/Christian Kleiber, Samuel Kotz.
p. cm.—(Wiley series in probability and statistics)
Includes bibliographical references and index.
ISBN 0-471-15064-9 (cloth)
1. Distribution (Economic theory) 3. Economics, Mathematical.
4. Insurance–Mathematics. I. Kotz, Samuel. II. Title. III. Series.

HB523.K55 2003
339.2′2–dc21 2003041140

Printed in the United States of America

10 9 8 7 6 5 4 3 2 1

Contents

Preface

This is a book about money, but it will not help you very much in learning how to make money. Rather, it will instruct you about the distribution of various kinds of income and their related economic size distributions. Specifically, we have painstakingly traced the numerous statistical models of income distribution, from the late nineteenth century when Vilfredo Pareto developed a bold and astonishing model for the distribution of personal income until the latest models developed some 100 years later. Our goal was to review, compare, and somehow connect all these models and to pinpoint the unfortunate lack of coordination among various researchers, which has resulted in the duplication of effort and waste of talent and to some extent has reduced the value of their contributions. We also discuss the size distributions of loss in actuarial applications that involve a number of distributions used for income purposes. An impatient reader may wish to consult the list of distributions covered in this book and their basic properties presented in Appendix C.

The task of compiling this interdisciplinary book took longer and was more arduous than originally anticipated. We have tried to describe the distributions outlined here within the context of the personalities of their originators since in our opinion the personality, temperament, and background of the authors cited did affect to some extent the nature and scope of their discoveries and contributions.

We hope that our readers come to regard this book as a reliable source of information and we gladly welcome all efforts to bring any remaining errors to our attention.

CHRISTIAN KLEIBER
Dortmund, Germany

SAMUEL KOTZ
Washington, D.C.

Acknowledgments

The authors are indebted to various researchers around the globe—too numerous to be mentioned individually—for generously providing us with preprints, reprints, and useful advice.

Special thanks are due to Professor Giovanni Maria Giorgi for writing four biographies of leading contributors to the field, to Professors Camilo Dagum and Gabriele Stoppa for reading parts of the original manuscript and offering us the most valuable suggestions and comments, to Professor Constance van Eeden and Meike Gebel for translations from the Dutch and Italian, respectively, and to Professor Fiorenzo Mornati for supplying important not easily accessible information about Vilfredo Pareto. The first author would also like to thank Professor Walter Krämer for his support over (by now) many years.

All of the graphs in this book were generated using the R statistical software package (http://www.r-project.org/), the GNU implementation of the S language.

CHAPTER ONE

Introduction

Certum est quia impossibile est. TERTULLIAN, 155/160 A.D.—after 220 A.D.

This book is devoted to the parametric statistical distributions of economic size phenomena of various types—a subject that has been explored in both statistical and economic literature for over 100 years since the publication of V. Pareto's famous breakthrough volume Cours d'économie politique *in 1897. To the best of our knowledge, this is the first collection that systematically investigates various parametric models—a more respectful term for distributions—dealing with income, wealth, and related notions. Our aim is marshaling and knitting together the immense body of information scattered in diverse sources in at least eight languages. We present empirical studies from all continents, spanning a period of more than 100 years.*

We realize that a useful book on this subject matter should be interesting, a task that appears to be, in T. S. Eliot's words, "not one of the least difficult." We have tried to avoid reducing our exposition to a box of disconnected facts or to an information storage or retrieval system. We also tried to avoid easy armchair research that involves computerized records and heavy reliance on the Web.

Unfortunately, the introduction by its very nature is always somewhat fragmentary since it surveys, in our case rather extensively, the content of the volume. After reading this introduction, the reader could decide whether continuing further study of the book is worthwhile for his or her purposes. It is our hope that the decision will be positive. To provide a better panorama, we have included in the Appendix brief biographies of the leading players.

1.1 OUR AIMS

The modeling of economic size distributions originated over 100 years ago with the work of Vilfredo Pareto on the distribution of income. He apparently was the first to

Statistical Size Distributions in Economics and Actuarial Sciences, By Christian Kleiber and Samuel Kotz.
ISBN 0-471-15064-9

observe that, for many populations, a plot of the logarithm of the number of incomes N_x above a level x against the logarithm of x yields points close to a straight line of slope $-\alpha$ for some $\alpha > 0$. This suggests a distribution with a survival function proportional to $x^{-\alpha}$, nowadays known as the Pareto distribution.

"Economic size distributions" comprise the distributions of personal incomes of various types, the distribution of wealth, and the distribution of firm sizes. We also include work on the distribution of actuarial losses for which similar models have been in use at least since Scandinavian actuaries (Meidell, 1912; Hagstrœm, 1925) observed that—initially in life insurance—the sum insured is likely to be proportional to the incomes of the policy holders, although subsequently there appears to have been hardly any coordination between the two areas. Since the lion's share of the available literature comprises work on the distribution of income, we shall often speak of income distributions, although most results apply with minor modifications to the other size variables mentioned above.

Zipf (1949) in his monograph *Human Behavior and the Principle of Least Effort* and Simon (1955) in his article "On a class of skew distribution functions" suggest that Pareto-type distributions are appropriate to model such different variables as city sizes, geological site sizes, the number of scientific publications by a certain author, and also the word frequencies in a given text. Since the early 1990s, there has been an explosion of work on economic size phenomena in the physics literature, leading to an emerging new field called *econophysics* (e.g., Takayasu, 2002). In addition, computer scientists are nowadays studying file size distributions in the World Wide Web (e.g., Crovella, Taqqu, and Bestavros, 1998), but these works are not covered in this volume. We also exclude discrete Pareto-type distributions such as the Yule distribution that have been utilized in connection with the size distribution of firms by Simon and his co-authors (see Ijiri and Simon, 1977).

Regarding the distribution of income, the twentieth century witnessed unprecedented attempts by powerful nations such as Russia (in 1917) and China (in 1949) and almost all Eastern European countries (around the same time) to carry out far-reaching economic reforms and establish economic regimes that will reduce drastically income inequality and result in something approaching the single-point distribution of income when everyone is paid the same wages.

The most radical example is, of course, the blueprint for the economy of the Peoples' Republic of China (PRC) proclaimed by Mao Tse Tung on October 1, 1949 (his delivery of this plan was witnessed by one of the authors of this book in his youth). Mao's daring and possibly utopian promise of total economic equality for close to 1 billion Chinese and a guaranteed "iron bowl" of rice for every citizen totally receded into the background over the next 30 years due partially to blunders, unfavorable weather conditions, fanaticism, and cruelty, but perhaps mainly because of—as claimed by Pareto in the 1890s— the inability to change human nature and to suppress the natural instinct for economic betterment each human seems to possess.

It is remarkable that only eight years after Mao's death, the following appeared in the Declaration of the Central Committee of the Communist Party of China in October 1984:

> There has long been a misunderstanding about the distribution of consumer goods under socialism as if it meant egalitarianism.

> If some members of society got higher wages through their labor, resulting in a wide gap in income it was considered polarization and a deviation from socialism. This egalitarian thinking is utterly incompatible with scientific Marxist views of socialism.

In modern terminology, this translates to "wealth creation seems to be more important than wealth redistribution." Even the rigid Stalinist regime in North Korea began flirting with capitalism after May 2002, triggering income inequality.

Much milder attempts at socialism (practiced, e.g., in Scandinavian countries in the early years of the second part of the twentieth century) to reduce inequality by government regulators, especially by substantial taxation on the rich, were a colossal failure, as we are witnessing now in the early years of the twenty-first century. Almost the entire world is fully entrenched in a capitalistic market economy that appears to lead to a mathematical expression of the income distribution close to the one discovered by Pareto, with possibly different α and insignificant modifications. In fact, in our opinion the bulk of this book is devoted to an analysis of Pareto-type distributions, some of them in a heavily disguised form, leading sometimes to unrecognizable mathematical expressions.

It was therefore encouraging for us to read a recent book review of Champernowne and Cowell's *Economic Inequality and Income Distribution* (1998) written by Thomas Piketty, in which Piketty defends the "old-fashionedness" of the authors in their frequent reference to Pareto coefficients and claims that due to the tremendous advances in computer calculations "at the age of SAS and STATA," young economists have never heard of "Pareto coefficients" and tend to assume "that serious research started in the 1980s or 1990s." We will attempt to provide the background on and hopefully a proper perspective of the area of parametric income distributions throughout its 100-year-plus history.

It should be admitted that research on income distribution was somewhat dormant during the period from 1910–1970 in Western countries, although periodically publications—mainly of a polemical nature—have appeared in basic statistical and economic journals (see the bibliography). (An exception is Italy where, possibly due to the influence of Pareto and Gini, the distribution of income has always been a favorite research topic among prominent Italian economists and statisticians.) This changed during the last 15 years with the rising inequality in Western economies over the 1980s and a surge in inequality in the transition economies of Eastern Europe in the 1990s. Both called for an explanation and prompted novel empirical research. Indeed, as indicated on a recent Web page of the Distributional Analysis Research Programme (DARP) at the London School of Economics (http://darp.lse.ac.uk/),

> the study of income distribution is enjoying an extraordinary renaissance: interest in the history of the eighties, the recent development of theoretical models of economic growth that persistent wealth inequalities, and the contemporary policy focus upon the concept of social exclusion are evidence of new found concern with distributional issues.

Readers are referred to the recently published 1,000-page *Handbook of Income Distribution* edited by Atkinson and Bourguignon (2000) for a comprehensive discussion of the economic aspects of income distributions. We shall concentrate on statistical issues here.

On the statistical side, methods can broadly be classified as parametric and nonparametric. The availability of ever more powerful computer resources during recent decades gave rise to various nonparametric methods of density estimation, the most popular being probably the kernel density approach. Their main attraction is that they do not impose any distributional assumptions; however, with small data sets—not uncommon in actuarial science—they might result in imprecise estimates. These inaccuracies may be reduced by applying parametric models.

A recent comment by Cowell (2000, p. 145) seems to capture lucidly and succinctly the controversy existing between the proponents and opponents to the parametric approach in the analysis of size distributions:

> The use of the parametric approach to distributional analysis runs counter to the general trend towards the pursuit of non-parametric methods, although [it] is extensively applied in the statistical literature. Perhaps it is because some versions of the parametric approach have had bad press: Pareto's seminal works led to some fanciful interpretations of "laws" of income distribution (Davis, 1941), perhaps it is because the non-parametric method seems to be more general in its approach.
>
> Nevertheless, a parametric approach can be particularly useful for estimation of indices or other statistics in cases where information is sparse [such as given in the form of grouped data, *our addition*].... Furthermore, some standard functional forms claim attention, not only for their suitability in modelling some features of many empirical income distributions but also because of their role as equilibrium distributions in economic processes.

We are not concerned with economic/empirical issues in this book that involve the choice of a type of data such as labor of nonlabor earnings, incomes before or after taxes, individual or household incomes. These are, of course, of great importance in empirical economic studies. Nor are we dealing with the equally important aspects of data quality; we refer interested readers to van Praag, Hagenaars, and van Eck (1983); Lillard, Smith, and Welch (1986); or Angle (1994) in this connection. This problem is becoming more prominent as more data become available and new techniques to cope with the incompleteness of data such as "top-coding" and outliers are receiving significant attention. In the latter part of the twentieth century, the works of Victoria-Feser and Ronchetti (1994, 1997), Cowell and Victoria-Feser (1996), and more recently Victoria-Feser (2000), provided a number of new tools for the application of parametric models of income distributions, among them robust estimators and related diagnostic tools. They can protect the researcher against model deviations such as gross errors in the data or grouping effects and therefore allow for more reliable estimation of, for example, income distributions and inequality indices. (The latter task has occupied numerous researchers for over half a century.)

1.2 TYPES OF ECONOMIC SIZE DISTRIBUTIONS

In this short section we shall enumerate for completeness the types of size distributions studied in this book. Readers who are interested solely in statistical aspects may wish to skip this section. Those inclined toward broader economic-statistical issues may wish to supplement our brief exposition by referring to numerous books and sources, such as Atkinson and Harrison (1978), Champernowne and Cowell (1998), Ijiri and Simon (1977), Sen (1997), or Wolff (1987) and books on actuarial economics and statistics.

Distributions of Income and Wealth

As Okun (1975, p. 65) put it, “income and wealth are the two box scores in the record book of people’s economic positions.” It is undoubtedly true that the size distribution of income is of vital interest to all (market) economies with respect to social and economic policy-making. In economic and social statistics, the size distribution of income is the basis of concentration and Lorenz curves and thus at the heart of the measurement of inequality and more general social welfare evaluations. From here, it takes only a few steps to grasp its importance for further economic issues such as the development of adequate taxation schemes or the evaluation of effectiveness of tax reforms. Income distribution also affects market demand and its elasticity, and consequently the behavior of firms and a fortiori market equilibrium. It is often mentioned that income distribution is an important factor in determining the amount of saving in a society; it is also a factor influencing the productive effort made by various groups in the society.

Distributions of Firm Sizes

Knowledge of the size distribution of firms is important to economists studying industrial organization, to government regulators, as well as to courts. For example, courts use firm and industry measures of market share in a variety of antitrust cases. Under the merger guidelines of the U.S. Department of Justice and the Federal Trade Commission, whether mergers are challenged depends on the relative sizes of the firms involved and the degree of concentration in the industry. In recent years, for example, the Department of Justice challenged mergers in railroads, banks, soft drink, and airline industries using data on concentration and relative firm size.

As of 2002 tremendous upheavals in corporate institutions that involve great firms are taking place throughout the world especially in the United States and Germany. This will no doubt result in drastic changes in the near future in the size distribution of firms, and the recent frequent mergers and occasional breakdowns of firms may even require a new methodology. We will not address these aspects, but it is safe to predict new theoretical and empirical research along these lines.

Distributions of Actuarial Losses

Coincidentally, the unprecedented forest fires that recently occurred in the western United States (especially in Colorado and Arizona) may challenge the conventional wisdom that “fire liabilities are rare.” The model of the total amount of losses in a given period presented below may undergo substantial changes: In particular, the existing probability distributions of an individual loss amount $F(x)$ will no doubt be reexamined and reevaluated.

In actuarial sciences, the total amount of losses in a given period is usually modeled as a risk process characterized by two (independent) random variables: the number of losses and the amount of individual losses. If

- $p_n(t)$ is the probability of exactly n losses in the observed period $[0, t]$,
- $F(x)$ is the probability that, given a loss, its amount is $\leq x$,
- $F^{*n}(x)$ is the nth convolution of the c.d.f. of loss amount $F(x)$,

then the probability that the total loss in a period of length t is $\leq x$ can be expressed as the compound distribution

$$G(x, t) = \sum_{n=0}^{\infty} p_n(t)F^{*n}(x).$$

Although the total loss distribution $G(x, t)$ is of great importance for insurers in their task of determining appropriate premiums or reinsurance policies, it is the probability distribution of an *individual loss amount*, $F(x)$, that is relevant when a property owner has to decide whether to purchase insurance or when an insurer designs deductible schedules. Here we are solely concerned with the distributions of individual losses.

1.3 BRIEF HISTORY OF THE MODELS FOR STUDYING ECONOMIC SIZE DISTRIBUTIONS

A statistical study of personal income distributions originated with Pareto's formulation of "laws" of income distribution in his famous *Cours d'économie politique* (1897) that is discussed in detail in this book and in Arnold's (1983) book *Pareto Distributions*.

Pareto was well aware of the imperfections of statistical data, insufficient reliability of the sources, and lack of veracity of income tax statements. Nonetheless, he boldly analyzed the data using his extensive engineering and mathematical training and succeeded in showing that there is a relation between N_x—the number of taxpayers with personal income greater or equal to x—and the value of the income x given by a downward sloping line

$$\log N_x = \log A - \alpha \log x \tag{1.1}$$

or equivalently,

$$N_x = \frac{A}{x^{\alpha}}, \quad A > 0, \alpha > 0, x > x_0, \tag{1.2}$$

x_0 being the minimum income (Pareto, 1895). Economists and economic statisticians (e.g., Brambilla, 1960; Dagum, 1977) often refer to α (or rather $-\alpha$) as the elasticity of the survival function with respect to income x

$$\frac{d \log \{1 - F(x)\}}{d \log x} = -\alpha.$$

Thus, α is the elasticity of a reduction in the number of income-receiving units when moving to a higher income class. The graph with coordinates $(\log x,\ \log N_x)$ is often referred to as the *Pareto diagram*. An exact straight line in this display defines the Pareto distribution.

Pareto (1896, 1897a) also suggested the second and third approximation equations

$$N_x = \frac{A}{(x + x_0)^{\alpha}}, \quad A > 0,\ \alpha > 0,\ x_0 + x > 0, \tag{1.3}$$

and

$$N_x = \frac{A}{(x + x_0)^{\alpha}} e^{-\beta x}, \quad A > 0,\ \alpha > 0,\ x > x_0,\ \beta > 0. \tag{1.4}$$

Interestingly enough, equation (1.2) provided the most adequate fit for the income distribution in the African nation of Botswana, a republic in South Central Africa, in 1974 (Arnold, 1985).

The fact that empirically the values of parameter α remain "stable" if not constant (see Table 1.1—based on the fitting of his equations for widely diverse economies such as semifeudal Prussia, Victorian England, capitalist but highly diversified Italian cities circa 1887, and the Communist-like regime of the Jesuits in Peru during Spanish rule (1556–1821)—caused Pareto to conclude that human nature, that is, humankind's varying capabilities, is the main cause of income inequality, rather than the organization of the economy and society. If we were to examine a community of thieves, Pareto wrote (1897a, p. 371), we might well find an income distribution similar to that which experience has shown is generally obtained. In this case, the determinant of the distribution of income "earners" would be their *aptitude for theft*. What presumably determines the distribution in a community in which the production of wealth is the only way to gain an income is the aptitude for work and saving, steadiness and good conduct. This prevents necessity or desirability of legislative redistribution of income. Pareto asserted (1897a, p. 360),

> This curve gives an equilibrium position and if one diverts society from this position automatic forces develop which lead it back there.

In the subsequent version of his *Cours*, Pareto slightly modified his position by asserting that "we cannot state that the shape of the income curve would not change

Table 1.1 Pareto's Estimates of α

Country	Date	α
England	1843	1.50
	1879–1880	1.35
Prussia	1852	1.89
	1876	1.72
	1881	1.73
	1886	1.68
	1890	1.60
	1894	1.60
Saxony	1880	1.58
	1886	1.51
Florence	1887	1.41
Perugia (city)	1889	1.69
Perugia (countryside)	1889	1.37
Ancona, Arezzo, Parma, Pisa (total)	1889	1.32
Italian cities (total)	1889	1.45
Basle	1887	1.24
Paris (rents)	1887	1.57
Augsburg	1471	1.43
	1498	1.47
	1512	1.26
	1526	1.13
Peru	ca. 1800	1.79

Source: Pareto, 1897a, Tome II, p. 312.

if the social constitution were to radically change; were, for example, collectivism to replace private property" (p. 376). He also admitted that "during the course of the 19th century there are cases when the curve (of income) has slightly changed form, the type of curve remaining the same, but the constants changing." [See, e.g., Bresciani Turroni (1905) for empirical evidence using German data from the nineteenth century.]

However, Pareto still maintained that "statistics tells us that the curve varies very little in time and space: different peoples, and at different times, give very similar curves. There is therefore a notable stability in the figure of this curve."

The first fact discovered by Ammon (1895, 1898) and Pareto at the end of the nineteenth century was that "the distribution of income is highly skewed." It was a somewhat uneasy discovery since several decades earlier the leading statistician Quetelet and the father of biometrics Galton emphasized that many human characteristics including mental abilities were normally distributed.

Numerous attempts have been made in the last 100 years to explain this paradox.

Firstly it was soon discovered that the original Pareto function describes only a portion of the reported income distribution. It was originally recognized by Pareto but apparently this point was later underemphasized.

Pareto's work has been developed by a number of Italian economists and statisticians. Statisticians concentrated on the meaning and significance of the parameter α and suggested alternative indices. Most notable is the work of Gini (1909a,b) who introduced a measure of inequality commonly denoted as δ. [See also Gini's (1936) Cowles Commission paper: *On the Measurement of Concentration with Special Reference to Income and Wealth*.] This quantity describes to which power one must raise the fraction of total income composed of incomes *above a given level* to obtain the fraction of all income earners composed of high-income earners.

If we let $x_1, x_2, \ldots, x_n$ indicate incomes of progressively increasing amounts and r the number of income earners, out of the totality of n income earners, with incomes of x_{n-r+1} and up, the distribution of incomes satisfies the following simple equation:

$$\left(\frac{x_{n-r+1} + x_{n-r+2} + \cdots + x_n}{x_1 + x_2 + \cdots + x_n}\right)^{\delta} = \frac{r}{n}. \tag{1.5}$$

If the incomes are equally distributed, then $\delta = 1$. Also, δ varies with changes in the selected limit (x_{n-r+1}) chosen and increases as the concentration of incomes increases. Nevertheless, despite its variation with the selected limit, in applications to the incomes in many countries, the δ index does not vary substantially.

Analytically, for a Pareto type I distribution (1.2)

$$\delta = \frac{\alpha}{\alpha - 1}, \tag{1.6}$$

however, repeated testing on empirical income data shows that calculated δ often appreciably differs from the theoretical values derived (for a known α) from this equation.

As early as 1905 Benini in his paper "I diagramma a scala logarithmica," and 1906 in his *Principii de Statistica Metodologica*, noted that many economic phenomena such as savings accounts and the division of bequests when graphed on a double logarithmic scale generate a parabolic curve

$$\log N_x = \log A - \alpha \log x + \beta(\log x)^2, \tag{1.7}$$

which provides a good fit to the distributions of legacies in Italy (1901–1902), France (1902), and England (1901–1902). This equation, however, contains two constants that may render comparisons between countries somewhat dubious. Benini thus finally proposes the "quadratic relation"

$$\log N_x = \log A + \beta(\log x)^2. \tag{1.8}$$

Mortara (1917) concurred with Benini's conclusions that the graph with the coordinates ($\log x$, $\log N_x$) is more likely to be an upward convex curve and suggested an equation of the type

$$\log N_x = a_0 + a_1 \log x + a_2(\log x)^2 + a_3(\log x)^3 + \cdots.$$

In his study of the income distribution in Saxony in 1908, he included the first four terms, whereas in a much later publication (1949) he used only the first three terms for the distribution of the total revenue in Brazil in the years 1945–1946. Bresciani Turroni (1914) used the same function in his investigation of the distribution of wealth in Prussia in 1905.

Observing the fragmentary form of the part of the curve representing lower incomes (which presumably must slope sharply upward), Vinci (1921, pp. 230–231) suggests that the complete income curve should be a Pearson's type V distribution with density

$$f(x) = Ce^{-b/x}x^{-p-1}, \quad x > 0, \tag{1.9}$$

or more generally,

$$f(x) = Ce^{-b/(x-x_0)}(x - x_0)^{-p-1}, \quad x > x_0, \tag{1.10}$$

where $b, p > 0$, x_0 denotes as above the minimum income, and C is the normalizing constant.

Cantelli (1921, 1929) provided a probabilistic derivation of "Pareto's second approximation" (1.3), and similarly D'Addario (1934, 1939) carried out a detailed investigation of this distribution that (together with the initial first approximation) has the following property: The average income $\zeta(x)$ of earners above a certain level x is an increasing linear function of the variable x. However, this is not a characterization of the Pareto distribution(s). D'Addario proposed an ingenious *average excess value* method that involves indirect determination of the graph of the function $f(x)$ by means of $\zeta(x)$ utilizing the formula

$$f(x) = \frac{\alpha\zeta'(x)}{x - \zeta(x)} \exp\left\{\int_x^\infty \frac{\zeta(z)}{z - \zeta(z)}\, dz\right\}.$$

This approach requires selecting the average $\zeta(x)$ and its parameters based on the empirical data. The method was later refined by D'Addario (1969) and rechecked by Guerrieri (1969–1970) for the lognormal and Pearson's distributions of type III and V.

For a complete income curve, Amoroso (1924–1925) provided the density function

$$f(x) = Ce^{-b(x-x_0)^{1/s}}(x - x_0)^{(p-s)/s}, \quad x > x_0, \tag{1.11}$$

x_0 being the minimum income, C, b, $p > 0$, and s a nonzero constant such that $p + s > 0$ and fit it to Prussian data. This distribution is well known in the English language statistical literature as the generalized gamma distribution introduced by Stacy in 1962 in the *Annals of Mathematical Statistics*—which is an indication of lack of coordination between the European Continental and Anglo-American statistical literature as late as the sixties of the twentieth century. The cases $s = 1$ and $s = -1$ correspond to Pearson's type III and type V distributions, respectively.

Rhodes (1944), in a neglected work, succeeded in showing that the Pareto distribution can be derived from comparatively simple hypotheses. These involve constancy of the coefficient of variation and constancy of the type of distribution of income of those in the same "talent" group, and require that, on average, the consequent income increases with the possession of more talents.

D'Addario—like many other investigators of income distributions—was concerned with the multitude of disconnected forms proposed by various researchers. He attempted to obtain a general, relatively simply structured formula that would incorporate numerous special forms. In his seminal contribution *La Trasformate Euleriane*, he showed how transforming variables in several expressions for the density of the income distribution lead to the general equation

$$f(x) = \frac{1}{\Gamma(p)} e^{-w(x)} [w(x)]^{p-1} |w'(x)| \tag{1.12}$$

[here $\Gamma(p)$ is the gamma function]. Given a density $g(z)$, transforming the variable $x = u(z)$ and obtaining its inverse $z = w(x)$, we calculate the density of the transformed variable, $f(x)$, say, by the formula

$$f(x) = g[w(x)]|w'(x)|.$$

Here, if we use D'Addario's terminology, $g(z)$ is the *generating* function, $z = w(x)$ the *transforming* function, and $f(x)$ the *transformed* function. If the generating function is the gamma distribution

$$g(z) = \frac{1}{\Gamma(p)} e^{-z} z^{p-1}, \quad z \geq 0,$$

then the *Eulerian transform* is given by (1.12). This approach was earlier suggested by Edgeworth (1898), Kapteyn (1903) in his *Skew Frequency Curves in Biology and Statistics*, and van Uven (1917) in his *Logarithmic Frequency Distributions*, but D'Addario applied it skillfully to income distributions. More details are provided in Section 2.4.

In 1931 Gibrat, a French engineer and economist, developed a widely used lognormal model for the size distributions of income and of firms based on Kapteyn's (1903) idea of the proportional effect (by adding increments of income to an initial income distribution in proportion to the level already achieved). Champernowne (1952, 1953) refined Gibrat's approach and developed formulas that often fit better than Gibrat's lognormal distribution. However, when applied to U.S. income data of 1947 that incorporate low-income recipients, his results are not totally satisfactory. Even his four-parameter model gives unacceptable, gross errors. Somewhat earlier Kalecki (1945) modified Gibrat's approach by assuming that the increments of the income are proportional to the excess in ability of given members of the distribution over the lowest (or median) member. (A thoughtful observation by Tinbergen, made as early as 1956, prompts to distinguish between two underlying causes for income distribution. One is dealing here simultaneously with the distribution of abilities to earn income as well as with a distribution of preferences for income.)

A somewhat neglected (in the English literature) contribution is the so-called *van der Wijk's law* (1939). Here it is assumed that the average income above a limit x, $\sum_{x_i>x} x_i/N_x$, is proportional to the selected income level x, leading to the "law"

$$\frac{\sum_{x_i>x} x_i}{N_x} = \eta x, \tag{1.13}$$

where η is a constant of proportionality. For instance, if $\eta = 2$, then the average income of people with at least \$20,000 must be in the vicinity of \$40,000 and so on. Bresciani Turroni proposed a similar relationship in 1910, but it was not widely noticed in the subsequent literature.

Van der Wijk in his rather obscure volume *Inkomens- en Vermogensverdeling* (1939) also provided an interpretation of Gibrat's equation by involving the concept of *psychic income*. This was in accordance with the original discovery of the lognormal distribution inspired by the Weber–Fechner law in psychology (Fechner, 1860), quite unrelated to income distributions.

Pareto's contribution stimulated further research in the specification of new models to fit the whole range of income. One of the earliest may be traced to the French statistician Lucien March who as early as 1898 proposed using the gamma distribution and fitted it to the distribution of wages in France, Germany, and the United States. March claimed that the suggestion of employing the gamma distribution was due to the work of German social anthropologist Otto Ammon (1842–1916) in his book *Die Gesellschaftsordnung und ihre natürlichen Grundlagen* (1896 [second edition]), but we were unable to find this reference in any one of the three editions of Ammon's text. Some 75 years later Salem and Mount (1974) fit the gamma distribution to U.S. income data (presumably unaware of March's priority).

Champernowne (1952) specified versions of the log-logistic distribution with two, three, and four parameters. Fisk (1961a,b) studied the two-parameter version in detail.

Mandelbrot (1960, p. 79) observed that

> over a certain range of values of income, its distribution is not markedly influenced either by the socio-economic structure of the community under study, or by the definition chosen for "income." That is, these two elements may at most influence the values taken by certain parameters of an apparently universal distribution law.

and proposed nonnormal stable distributions as appropriate models for the size distribution of incomes.

Metcalf (1969) used a three-parameter lognormal distribution. Thurow (1970) and McDonald and Ransom (1979a) dealt with the beta type I distribution.

Dagum in 1977 devised two categories of properties for a p.d.f. to be specified as a model of income or wealth distribution: The first category includes essential properties, the second category important (but not necessary) properties. The essential properties are

- Model foundations
- Convergence to the Pareto law
- Existence of only a small number of finite moments
- Economic significance of the parameters
- Model flexibility to fit both unimodal and zeromodal distributions

(It seems to us that property 3 is implied by property 2.) Among the important properties are

- Good fit of the whole range of income
- Good fit of distributions with null and negative incomes
- Good fit of the whole income range of distributions starting from an unknown positive origin
- Derivation of an explicit mathematical form of the Lorenz curve from the specified model of income distributions and conversely

Dagum attributed special importance to the concept of income elasticity

$$\eta(x, F) = \frac{x}{F(x)} \frac{dF(x)}{dx} = \frac{d \log F(x)}{d \log x}$$

of a distribution function as a criterion for an income distribution.

He noted that the observed income elasticity of a c.d.f. behaves as a nonlinear and decreasing function of F. To represent this characteristic of the income elasticity, Dagum specified (in the simplest case) the differential equation

$$\eta(x, F) = ap\{1 - [F(x)]^{1/p}\}, \quad x \geq 0,$$

subject to $p > 0$ and $ap > 0$, which leads to the Dagum type I distribution

$$F(x) = \left[1 + \left(\frac{x}{b}\right)^{-a}\right]^{-p}, \quad x > 0,$$

where $a, b, p > 0$.

It was noted by Dagum (1980c, 1983) [see also Dagum (1990a, 1996)] that it is appropriate to classify the income distributions based on three generating systems:

- Pearson system
- D'Addario's system
- Generalized logistic (or Burr logistic) system

Only Champernowne's model does not belong to any of the three systems.

The pioneering work of McDonald (1984) and Venter (1983) led to the generalized beta (or transformed beta) distribution given by

$$f(x) = \frac{ax^{ap-1}}{b^{ap}B(p, q)[1 + (x/b)^a]^{p+q}}, \quad x > 0. \tag{1.14}$$

It is also known in the statistical literature as the generalized F (see, e.g., Kalbfleisch and Prentice, 1980) and was rediscovered in a slightly different parameterization by Majumder and Chakravarty (1990) a few years later. This family includes numerous models used as income and size distributions, in particular the Singh and Maddala (1976) model, the Dagum type I model (Dagum, 1977), the Fisk model, and evidently the beta distribution of the second kind. In actuarial science the Singh–Maddala and Dagum models are usually referred to as the Burr and inverse Burr distributions, respectively, since they are members of the Burr (1942) system of distributions.

We also mention the natural generalization of the Pareto distribution proposed by Stoppa in 1990b,c. It is given by

$$F(x) = \left[1 - \left(\frac{x}{x_0}\right)^{-\alpha}\right]^{\varphi}, \quad 0 < x_0 \leq x. \tag{1.15}$$

This book is devoted to a detailed study of the distributions surveyed in this section and their interrelations. The literature is immense and omissions are unavoidable although we tried to utilize all the references collected during a six-month extensive search. Due to the rather sporadic developments in that area, only some isolated multivariate distributions are included.

1.4 STOCHASTIC PROCESS MODELS FOR SIZE DISTRIBUTIONS

Interestingly enough, income and wealth distributions of various types can be obtained as steady-state solutions of stochastic processes.

The first example is Gibrat's (1931) model leading to the lognormal distribution. He views income dynamics as a multiplicative random process in which the product of a large number of individual random variables tends to the lognormal distribution. This multiplicative central limit theorem leads to a simple Markov model of the "law of proportionate effect." Let X_t denote the income in period t. It is generated by a first-order Markov process, depending only on X_{t-1} and a stochastic influence

$$X_t = R_t X_{t-1}.$$

Here $\{R_t\}$ is a sequence of independent and identically distributed random variables that are independent of X_{t-1} as well. X_0 is the income in the initial period. Substituting backward, we see that

$$X_t = X_0 \cdot R_0 \cdot R_1 \cdot R_2 \cdot \ldots R_{t-1},$$

and as t increases, the distribution of X_t tends to a lognormal distribution provided $\mathrm{var}(\log R_t) < \infty$.

In the Gibrat model we assume the independence of R_t, which may not be realistic. Moreover, the variance of $\log X_t$ is an increasing function of t and this often contradicts the data. Kalecki (1945), in a paper already mentioned, modified the model by introducing a negative correlation between X_{t-1} and R_t that *prevents* $\mathrm{var}(\log X_t)$ from growing. Economically, it means that the probability that income will rise by a given percentage is lower for the rich than for the poor. (The modification is an example of an ingenious but possibly ad hoc assumption.)

Champernowne (1953) demonstrated that under certain assumptions the stationary income distribution will approximate the Pareto distribution irrespectively of the initial distribution. He also viewed income determination as a Markov process (income for the current period depends only on one's income for the last period and random influence). He subdivided the income into a finite number of classes and defined p_{ij} as the probability of being in class j at time $t+1$ given that one was in class i at time t. The income intervals defining each class are assumed (1) to form a *geometric* (not arithmetic) progression. The limits of class j are higher than those of class $j-1$ by a certain *percentage* rather than a certain absolute amount of income and the transitional probabilities p_{ij} depend only on the differences $j-i$. (2) Income cannot *move up* more than one interval nor down more than n intervals in any one period; (3) there is a lowest interval beneath which no income can fall, and (4) the average number of intervals shifted in a period is negative in each income bracket. Under these assumptions, Champernowne proved that the distribution eventually behaves like the Pareto law.

The assumptions of the Champernowne model can be relaxed by allowing for groups of people (classified by age, occupation, etc.) and permitting movement from one group to another. However, constancy of the transition matrix is essential; otherwise, no stationary distribution will emerge from the Markov process. Moreover, probabilities of advancing or declining ought to be independent of the amount of income. Many would doubt the existence of a society whose institutional framework is so static, noting that such phenomena as "inherited privilege," and cycles of poverty or prosperity are part and parcel of all viable societies.

To complicate the matter with the applicability of Champernowne's model, it was shown by Aitchison and Brown (1954) that if the transition probabilities p_{ij} depend on j/i (rather than $j - i$, as is the case in Champernowne's model) and further that the income brackets form an arithmetic (rather than geometric) progression, then the limiting distribution is lognormal rather than Pareto. In our opinion the dependence on $j - i$ may seem to be more natural, but it is a matter of subjective opinion.

It should also be noted that the Champernowne and Gibrat models and some others require long durations of time until the approach to stationarity is obtained. This point has been emphasized by Shorrocks (1975).

Rutherford (1955) incorporated birth–death considerations into a Markov model. His assumptions were as follows:

- The supply of new entrants grows at a constant rate.
- These people enter the labor force with a lognormal distribution of income.
- The number of survivors in each cohort declines exponentially with age.

Under these assumptions, the data eventually approximate the Gram–Charlier type A distribution, which often provides a better fit than the lognormal. In Rutherford's model the overall variance remains *constant* over time.

Mandelbrot (1961) constructed a Markov model that approximates the Pareto distribution similarly to Champernowne's model, but does not require the strict law of proportionate effect (a random walk in logarithms).

Wold and Whittle (1957) offered a rather general continuous-time model that also generates the Pareto distribution: It is applied to stocks of wealth that grow at a compound interest rate during the lifetime of a wealth-holder and are then divided among his heirs. Deaths occur randomly with a known mortality rate per unit time. Applying the model to wealth above a certain minimum (this is necessary because the Pareto distribution only applies above some positive minimum wealth), Wold and Whittle derived the Pareto law and expressed the exponent α as a function of (1) the number of heirs per person, (2) the growth rate of wealth, and (3) the mortality rate of the wealth owners.

The most complicated model known to us seems to be due to Sargan (1957). It is a continuous-time Markov process: The ways in which transitions occur are explicitly spelled out. His approach is quite general; it accommodates

- Setting of new households and dissolving of old ones
- Gifts between households
- Savings and capital gains
- Inheritance and death

It is its generality that makes it unwieldy and unintelligible.

As an alternative to the use of ergodic Markov processes, one can also explain wealth or income distributions by means of branching processes. Steindl (1972), building on the model of Wold and Whittle (1957) mentioned above, showed in this way that the distribution of wealth can be regarded as a certain transformation of an age distribution. Shorrocks (1975) explained wealth accumulations using the theory of queues. He criticized previously developed stochastic models for concentrating on equilibrium distributions and proposed a model in which the transition probabilities or parameters of the distribution are allowed to change over time.

These models were often criticized by applied economists who favor models based on human capital and the concept of economic man (Mincer, 1958; Becker, 1962, 1964). Some of them scorn size distribution of income and refer to them as antitheories. Their criticism often goes like this:

> Allowing a stochastic mechanism to be the sole determinant of the income distribution is TO GIVE UP BEFORE YOU START. The deterministic part of a model (in econometrics) is "what we think we know," the disturbance term is "what we don't know." The probabilistic approach allocates 100% variance in income to the latter.

In our opinion this type of argument shows a lack of understanding of the concept of stochastic model and by extension of the probabilistic-statistical approach.

CHAPTER TWO

General Principles

Before embarking on a detailed discussion of the models for economic and actuarial size phenomena, we will discuss a number of unifying themes along with several tools that are required in the sequel. These include, among others, the ubiquitous Lorenz curve and associated inequality measures. In addition, we present some concepts usually associated with reliability and engineering statistics such as the hazard rate and the mean residual life function that are known in actuarial science under different names. Here these functions are often used for preliminary model selection because they highlight the area of a distribution that is of central interest in these applications, the extreme right tail.

We also briefly discuss systems of distributions in order to facilitate subsequent classifications, namely, the Pearson and Burr systems and the less widely known Stoppa system. The largest branch of the size distributions literature, dealing with the size distribution of personal income, has developed its own systems for the generation *of distributions; these we survey in Section 2.4.*

Unless explicitly stated otherwise, we assume throughout this chapter that the underlying c.d.f. F *is continuous and supported on an interval.*

2.1 SOME CONCEPTS FROM ECONOMICS

The literature on Lorenz curves, inequality measures, and related notions is by now so substantial that it would be easy to write a 500-page volume dealing exclusively with these concepts and their ramifications. We shall be rather brief and only present the basic results. For refinements and further developments, we refer the interested reader to Kakwani (1980b), Arnold (1987), Chakravarty (1990), Mosler (1994), or Cowell (2000) and the references therein.

Statistical Size Distributions in Economics and Actuarial Sciences, By Christian Kleiber and Samuel Kotz.
ISBN 0-471-15064-9

2.1.1 Lorenz Curves and the Lorenz Order

In June 1905 a paper entitled "Methods of measuring the concentration of wealth," written by Max Otto Lorenz (who was completing at that time his Ph.D. dissertation at the University of Wisconsin and destined to become an important U.S. Government statistician), appeared in the *Journal of the American Statistical Association.*

It truly revolutionized the economic and statistical studies of income distributions, and even today it generates a fertile field of investigation into the bordering area between statistics and economics. The *Current Index of Statistics* (for the year 1999) lists 13 papers with the titles Lorenz curve and Lorenz ordering. It would not be an exaggeration to estimate that several hundred papers have been written in the last 50 years in statistical journals and at least the same number in econometric literature. It should be acknowledged that Lorenz's pioneering work lay somewhat dormant for a number of decades in the English statistical literature until it was resurrected by Gastwirth in 1971.

To draw the Lorenz curve of an n-point empirical distribution, say, of household income, one plots the share $L(k/n)$ of total income received by the $k/n \cdot 100\%$ of the lower-income households, $k = 0, 1, 2, \ldots, n$, and interpolates linearly.

In the discrete (or empirical) case, the Lorenz curve is thus defined in terms of the $n + 1$ points

$$L\left(\frac{k}{n}\right) = \frac{\sum_{i=1}^{k} x_{i:n}}{\sum_{i=1}^{n} x_{i:n}}, \quad k = 0, 1, \ldots, n, \tag{2.1}$$

where $x_{i:n}$ denotes the ith smallest income, and a continuous curve $L(u)$, $u \in [0, 1]$, is given by

$$L(u) = \frac{1}{n\bar{x}} \left\{ \sum_{i=1}^{\lfloor un \rfloor} x_{i:n} + (un - \lfloor un \rfloor) x_{\lfloor un \rfloor + 1:n} \right\}, \quad 0 \leq u \leq 1, \tag{2.1a}$$

where $\lfloor un \rfloor$ denotes the largest integer not exceeding un.

Figure 2.1 depicts the Lorenz curve for the (income) vector $\mathbf{x} = (1, 3, 5, 11)$. By definition, the diagonal of the unit square corresponds to the Lorenz curve of a society in which everybody receives the same income and thus serves as a benchmark case against which actual income distributions may be measured.

The appropriate definition of the Lorenz curve for a general distribution follows easily by recognizing the expression (2.1) as a sequence of standardized empirical incomplete first moments. In view of $E(X) = \int_0^1 F^{-1}(t)\, dt$, where the quantile function F^{-1} is defined as

$$F^{-1}(t) = \sup\{x \mid F(x) \leq t\}, \quad t \in [0, 1], \tag{2.2}$$

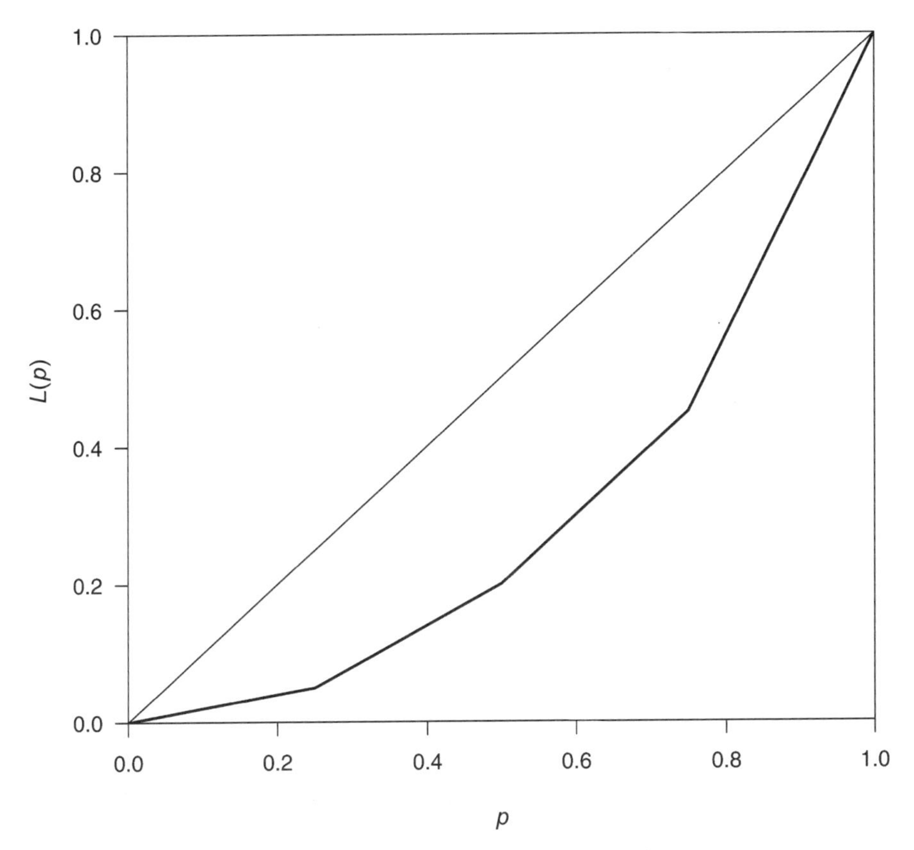

Figure 2.1 Lorenz curve of $\mathbf{x} = (1, 3, 5, 11)$.

equation (2.1a) may be rewritten as

$$L(u) = \frac{1}{E(X)} \int_0^u F^{-1}(t)\, dt, \quad u \in [0, 1]. \tag{2.3}$$

It follows that any distribution supported on the nonnegative halfline with a finite and positive first moment admits a Lorenz curve. Following Arnold (1987), we shall occasionally denote the set of all random variables with distributions satisfying these conditions by $\mathcal{L}$. Clearly, the empirical Lorenz curve can now be rewritten in the form

$$L_n(u) = \frac{1}{\bar{x}} \int_0^u F_n^{-1}(t)\, dt, \quad u \in [0, 1], \tag{2.4}$$

an expression that is useful for the derivation of the sampling properties of the Lorenz curve.

In the Italian literature the representation (2.3) in terms of the quantile function was used as early as 1915 by Pietra who obviously was not aware of Lorenz's contribution. It has also been popularized by Piesch (1967, 1971) in the German literature.

In the era preceding Gastwirth's (1971) influential article (reviving the interest in Lorenz curves in the English statistical literature), the Lorenz curve was commonly defined in terms of the *first-moment distribution*. The moment distributions are defined by

$$F_{(k)}(x) = \frac{1}{E(X^k)} \int_0^x t^k f(t)\, dt, \quad x \geq 0, k = 0, 1, 2, \ldots, \tag{2.5}$$

provided $E(X^k) < \infty$. Hence, they are merely normalized partial moments. Like the higher-order moments themselves, the higher-order moment distributions are difficult to interpret; however, the c.d.f. $F_{(1)}(x)$ of the first-moment distribution simply gives the share of the variable X accruing to the population below x. In the context of income distributions, Champernowne (1974) refers to $F_{(0)}$, that is, the underlying c.d.f. F, as the *people curve* and to $F_{(1)}$ as the *income curve*.

It is thus not difficult to see that the Lorenz curve can alternatively be expressed as

$$\{(u, L(u))\} = \{(u, v)|u = F(x), v = F_{(1)}(x); x \geq 0\}. \tag{2.6}$$

Although the representation (2.3) is often more convenient for theoretical considerations, the "moment distribution form" (2.6) also has its moments, especially for parametric families that do not admit a quantile function expressed in terms of elementary functions. In the following chapters, we shall therefore use whatever form is more tractable in a given context. It is also worth noting that several of the distributions considered in this book are closed with respect to the formation of moment distributions, that is, $F_{(k)}$ is of the same type as F but with a different set of parameters (Butler and McDonald, 1989). Examples include the Pareto and lognormal distributions and the generalized beta distribution of the second kind discussed in Chapter 6.

It follows directly from (2.3) that the Lorenz curve has the following properties:

- L is continuous on $[0, 1]$, with $L(0) = 0$ and $L(1) = 1$.
- L is increasing.
- L is convex.

Conversely, any function possessing these properties is the Lorenz curve of a certain statistical distribution (Thompson, 1976).

Since any distribution is characterized by its quantile function, it is clear from (2.3) that the Lorenz curve characterizes a distribution in $\mathcal{L}$ up to a scale parameter (e.g., Iritani and Kuga, 1983). It is also worth noting that the Lorenz curve itself may be considered a c.d.f. on the unit interval. This implies, among other things, that this

"Lorenz curve distribution"—having bounded support—can be characterized in terms of its moments, and moreover that these "Lorenz curve moments" characterize the underlying income distribution up to a scale, even if this distribution is of the Pareto type and only a few of the moments exist (Aaberge, 2000).

By construction, the quantile function associated with the "Lorenz curve distribution" is also a c.d.f. It is sometimes referred to as the Goldie curve, after Goldie (1977) who studied its asymptotic properties.

Although the Lorenz curve has been used mainly as a convenient graphical tool for representing distributions of income or wealth, it can be used in all contexts where "size" plays a role. As recently as 1992, Aebi, Embrechts, and Mikosch have used Lorenz curves under the name of *large claim index* in actuarial sciences. Also, the Lorenz curve is intimately related to several concepts from engineering statistics such as the so-called total-time-on-test transform (TTT) (Chandra and Singpurwalla, 1981; Klefsjö, 1984; Heilmann, 1985; Pham and Turkkan, 1994). It continues to find new applications in many branches of statistics; recently, Zenga (1996) introduced a new concept of kurtosis based on the Lorenz curve (see also Polisicchio and Zenga, 1996).

As an example of Lorenz curves, consider the classical Pareto distribution (see Chapter 3) with c.d.f. $F(x) = 1 - (x/x_0)^{-\alpha}, x \geq x_0 > 0$, and quantile function $F^{-1}(u) = x_0(1-u)^{-1/\alpha}, 0 < u < 1$. The mean $E(X) = \alpha x_0/(\alpha - 1)$ exists if and only if $\alpha > 1$. This yields

$$L(u) = 1 - (1-u)^{1-1/\alpha}, \quad 0 < u < 1, \tag{2.7}$$

provided $\alpha > 1$. We see that Lorenz curves from Pareto distributions with a different α never intersect. Empirical Lorenz curves occasionally do intersect, so Pareto distributions may not be useful in these situations. Figure 2.2 depicts the Lorenz curves of two Pareto distributions, with $\alpha = 1.5$ and $\alpha = 2.5$.

It is natural to study the geometric aspects of Lorenz curves, for example, their symmetry (or lack thereof) with respect to the alternate diagonal $\{(x, 1-x)|x \in [0, 1]\}$, the line perpendicular to the line of equal distribution. A general condition for self-symmetry was given by Kendall (1956) in the form of a functional equation for the density

$$f(x) = \left[\frac{E(X)}{x}\right]^{3/2} g\left[\log \frac{x}{E(X)}\right], \tag{2.8}$$

where $g(y)$ is an even function of y.

Clearly, the Lorenz curve of the Pareto distribution (2.7) does not possess this symmetry property. The best known example of a distribution with self-symmetric Lorenz curves is the lognormal; see Figure 4.3 in Chapter 4. See also Champernowne (1956), Taguchi (1968), and especially Piesch (1975) for further details on the geometry of Lorenz curves.

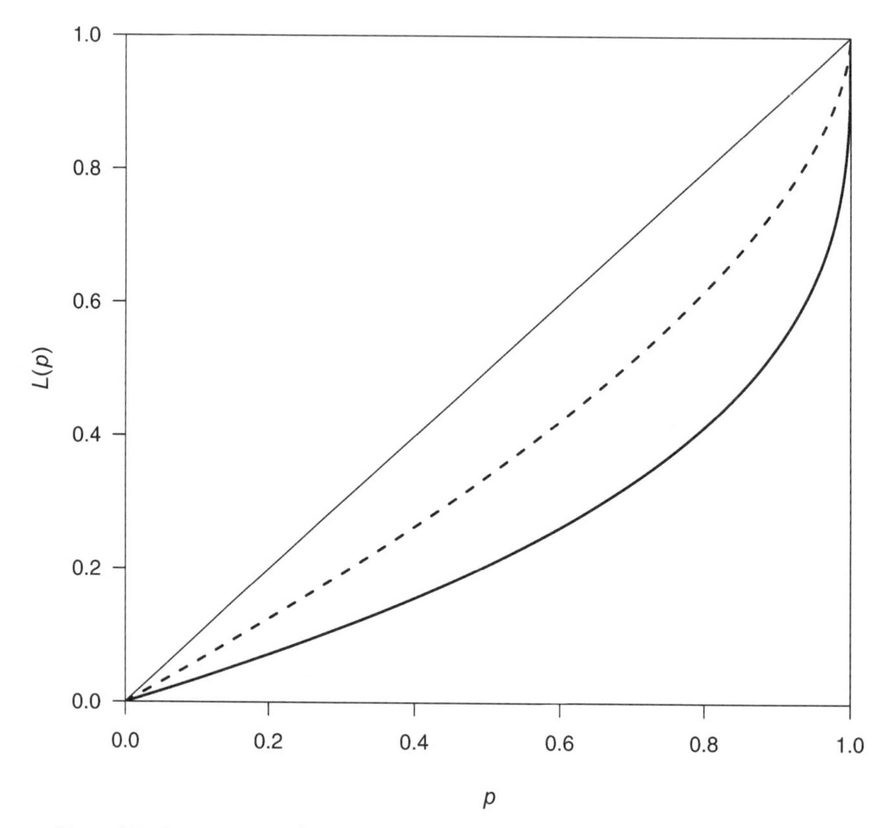

Figure 2.2 Lorenz curves of two Pareto distributions: $\alpha = 1.5$ (solid) and $\alpha = 2.5$ (dashed).

Arnold et al. (1987) observed that every distribution F which is strongly unimodal and symmetric about 0 leads to a self-symmetric Lorenz curve representable as $L_\tau(u) = F[F^{-1}(u) - \tau]$, $u \in (0, 1)$, $\tau \geq 0$. The prime example is the normal distribution that leads to a lognormal Lorenz curve.

Figure 2.2 also prompts us to compare two distributions, in a global sense, by comparing their corresponding Lorenz curves. If two Lorenz curves do not intersect, it may perhaps be appropriate to call the distribution with the lower curve "more unequal" or "more variable," and indeed a stochastic ordering based on this notion, the Lorenz partial ordering, was found to be a useful tool in many applications. For $X_1, X_2 \in \mathcal{L}$, the Lorenz ordering is defined as

$$X_1 \geq_L X_2 :\Longleftrightarrow F_1 \geq_L F_2 :\Longleftrightarrow L_{X_1}(u) \leq L_{X_2}(u), \quad \text{for all } u \in [0, 1]. \tag{2.9}$$

Here X_1 is said to be larger than X_2 (or more unequal) in the Lorenz sense. From (2.3) it is evident that the Lorenz ordering is scale-free; hence,

$$X_1 \geq_L X_2 \Longleftrightarrow a \cdot X_1 \geq_L b \cdot X_2, \quad a, b > 0. \tag{2.10}$$

Economists usually prefer to denote the situation where $L_{X_1} \leq L_{X_2}$ as $X_2 \geq_L X_1$, because F_2 is, in a certain sense, associated with a higher level of economic welfare (Atkinson, 1970). We shall use the notation (2.9) that appears to be the common one in the statistical literature, employed by Arnold (1987) or Shaked and Shanthikumar (1994), among others.

Among the methods for verifying Lorenz ordering relationships, we mention that if $X_2 \stackrel{d}{=} g(X_1)$, then (Fellman, 1976)

$$\frac{g(x)}{x} \text{ is nonincreasing on } (0, \infty) \Longrightarrow X_1 \geq_L g(X_1). \tag{2.11}$$

Under the additional assumption that g be increasing on $[0, \infty)$, the condition is also necessary (Arnold, 1987). This result is useful, among other things, in connection with the lognormal distribution; see Section 4.5.

Closely connected conditions are in terms of two stronger concepts of stochastic ordering, the convex and star-shaped orderings. For two distributions F_i, $i = 1, 2$, supported on $[0, \infty)$ or a subinterval thereof, distribution F_1 is said to be convex (star-shaped) ordered with respect to a distribution F_2, denoted as $F_1 \geq_{\text{conv}} F_2 (F_1 \geq_* F_2)$, if $F_1^{-1} F_2$ is convex $[F_1^{-1} F_2(x)/x$ is nonincreasing$]$ on the support of F_2. It can be shown that the convex ordering implies the star-shaped ordering, which in turn implies the Lorenz ordering (Chandra and Singpurwalla, 1981; Taillie, 1981).

These criteria are useful when the quantile function is available in a simple closed form, as is the case with, among others, the Weibull distribution; see Chapter 5. Several further methods for verifying Lorenz dominance were discussed by Arnold (1987) or Kleiber (2000a).

Various suggestions have been made as to how to proceed when two Lorenz curves intersect. In international comparisons of income distributions, this is particularly common for countries on different economic levels, for example, industrialized and developing countries. This suggests that the problem can be resolved by scaling up the Lorenz curves by the first moment, leading to the so-called *generalized Lorenz curve* (Shorrocks, 1983; Kakwani, 1984)

$$\text{GL}(u) = E(X) \cdot L(u) = \int_0^u F^{-1}(t)\, dt, \quad 0 < u < 1. \tag{2.12}$$

In contrast to the classical Lorenz curve, the generalized Lorenz curve is no longer scale-free and so it completely determines any distribution with a finite mean (Thistle, 1989a). The associated ordering concept is the generalized Lorenz ordering, denoted as $X_1 \geq_{\text{GL}} X_2$. In economic parlance, the generalized Lorenz ordering is a welfare order, since it takes not only distributional aspects into account (as does the Lorenz ordering) but also size-related aspects such as the first moment. In statistical terms it is simply second-order stochastic dominance (SSD), since (e.g., Thistle, 1989b)

$$X_1 \geq_{\text{GL}} X_2 \Longleftrightarrow \int_0^x F_1(t)\, dt \geq \int_0^x F_2(t)\, dt, \quad \text{for all} \quad x \geq 0.$$

Shorrocks and others have provided many empirical examples for which generalized Lorenz dominance applies; hence, this extension appears to be of considerable practical importance.

Another variation on this theme is the absolute Lorenz ordering introduced by Moyes (1987); it replaces scale invariance with location invariance and is defined in terms of the *absolute Lorenz curve*

$$AL(u) = E(X) \cdot \{L(u) - u\} = \int_0^u \{F^{-1}(t) - E(X)\}\, dt, \quad 0 < u < 1. \tag{2.13}$$

However, these proposals clearly do not exhaust the possibilities. See Alzaid (1990) for additional Lorenz-type orderings defined via weighting functions that emphasize certain parts of the Lorenz curves.

2.1.2 Parametric Families of Lorenz Curves

In view of the importance of the Lorenz curve in statistical and economic analyses of income inequality, it should not come as a surprise that a plethora of parametric models for approximating empirical Lorenz curves has been suggested. Since the Lorenz curve characterizes a distribution up to scale, it is indeed quite natural to start directly from the Lorenz curve (or the quantile function), especially since many statistical offices report distributional data in the form of quintiles or deciles, occasionally even in the form of percentiles. In these cases the shape of the income distribution is only indirectly available and perhaps not even required if an assessment of inequality associated with the distribution is all that is desired. In short, does one fit a distribution function to the data and obtain the implied Lorenz curve (and Gini coefficient), or does one fit a Lorenz curve and obtain the implied distribution function (and Gini coefficient)?

The pioneering effort of Kakwani and Podder (1973) triggered a veritable avalanche of papers concerned with the direct modeling of the Lorenz curve, of which we shall only present a brief account. Since any function that passes through (0, 0) and (1, 1) and that is monotonically increasing and convex in between is a bona fide Lorenz curve, the possibilities are virtually endless. Kakwani and Podder (1973, 1976) suggested two forms. Their 1973 form is

$$L(u) = u^{\delta} e^{-\eta(1-u)}, \quad 0 < u < 1, \tag{2.14}$$

where $\eta > 0$ and $1 < \delta < 2$, whereas the more widely known second form (Kakwani and Podder, 1976) has a geometric motivation. Introducing a new coordinate system defined in terms of

$$\eta = \frac{1}{\sqrt{2}}(u + v) \quad \text{and} \quad \pi = \frac{1}{\sqrt{2}}(u - v), \tag{2.15}$$

where $0 < u < 1$ and $v = L(u)$, this form is given by

$$\eta = a\pi^{\alpha}(\sqrt{2} - \pi)^{\beta}. \tag{2.16}$$

Here $a \geq 0$, $0 \leq \alpha \leq 1$ and $0 < \beta \leq 1$. This model amounts to expressing a point $[F(x), F_{(1)}(x)]$ on the Lorenz curve in the form (π, η), where η is the length of the ordinate from $[F(x), F_{(1)}(x)]$ on the egalitarian line and π is the distance of the ordinate from the origin along the egalitarian line. [As pointed out by Dagum (1986), the new coordinate system (2.15) was initially introduced by Gini (1932) in the Italian literature.]

Other geometrically motivated specifications include several models based on conic sections: Ogwang and Rao (1996) suggested the use of a circle's arc, Arnold (1986) employed a hyperbolic model, whereas Villaseñor and Arnold (1989) used a segment of an ellipse. Although the resulting fit is sometimes excellent, all these models have the drawbacks that their parameters must satisfy certain constraints which are not easily implemented in the estimation process and also that the expressions for the Gini coefficients are somewhat formidable (an exception is the Ogwang–Rao specification).

A further well-known functional form is the one proposed by Rasche et al. (1980) who suggested

$$L(u) = [1 - (1 - u)^{\alpha}]^{1/\beta}, \quad 0 < u < 1, \tag{2.17}$$

where $0 < \alpha, \beta \leq 1$. This is a direct generalization of the Lorenz curve of the Pareto distribution (2.7) obtained for $\beta = 1$ and $\alpha < 1$. For $\alpha = \beta$ the curve is self-symmetric (in the sense of Section 2.1.1), as pointed out by Anstis (1978).

In order to overcome the drawback of many previously considered functional forms, namely, a lack of fit over the entire range of income, several authors have proposed generalizations or combinations of the previously considered functions. Quite recently, Sarabia, Castillo, and Slottje (1999) have suggested a family of parametric Lorenz curves that synthetizes and unifies some of the previously considered functions. They point out that for any Lorenz curve L_0 the following curves are also Lorenz curves that generalize the initial model L_0:

- $L_1(u) = u^{\alpha}L_0(u)$, $0 < u < 1$, where either $\alpha \geq 1$ or $0 \leq \alpha < 1$ and $L_0'''(u) \geq 0$.
- $L_2(u) = \{L_0(u)\}^{\gamma}$, $0 < u < 1$, where $\gamma \geq 1$.
- $L_3(u) = u^{\alpha}\{L_0(u)\}^{\gamma}$, $0 < u < 1$, and $\alpha, \gamma \geq 1$.

An advantage of this approach is that Lorenz ordering results are easily obtained, in particular

- $L_1(u; \alpha_1) \geq_L L_1(u; \alpha_2)$, if and only if $\alpha_1 \geq \alpha_2 > 0$.
- $L_2(u; \gamma_1) \geq_L L_2(u; \gamma_2)$, if and only if $\gamma_1 \geq \gamma_2 > 0$.
- A combination of the preceding two cases yields results for L_3.

For the particular choice $L_0(u) = 1 - (1-u)^k$, $k \leq 1$, the Lorenz curve of the Pareto distribution (2.7), Sarabia, Castillo, and Slottje obtained a class of parametric Lorenz curves comprising two previously considered functions: the model proposed by Rasche et al. (1980), cf. (2.17), and a proposal due to Ortega et al. (1991). The resulting family is baptized the *Pareto family*; its members are listed in Table 2.1.

Sarabia, Castillo, and Slottje obtained excellent results when fitting this family to Swedish and Brazilian data given in Shorrocks (1983) and concluded that the Pareto distribution, which has the disadvantage not to fit the entire income range, does much better when serving as a generator of parametric Lorenz curves.

The process may be repeated using other Lorenz curves as the generating function L_0. Specifically, for the model proposed by Chotikapanich (1993),

$$L_0(u) = \frac{e^{ku} - 1}{e^k - 1}, \quad 0 < u < 1, \tag{2.18}$$

where $k > 0$, one obtains a family that is called the *exponential family of Lorenz curves* by Sarabia, Castillo, and Slottje (2001), whereas for

$$L_0(u) = u\beta^{u-1}, \quad 0 < u < 1, \tag{2.19}$$

where $\beta > 0$, a specification proposed by Gupta (1984), one obtains the Rao–Tam (1987) curve as the L_1-type curve, namely

$$L_1(u) = u^{\alpha}\beta^{u-1}, \quad 0 < u < 1. \tag{2.20}$$

A further method for generating new parametric Lorenz curves from previously considered ones has been suggested by Ogwang and Rao (2000) who employed convex combinations—their "additive model"—and weighted products—their "multiplicative model"—of two constituent Lorenz curves. Employing combinations

Table 2.1 Parametric Lorenz Curves: The Pareto Family

Lorenz Curve	Gini Coefficient	Source
$L_0(u) = 1 - (1-u)^k$	$\dfrac{1-k}{1+k}$	
$L_1(u) = u^{\alpha}[1 - (1-u)^k]$	$1 - 2[B(\alpha+1, 1) - B(\alpha+1, k+1)]$	Ortega et al. (1991)
$L_2(u) = [1 - (1-u)^k]^{\gamma}$	$1 - \frac{2}{k}B(k^{-1}, \gamma+1)$	Rasche et al. (1980)
$L_3(u) = u^{\alpha}[1 - (1-u)^k]^{\gamma}$	$1 - 2\sum_{i=0}^{\infty} \frac{\Gamma(i-\gamma)}{\Gamma(i+1)\Gamma(-\gamma)} \times B(\alpha+1, ki+1)$	Sarabia, Castillo, and Slottje (1999)

of the Ortega et al. (1991) and Chotikapanich (1993) as well as the Rao–Tam (1987) and Chotikapanich (1993) models, they concluded that the additive models perform distinctly better than either constituent model and moreover yield a satisfactory fit over the entire range of income.

Among the many further proposals, we should mention the work of Maddala and Singh (1977b) who suggested expressing the Lorenz curve as a sum of powers of u and $1-u$. Holm (1993) proposed a maximum entropy approach using side conditions on the Gini coefficient and the distance of the mean income from the minimum income. For the resulting maximum entropy Lorenz curve, the fit is often excellent. Ryu and Slottje (1996) considered nonparametric series estimators based on Bernstein or exponential polynomials and Sarabia (1997) took one of the few distributions parameterized in terms of their associated quantile function, the Tukey's lambda distribution, as the starting point.

For any given parametric Lorenz curve, it is natural to inquire about the form of the implied p.d.f. of the income distribution. Not surprisingly, the resulting expressions are often rather involved, yielding, for example, distributions with bounded support—this being the case, for instance, for the Chotikapanich (1993) model—or a severely restricted behavior in the upper tail. For example, the elliptical model of Villaseñor and Arnold (1989) implies that $f(x) \sim x^{-3}$, for large x, irrespective of the parameters. This does however not diminish the usefulness of these models for approximating the Lorenz curve. It is also interesting that a subclass of the Rasche et al. curve corresponds to a subclass of Lorenz curves implied by the Singh–Maddala income distribution (cf. Section 6.2). See Chotikapanich (1994) for a discussion of the general form of the p.d.f. implied by the Rasche et al. curves.

All the specifications considered are usually estimated by a nonlinear (generalized) least-squares procedure, possibly after a logarithmic transformation.

In a comprehensive study fitting 13 parametric Lorenz curves to 16 data sets describing the disposable household incomes in the Federal Republic of Germany for several nonconsecutive years between 1950 and 1988, Schader and Schmid (1994) concluded that one- and two-parameter models are often inappropriate for this purpose. (The investigation also include several curves obtained from parametric models of the income density, namely, the lognormal, Singh–Maddala, and Dagum type I models.) Their criterion is a comparison with Gastwirth's (1972) nonparametric bounds for the Gini coefficient that are violated by a number of curves. In particular, the Kakwani–Podder (1973) and Gupta (1984) models perform rather poorly, whereas the Kakwani–Podder (1976) model does very well.

A popular benchmark data set in this line of research consists of the deciles of the income distributions of 19 countries derived from Jain (1975) and later published by Shorrocks (1983).

2.1.3 Inequality Measures

Inequality measures are an immensely popular and favorite topic in the modern statistical and econometric literature, especially among "progressive" researchers.

Oceans of ink and tons of computer software have been used to analyze this somewhat controversial and touchy topic, and numerous books, theses, pamphlets, technical and research reports, memoranda, etc. are devoted to this subject matter (see, e.g., Chakravarty, 1990; Sen, 1997; or Cowell, 1995, 2000). It is not our aim to analyze these works and we shall take it for granted that the reader is familiar with the structure of the basic time-honored inequality measure, the Gini coefficient. However, in the authors' opinion the overemphasis—bordering on obsession—on the Gini coefficient as *the* measure of income inequality that permeates the relevant publications of research staffs and their consultants in the IMF and World Bank is an unhealthy and possibly misleading development.

One of the numerous definitions of the Gini index is given as twice the area between the Lorenz curve and the "equality line":

$$G = 2\int_0^1 [u - L(u)]\,du = 1 - 2\int_0^1 L(u)\,du. \tag{2.21}$$

Clearly, the Gini coefficient satisfies the Lorenz order; in economic parlance: It is "Lorenz consistent." Alternative representations are too numerous to mention here; see Giorgi (1990) for a partial bibliography with 385 mainly Italian sources. We do however require a formula in terms of the expectations of order statistics

$$G = 1 - \frac{E(X_{1:2})}{E(X)} = 1 - \frac{\int_0^\infty [1 - F(x)]^2\,dx}{E(X)}, \tag{2.22}$$

which is presumably due to Arnold and Laguna (1977), at least in the non-Italian literature. In the economics literature it has been independently rediscovered by Dorfman (1979).

It should not come as a surprise that various generalizations of the Gini coefficient have also been suggested. Kakwani (1980a), Donaldson and Weymark (1980, 1983), and Yitzhaki (1983) proposed a one-parameter family of generalized Gini indices by introducing different weighting functions for the area under the Lorenz curve

$$G_\nu = 1 - \nu(\nu - 1)\int_0^1 L(u)(1 - u)^{\nu-2}\,du, \tag{2.23}$$

where $\nu > 1$. Muliere and Scarsini (1989) observed that

$$G_\nu = 1 - \frac{E(X_{1:\nu})}{E(X)}. \tag{2.24}$$

Equation (2.24) is a direct generalization of (2.22).

It is of some at least theoretical interest that income distributions can be characterized in terms of these generalized Gini coefficients, which is equivalent—in view of (2.22)—to a characterization in terms of the first moments of the order statistics. As we shall see in the following chapters, most parametric models for the size distribution of incomes possess heavy (polynomial) tails, so only a few of the moments exist. However, these distributions are determined by the sequence of the associated generalized Gini coefficients, provided the mean is finite (Kleiber and Kotz, 2002).

Another classical index is the Pietra coefficient or relative mean deviation, defined as

$$P = \frac{E|X - E(X)|}{2E(X)}. \tag{2.25}$$

It has the interesting geometrical property of being equal to the maximum distance between the Lorenz curve and the equality line.

Being a natural scale-free index, the variance of logarithms

$$\mathrm{VL}(X) = \mathrm{var}(\log X) \tag{2.26}$$

has also attracted some attention, especially in applied work, apparently because of its simple interpretation in connection with the popular lognormal distribution. However, Foster and Ok (1999) showed that it can grossly violate the Lorenz ordering, thus casting some doubt on its usefulness. [Earlier although less extreme results in this regard were obtained by Creedy (1977).]

Among the many further inequality coefficients proposed over the last 100 years, we shall confine ourselves to two one-parameter families that are widely used in applied work. These are the Atkinson (1970) measures

$$A_\epsilon = 1 - \frac{1}{E(X)}\left\{\int_0^\infty x^{1-\epsilon} dF(x)\right\}^{1/(1-\epsilon)}, \tag{2.27}$$

where $\epsilon > 0$ is a sensitivity parameter giving more and more weight to the small incomes as it increases, and the so-called generalized entropy measures (Cowell and Kuga, 1981)

$$\mathrm{GE}_\theta = \frac{1}{\theta(\theta-1)}\int_0^\infty \left[\left\{\frac{x}{E(X)}\right\}^\theta - 1\right] dF(x), \tag{2.28}$$

where $\theta \in \mathbb{R}\backslash\{0, 1\}$. Again, θ is a sensitivity parameter emphasizing the upper tail for $\theta > 0$ and the lower tail for $\theta < 0$. It is worth noting that for $\theta = 2$ we essentially obtain the squared coefficient of variation; hence, this widely used characteristic of a distribution is of special significance in connection with size phenomena. For $\theta = 0, 1$ the generalized entropy coefficients are defined via a limiting argument, yielding

$$T_1 := \mathrm{GE}_1 = \int_0^\infty \frac{x}{E(X)} \log\left\{\frac{x}{E(X)}\right\} dF(x) \tag{2.29}$$

and

$$T_2 := \mathrm{GE}_0 = \int_0^\infty \log\left\{\frac{E(X)}{x}\right\} dF(x). \qquad (2.30)$$

The latter two measures are known as the Theil coefficients—after Theil (1967) who derived them from information-theoretic considerations—T_1 being often referred to as the *Theil coefficient* and T_2 as *Theil's second measure* or the mean logarithmic deviation.

A drawback of both families of coefficients is that they are simple functions of the moments of the distributions; hence, they will only be meaningful for a limited range of the sensitivity parameters ϵ and θ if the underlying distribution possesses only a few finite moments (Kleiber, 1997). Unfortunately, this is precisely the type of distributions we shall encounter below.

2.1.4 Sampling Theory of Lorenz Curves and Inequality Measures

In view of the long history of inequality measurement, it is rather surprising that only comparatively recently the asymptotics of time-honored tools such as the Lorenz curve and associated inequality measures have been investigated. A possible explanation is that in earlier literature income data were often believed to come from censuses. Nowadays it is however generally acknowledged that most data are, in fact, obtained from surveys (although not necessarily from simple random samples). In addition, the applications of Lorenz curves extend to other areas such as actuarial science where samples may be much smaller. This creates the need for an adequate theory of sampling variation.

As mentioned above, it would by now be easy to write a 500-page monograph dealing exclusively with inequality measurement, parametric and nonparametric, classical and computer-intensive. Our following brief account presents the core results in the sampling theory of Lorenz curves and some popular inequality measures when raw microdata are available, and only in the case of complete data. [We do not even discuss the celebrated Gastwirth (1972) bounds on the Gini coefficient and its associated sampling theory.]

Lorenz Curves

The pointwise strong consistency of the empirical Lorenz curve was proved by Gail and Gastwirth (1978) and Sendler (1979) under the assumption of a finite mean and uniqueness of the quantile under consideration. The question arises if, and under what conditions, the entire empirical Lorenz curve can be considered a good estimator of its theoretical counterpart. This requires the use of more advanced probabilistic tools from the theory of weak convergence of stochastic processes. The first results in this area were due to Goldie (1977) who showed that (1) the empirical Lorenz curve L_n converges almost surely uniformly to the population Lorenz curve

$$\sup_{u\in[0,1]} |L_n(u) - L(u)| \xrightarrow{a.s.} 0$$

(a Glivenko–Cantelli-type result), and (2) a functional central limit theorem holds. For the latter one must consider the normed difference between the empirical and theoretical Lorenz curves—the *empirical Lorenz process*

$$\sqrt{n}\{L_n - L\};$$

under appropriate regularity conditions it converges weakly (in the space $C[0, 1]$ of continuous functions on $[0, 1]$) to a Gaussian process that is related to (without being identical to) the familiar Brownian bridge process.

More recent results along these lines include works on the rate of convergence of the empirical Lorenz process (a function-space law of the iterated logarithm) due to Rao and Zhao (1995), and subsequently refined by Csörgő and Zitikis (1996, 1997). These results allow for the construction of asymptotic confidence bands for the entire Lorenz curves; see Csörgő, Gastwirth, and Zitikis (1998). [Confidence intervals for single points on the Lorenz curves—a much easier problem—were obtained by Sendler (1979).]

However, although the empirical process approach considering the entire Lorenz curve as the random element of interest is perhaps the most appropriate setup from a theoretical point of view, in applied work it is often sufficient to consider the Lorenz curve at a finite set of points (the deciles, say). A line of research dealing with inference based on a vector of Lorenz curve ordinates was initiated by Beach and Davidson (1983). Consider the k points $0 < u_1 < u_2 < \cdots < u_k < 1$, with corresponding quantiles $F^{-1}(u_i)$, $i = 1, \ldots, k$. (Note that under the general assumption of this chapter—namely, that F be supported on an interval—these quantiles are unique.) Writing the conditional mean of incomes less than or equal to $F^{-1}(u_i)$ as $\gamma_i := E[X|X \leq F^{-1}(u_i)]$, $i = 1, \ldots, k$, and $\mu = E(X)$ we can express the corresponding Lorenz curve ordinates in the form

$$L(u_i) = \frac{1}{\mu}\int_0^{F^{-1}(u_i)} x\,dF(x) = \frac{u_i}{\mu}\int_0^{F^{-1}(u_i)} \frac{x\,dF(x)}{u_i} = u_i \cdot \frac{\gamma_i}{\mu}. \tag{2.31}$$

Natural estimates of these quantities are computed as $\hat{u}_i = X_{r_i:n}$, where $r_i = \lfloor nu_i \rfloor$, and

$$\hat{L}(u_i) = \frac{\sum_{j=1}^{r_i} X_{j:n}}{\sum_{j=1}^{n} X_{j:n}} \approx u_i \frac{\hat{\gamma}_i}{\hat{\mu}},$$

with $\hat{\gamma}_i = \sum_{j=1}^{r_i} X_{j:n}/r_i$ and $\hat{\mu} = \bar{X}_n$. This shows that the asymptotics of sample Lorenz curve ordinates can be reduced to the asymptotics of the sample quantiles for which there is a well-developed theory (e.g., David, 1981, pp. 254–258).

Under the assumption that the observations form a simple random sample from an underlying distribution with finite variance (σ^2, say), Beach and Davidson showed that the $(k+1)$-dimensional random vector $\hat{\theta} := (u_1\hat{\gamma}_1, \ldots, u_k\hat{\gamma}_k, u_{k+1}\hat{\gamma}_{k+1})^\top$, with $u_{k+1} = 1$ and $\hat{\gamma}_{k+1} = \bar{X}_n$, is asymptotically jointly multivariate normal

$$\sqrt{n}(\hat{\theta} - \theta) \xrightarrow{d} N(0, \Omega), \tag{2.32}$$

where the asymptotic covariance matrix $\Omega = (\omega_{ij})$ is given by

$$\omega_{ij} = u_i \Big[\lambda_i^2 + (1-u_i)\{F^{-1}(u_i) - \gamma_i\}\{F^{-1}(u_j) - \gamma_j\} + \{F^{-1}(u_i) - \gamma_i\}(\gamma_j - \gamma_i) \Big], \quad i \le j. \tag{2.33}$$

Here $\lambda_i^2 := \mathrm{var}[X|X \le F^{-1}(u_i)]$. (Note that $\omega_{k+1,k+1}/n = \sigma^2/n$, the variance of the sample mean.) Writing

$$\hat{L}(u) := [\hat{L}(u_1), \ldots, \hat{L}(u_k)]^\top = [u_1\hat{\gamma}_1/(u_{k+1}\hat{\gamma}_{k+1}), \ldots, u_k\hat{\gamma}_k/(u_{k+1}\hat{\gamma}_{k+1})]^\top$$

an application of the delta method shows that the k-dimensional vector of sample Lorenz curve ordinates $\hat{L}(u)$ is also jointly multivariate normal, specifically

$$\sqrt{n}[\hat{L}(u) - L(u)] \xrightarrow{d} N(0, V), \tag{2.34}$$

where $V = (v_{ij})$ is given by

$$v_{ij} = \left(\frac{1}{\mu^2}\right)\omega_{ij} + \left(\frac{u_i\gamma_i}{\mu^2}\right)\left(\frac{u_j\gamma_j}{\mu^2}\right)\sigma^2 - \left(\frac{u_i\gamma_i}{\mu^3}\right)\omega_{j,k+1} - \left(\frac{u_j\gamma_j}{\mu^3}\right)\omega_{i,k+1}, \quad i \le j. \tag{2.35}$$

We see that V depends solely on the u_i, the unconditional mean and variance μ and σ^2, the income quantiles $F^{-1}(u_i)$, and the conditional means and variances γ_i and λ_i^2. All these quantities can be estimated consistently. For example, a natural consistent estimator of λ_i^2 is

$$\hat{\lambda}_i^2 = \frac{1}{r_i}\sum_{j=1}^{r_i}(X_{j:n} - \hat{\gamma}_i)^2. \tag{2.36}$$

Consequently, tests of joint hypotheses on Lorenz curve ordinates are now available in a straightforward manner. For example, if it is required to compare an estimated Lorenz curve $\hat{L}(u) = [\hat{L}(u_1), \ldots, \hat{L}(u_k)]$ against a theoretical Lorenz curve $L^0(u) = [L^0(u_1), \ldots, L^0(u_k)]$ in order to test $H_0 : L(u) = L^0(u)$, we may use the quadratic form

$$n[\hat{L}(u) - L^0(u)]^\top \hat{V}^{-1}[\hat{L}(u) - L^0(u)],$$

a statistic that is asymptotically distributed as a χ_k^2 under the null hypothesis. Similarly, to compare two estimated Lorenz curves from independent samples in order to test $H_0 : L^1(u) = L^2(u)$, we may use

$$n[\hat{L}^1(u) - \hat{L}^2(u)]^\top[\hat{V}_1/n_1 + \hat{V}_2/n_2]^{-1}[\hat{L}^1(u) - \hat{L}^2(u)],$$

which is also asymptotically distributed as a χ_k^2 under the null hypothesis.

These procedures are omnibus tests in that they have power against a wide variety of differences between two Lorenz curves. In particular, they are asymptotically distribution-free and consistent tests against the alternative of crossing Lorenz curves.

This line of research has been extended in several directions: Beach and Richmond (1985) provided asymptotically distribution-free simultaneous confidence intervals for Lorenz curve ordinates. More recently, Dardanoni and Forcina (1999) considered comparisons among more than two population distributions using methodology from the literature on order-restricted statistical inference. Davidson and Duclos (2000) derived asymptotic distributions for the ranking of distributions in terms of poverty, inequality, and stochastic dominance of arbitrary order (note that the generalized Lorenz order is equivalent to second-order stochastic dominance). Zheng (2002) extended the Beach–Davidson results to stratified, cluster, and multistage samples.

Inequality Measures

Several possibilities exist to study the asymptotic properties of inequality measures. Since the classical and generalized Gini coefficients are defined in terms of the Lorenz curve, a natural line of attack is to consider them as functionals of the empirical Lorenz process. This was the approach followed by Goldie (1977) for the case of the classical Gini coefficient and by Barrett and Donald (2000) for the generalized Gini. However, an alternative approach using somewhat simpler probabilistic tools (not requiring empirical process techniques) is also feasible. It was developed by Sendler (1979) for the classical Gini coefficient and more recently by Zitikis and Gastwirth (2002) for the generalized version. An additional benefit is that fewer assumptions—essentially only moment assumptions—are required.

Using general representations for the moments of order statistics (see David, 1981), the generalized Gini coefficient (2.24) can be expressed in the form

$$G_\nu = 1 - \frac{\nu}{\mu}\int_0^1 F^{-1}(t)(1-t)^{\nu-1}dt = 1 + \frac{1}{\mu}\int_0^1 F^{-1}(t)\, d\{(1-t)^\nu\}. \tag{2.37}$$

This suggests that G_ν can be estimated by

$$\hat{G}_{n,\nu} = 1 - \frac{1}{\bar{X}_n n^\nu}\sum_{i=1}^{n}\{(n-i+1)^\nu - (n-i)^\nu\}X_{i:n},$$

which is a ratio of two linear functions of the order statistics. In the case of $\nu = 2$ (the classical Gini coefficient), we obtain

$$\begin{aligned}\hat{G}_{n,2} &= 1 - \frac{1}{\bar{X}_n n^2}\sum_{i=1}^{n}\{(n-i+1)^2 - (n-i)^2\}X_{i:n} \\ &= \frac{1}{\bar{X}_n n^2}\sum_{i=1}^{n}\sum_{j=1}^{n}|X_i - X_j|,\end{aligned}$$

the numerator of which is recognized as a variant of a measure of spread commonly referred to as *Gini's mean difference*; see, for example, David (1968).

From (2.37) we moreover see that the weight function $(1-t)^{\nu}$ is smooth and bounded (for $\nu > 1$) on [0, 1]. Using limit theory for such functions of order statistics (see, e.g., Stigler, 1974), Zitikis and Gastwirth obtained for $\nu > 1$ under the sole assumption that $E(X^2)$ is finite

$$\sqrt{n}(\hat{G}_{n,\nu} - G_{\nu}) \xrightarrow{d} N(0, \sigma^2_{F,\nu}),$$

where

$$\sigma^2_{F,\nu} = \frac{1}{\mu^2}\left\{\sigma_F(\nu,\nu) + 2(G_\nu - 1)\sigma_F(1,\nu) + (G_\nu - 1)^2\sigma_F(1,1)\right\} \tag{2.38}$$

with, for $\alpha, \beta \in \{1, \nu\}$,

$$\sigma_F(\alpha,\beta) = \int_0^\infty \int_0^\infty \{F(s \wedge t) - F(s)F(t)\}\{1 - F(s)\}^{\alpha-1}\{1 - F(t)\}^{\beta-1}\, ds\, dt.$$

Alternative expressions for the asymptotic variance of the classical Gini index ($\nu = 2$) may be found in Goldie (1977) and Sendler (1979).

To apply this result for the construction of approximate confidence intervals, a consistent estimate of (2.38) is required. A simple nonparametric estimate of $\sigma^2_{F,\nu}$ is obtained by replacing F in (2.38) by its empirical counterpart F_n. This yields the estimator

$$s^2_{n,\nu} = \frac{1}{\bar{X}^2}\left\{s_n(\nu,\nu) + 2(G_{n,\nu} - 1)\, s_n(1,\nu) + (G_{n,\nu} - 1)^2\, s_n(1,1)\right\}$$

with, for $\alpha, \beta \in \{1, \nu\}$,

$$s_n(\alpha,\beta) = \sum_{i=1}^{n-1}\sum_{j=1}^{n-1} \phi_{ij}^{(n)}(\alpha,\beta)(X_{i+1:n} - X_{i:n})(X_{j+1:n} - X_{j:n}),$$

where

$$\phi_{ij}^{(n)}(\alpha,\beta) = \alpha\beta\left\{\left(\frac{i}{n} \wedge \frac{j}{n}\right) - \frac{i}{n}\frac{j}{n}\right\}\left(1 - \frac{i}{n}\right)^{\alpha-1}\left(1 - \frac{j}{n}\right)^{\beta-1}.$$

Under the previously stated assumptions, Zitikis and Gastwirth showed that this estimator is strongly consistent, implying that an approximate $100(1-\alpha)\%$ confidence interval for G_ν is given by

$$\hat{G}_{n,\nu} \pm z_{\alpha/2}s_{n,\nu}/\sqrt{n},$$

where $z_{\alpha/2}$ denotes the $\alpha/2$ fractile of the standard normal distribution. For the case where instead of complete data only a vector of Lorenz curve ordinates is available, Barrett and Pendakar (1995) provided the asymptotic results for the generalized Gini measures following the Beach and Davidson (1983) approach.

A similar derivation yields the asymptotic distribution of the Pietra coefficient, although the presence of absolute values provides some additional complications. A natural estimator of (2.25) is

$$\hat{P}_n = \frac{\sum_{i=1}^n |X_i - \bar{X}|}{2n\bar{X}} =: \frac{D}{\bar{X}}. \tag{2.39}$$

Writing

$$\sum_{i=1}^n |X_i - \bar{X}| = \sum_{X_i \leq \bar{X}} (\bar{X} - X_i) + \sum_{X_i > \bar{X}} (X_i - \bar{X}),$$

we see that $2nD$ may be expressed in the form

$$\sum_{i=1}^n |X_i - \bar{X}| = 2\left(N\bar{X} - \sum_{X_i \leq \bar{X}} X_i \right),$$

where N denotes the random number of observations less than $\bar{X}$.

Gastwirth (1974) showed that if $E(X^2) < \infty$ and the underlying density is continuous in the neighborhood of $\mu = E(X)$, the numerator and denominator of (2.39) are asymptotically jointly bivariate normally distributed. Hence, an application of the delta method yields

$$\sqrt{n}(\hat{P}_n - P) \xrightarrow{d} N(0, \sigma_P^2),$$

where the asymptotic variance σ_P^2 has the somewhat formidable representation

$$\sigma_P^2 = \frac{v^2}{\mu^2} + \frac{\delta^2 \sigma^2}{4\mu^2} - \frac{\delta}{\mu^3}\left[p\sigma^2 - \int_0^\mu (x-\mu)^2 \, dF(x) \right],$$

with $p = F(\mu)$, $\delta = E|X - E(X)|$ and

$$v^2 = p^2 \int_\mu^\infty (x-\mu)^2 \, dF(x) + (1-p)^2 \int_0^\mu (x-\mu)^2 \, dF(x) - \frac{\delta^2}{4}.$$

Although the Gini and Pietra coefficients are directly related to the Lorenz curve and can therefore be treated using results on the asymptotics of linear functions of order statistics, this is not true for the Atkinson and generalized entropy measures. However, these quantities are simple functions of the moments of the size distribution, so it is natural to estimate them by the method of moments. For complete random samples the large sample properties of the resulting estimators

$$\hat{A}_{n,\epsilon} = 1 - \frac{m(\epsilon)^{1/\epsilon}}{m(1)} \tag{2.40}$$

and

$$\widehat{\mathrm{GE}}_{n,\theta} = \frac{1}{\theta(\theta-1)}\left\{\frac{m(\theta)}{m(1)^{\theta}} - 1\right\}, \tag{2.41}$$

where $m(\epsilon)$ denotes the sample moment of order ϵ, were derived by Thistle (1990) and Kakwani (1990). It is also convenient to denote the variance of $\sqrt{n}m(\epsilon)$ by $\sigma^2(\epsilon)$ and the covariance of $\sqrt{n}m(\epsilon)$ and $\sqrt{n}m(\epsilon')$ by $\gamma(\epsilon, \epsilon')$.

Provided the required moments exist, a straightforward application of the delta method yields

$$\sqrt{n}\frac{\hat{A}_{n,\epsilon} - A_\epsilon}{\sqrt{v_A(\epsilon)}} \xrightarrow{d} N(0, 1), \tag{2.42}$$

where

$$v_A(\epsilon) = \left[\frac{1-A_\epsilon}{\epsilon\mu(\epsilon)}\right]^2\left\{\sigma^2(\epsilon) - \frac{2\epsilon\mu(\epsilon)}{\mu}\gamma(\epsilon, 1) + \left[\frac{\epsilon\mu(\epsilon)}{\mu}\right]^2\sigma^2\right\}, \tag{2.43}$$

with $\mu(\epsilon) := E(X^\epsilon)$ and $\sigma^2(\epsilon) := \mu(2\epsilon) - \mu(\epsilon)^2$ Thistle also showed that the estimator $\hat{A}_\epsilon$ is strongly consistent and that (2.43) can be consistently estimated by replacing the population moments with their sample counterparts, thus allowing for the construction of asymptotic confidence intervals and tests based on (2.42).

An analogous result is available for the generalized entropy measures

$$\sqrt{n}\frac{\widehat{\mathrm{GE}}_{n,\theta} - \mathrm{GE}_\theta}{\sqrt{v_{\mathrm{GE}}(\theta)}} \xrightarrow{d} N(0, 1), \tag{2.44}$$

where

$$v_{\mathrm{GE}}(\theta) = \frac{1}{\theta^2(\theta-1)^2\mu^{2(\theta+1)}}\{\mu^2\sigma^2(\theta) - 2\theta\mu\mu(\theta)\gamma(\theta, 1) + \theta^2\mu^2(\theta)\sigma^2\}, \tag{2.45}$$

Again, (2.45) can be estimated consistently so that asymptotic tests and confidence intervals based on (2.44) are feasible.

Alternatively, confidence intervals for inequality measures may be obtained via the bootstrap method; see Mills and Zandvakili (1997) for a bootstrap approach in the case of the Theil and classical Gini coefficients and Xu (2000) for the generalized Gini case.

2.1.5 Multivariate Lorenz Curves

With the increasing use of multivariate data, multivariate Lorenz curves are no doubt the wave of the future. So far only a limited number of results are available on this challenging concept. The following exposition draws heavily on Koshevoy and Mosler (1996).

Taguchi (1972a,b) suggested defining a bivariate Lorenz curve—or rather surface—as the set of points $[s, t, L^T(s, t)] \in \mathbb{R}^3_+$, where

$$s = \int_{\mathbb{R}^2_+} \chi_{\mathbf{a}}(\mathbf{x})\, dF(\mathbf{x}),$$

$$t = \int_{\mathbb{R}^2_+} \chi_{\mathbf{a}}(\mathbf{x})\tilde{x}_1\, dF(\mathbf{x}),$$

$$L^T(s, t) = \int_{\mathbb{R}^2_+} \chi_{\mathbf{a}}(\mathbf{x})\tilde{x}_2\, dF(\mathbf{x})$$

and

$$\chi_{\mathbf{a}}(\mathbf{x}) = \begin{cases} 1, & \mathbf{x} \le \mathbf{a}, \\ 0, & \text{otherwise}, \end{cases}$$

with $\mathbf{x} = (x_1, x_2)$, $\mathbf{a} = (a_1, a_2)$, $0 \le a_j \le \infty$, and $\tilde{x}_i = x_i/E(X_i)$, $i = 1, 2$. [$\mathbf{x} \le \mathbf{a}$ is defined in the componentwise sense.]

This may be called the *Lorenz–Taguchi surface*. A problem with Taguchi's proposal is that it does not treat the coordinate random variables in a symmetric fashion.

Arnold (1983, 1987) introduced the following definition which is somewhat easier to handle. The *Lorenz–Arnold surface* L^A is the graph of the function

$$L^A(u_1, u_2) = \frac{\int_0^{\xi_1}\int_0^{\xi_2} x_1 x_2\, dF(x_1, x_2)}{\int_0^{\infty}\int_0^{\infty} x_1 x_2\, dF(x_1, x_2)} \tag{2.46}$$

where

$$u_i = \int_0^{\xi_i} dF^{(i)}(x_i), \quad 0 \le u_i \le 1, \; i = 1, 2,$$

$F^{(i)}$, $i = 1, 2$, being the marginals of F and $\xi_i \in \mathbb{R}_+$. An appealing feature of this approach is that if F is a product distribution function, $F(x_1, x_2) = F^{(1)}(x_1) \cdot F^{(2)}(x_2)$, then $L^A(u_1, u_2)$ is just the product of the marginal Lorenz curves. Hence for two product distribution functions F and G, $L_F^A(u_1, u_2) \leq L_G^A(u_1, u_2)$ if and only if the two univariate marginals are ordered in the usual sense.

Unfortunately, apart from the case of independent marginals, neither the Lorenz–Taguchi nor the Lorenz–Arnold approach has a simple economic interpretation.

Koshevoy and Mosler (1996) took a different route. Their multivariate Lorenz "curve" is a manifold in $(d+1)$ space. They started from a special view of the univariate Lorenz curve: Defining a dual Lorenz curve via $\bar{L}(u) = 1 - L(1-u)$, $0 \leq u \leq 1$, they introduced the *Lorenz zonoid* as the convex region bordered by L and $\bar{L}$. Clearly, the area between these two curves is equal to the classical Gini coefficient.

The Koshevoy–Mosler multivariate Lorenz curve is now given by a generalization of this idea to $(d+1)$ space. Let $\mathcal{L}^\Gamma$ be the set of d-variate random variables that have finite and positive expectation vectors, $E(X_j) = \int_{\mathbb{R}_+^d} x_j \, dF(\mathbf{x}) > 0$, $j = 1, \ldots, d$, and set

$$\tilde{\mathbf{x}} = (\tilde{x}_1, \ldots, \tilde{x}_d)^\top, \quad \tilde{x}_j = x_j / E(X_j), \quad j = 1, \ldots, d.$$

Thus, $\tilde{\mathbf{X}}$ is the normalization of $\mathbf{X}$ with expectation $\mathbf{1}_d = (1, \ldots, 1)^\top$. If we set

$$\varsigma(h) = \left[\int_{R_+^d} h(\mathbf{x}) \, dF(\mathbf{x}), \int_{R_+^d} h(\mathbf{x}) \tilde{\mathbf{x}} \, dF(\mathbf{x}) \right],$$

for every (measurable) $h : \mathbb{R}_+^d \to [0, 1]$, the $d+1$ dimensional Lorenz zonoid $LZ(F)$ is defined as

$$LZ(F) = \{\mathbf{z} \in \mathbb{R}^{d+1} \,|\, \mathbf{z} = (z_0, z_1, \ldots, z_d)^\top = \varsigma(h)\}. \tag{2.47}$$

The Lorenz zonoid is therefore a convex compact subset of the unit (hyper) cube in $\mathbb{R}_+^{d+1}$—it may be shown to be strictly convex if F is an absolutely continuous distribution—containing the origin as well as the point $\mathbf{1}_{d+1} = (1, \ldots, 1)^\top$ in $\mathbb{R}^{d+1}$.

Now that we have defined a generalization of the area between the Lorenz and inverse Lorenz functions as a convex set in $d+1$ space, it remains to define a generalization of the curve itself. The solution is as follows. Consider the set

$$Z(F) = \left\{ \mathbf{y} \in \mathbb{R}_+^d \;\middle|\; \mathbf{y} = \int_{R_+^d} h(\mathbf{x}) \tilde{\mathbf{x}} \, dF(\mathbf{x}), \, h : \mathbb{R}_+^d \to [0, 1] \text{ measurable} \right\}. \tag{2.48}$$

$Z(F)$ is called the F zonoid. Note that if $(z_0, z_1, \ldots, z_d)^\top \in LZ(F)$, then $(z_1, \ldots, z_d)^\top \in Z(F)$. $Z(F)$ is contained in the unit cube of $\mathbb{R}_+^d$ and consists of all total portion vectors $\tilde{\mathbf{x}}$ held by subpopulations. In particular, for $d = 1$ the F zonoid $Z(F)$ is equal to the unit interval. Now for a given $(z_1, \ldots, z_d)^\top \in Z(F)$, we have $(z_0, z_1, \ldots, z_d)^\top \in LZ(F)$ if and only if z_0 is an element of the closed interval between the smallest and largest percentage of the population by which the portion vector $(z_1, \ldots, z_d)^\top$ is held.

The function $l_F Z(F) \to \mathbb{R}_+$ defined by

$$l_F(\mathbf{y}) = \max\{t \in \mathbb{R}_+ | (t, \mathbf{y}) \in LZ(F)\} \tag{2.49}$$

is now called the d-dimensional inverse Lorenz function and its graph is the Koshevoy–Mosler Lorenz surface of F. Because $LZ(F)$ and $Z(F)$ are convex sets, l_F is a concave function that is continuous on $Z(F) \cap \mathbb{R}^d_+$, with $l_F(\mathbf{1}_d) = 1$. If F has no mass at the origin, then $l_F(\mathbf{0}) = 0$. In this sense, it represents a natural generalization of the univariate Lorenz curve. It may also be shown that the Lorenz surface determines the distribution F uniquely up to a vector of scaling factors. [It should be noted that the Lorenz surface is not necessarily a surface in the usual sense but a manifold that may have any dimension between 1 and d. Its dimension equals the dimension of $Z(F)$.]

In terms of a distribution of commodities, $l_F(\mathbf{y})$ is equal to the maximum percentage of the population whose total portion amounts to $\mathbf{y}$. The Lorenz zonoid has the following economic interpretation: To every unit of a population the vector $\mathbf{X}$ of endowments in d commodities is assigned. This unit then holds the vector $\tilde{\mathbf{X}}$ of portions of the mean endowment. A given function h may now be considered a selection of a subpopulation. Of all those units that have endowment vector $\mathbf{x}$ or portion vector $\tilde{\mathbf{x}}$, the percentage $h(\mathbf{x})$ is selected. Thus, $\int h(\mathbf{x})\, dF(\mathbf{x})$ is the size of the subpopulation selected by h, and $\int h(\mathbf{x})\tilde{\mathbf{x}}\, dF(\mathbf{x})$ amounts to the total portion vector held by this subpopulation.

The Koshevoy–Mosler multivariate Lorenz order is now defined as the set inclusion ordering of Lorenz zonoids, that is,

$$F \geq_{LZ} G :\Longleftrightarrow LZ(F) \supseteq LZ(G). \tag{2.50}$$

It has the appealing property that it implies the classical Lorenz ordering of all univariate marginal distributions.

In view of the geometric motivation of the classical Gini coefficient (2.21), the question of what a multivariate Gini index might look like arises. Koshevoy and Mosler (1997) discussed a multivariate Gini index defined as the volume of their Lorenz zonoid $LZ(F)$, specifically

$$\mathbf{G} := \text{vol}[LZ(F)] = \frac{1}{(d+1)! \prod_{j=1}^{d} E(X_j)} E(|\det \mathbf{Q}_F|), \tag{2.51}$$

where $\mathbf{Q}_F$ is the $(d+1) \times (d+1)$ matrix with rows $(1, \mathbf{X}_i)$, $i = 1, \ldots, d+1$, and $\mathbf{X}_1, \mathbf{X}_2, \ldots, \mathbf{X}_{d+1}$ are i.i.d. with the c.d.f. F.

It follows that this multivariate Gini coefficient may be equal to zero *without* all commodities being equally distributed. In fact, it will be equal to zero if *at least one* of the commodities is equally distributed or if two commodities have the *same* distribution. (It will also be equal to zero if there are fewer income receiving units than commodities, that is, if $n < d$.)

2.1.6 Zenga Curves and Associated Inequality Measures

Fairly recently, an alternative to the Lorenz curve has received some attention in the Italian literature. Like the Lorenz curve, the Zenga curve (Zenga, 1984) is defined via the first-moment distribution, hence, we require $E(X) < \infty$. The Zenga curve is now defined in terms of the quantiles of the size distribution and the corresponding first-moment distribution: For

$$Z(u) = \frac{F_{(1)}^{-1}(u) - F^{-1}(u)}{F_{(1)}^{-1}(u)} = 1 - \frac{F^{-1}(u)}{F_{(1)}^{-1}(u)}, \quad 0 < u < 1, \tag{2.52}$$

the set

$$\{[u, Z(u)] \mid u \in (0, 1)\}$$

is the Zenga concentration curve. Note that $F_{(1)} \leq F$ implies $F^{-1} \leq F_{(1)}^{-1}$, so that the Zenga curve belongs to the unit square. It is also evident from (2.52) that the curve is scale-free.

It is instructive to compare the Zenga and Lorenz curves: For the Lorenz curve, the amount of concentration associated with the $u \cdot 100\%$ poorest of the population is described by the difference $u - L(u)$. Rewriting the normalized form $[u - L(u)]/u$ in the form

$$\frac{u - L(u)}{u} = 1 - \frac{F_{(1)}[F^{-1}(u)]}{F[F^{-1}(u)]},$$

we see that concentration measurement according to Lorenz amounts to comparing the c.d.f. F and the c.d.f. of the first-moment distribution $F_{(1)}$ at the *same abscissa* $F^{-1}(u)$. An alternative approach consists of comparing the abscissas at which F and $F_{(1)}$ take the *same value* u. This is the idea behind the Zenga curve. Zenga refers to $Z(u)$ as a point measure of inequality.

If the Zenga curve is an alternative to the Lorenz curve, the question of what corresponding summary measures look like arises. There are several possibilities to aggregate the information contained in the point measure (2.52) into a single coefficient. Zenga (1984, 1985) proposed two measures; the first suggested taking the arithmetic mean of the $Z(u)$, $u \in [0, 1]$,

$$\xi = \int_0^1 Z(u)\, du = 1 - \int_0^1 \frac{F^{-1}(u)}{F_{(1)}^{-1}(u)}\, du, \tag{2.53}$$

whereas the second utilizes the geometric mean of the ratios $F^{-1}(u)/F_{(1)}^{-1}(u)$ and is therefore given by

$$\xi_2 = 1 - \exp\left\{\int_0^1 \log\left[\frac{F^{-1}(u)}{F_{(1)}^{-1}(u)}\right] du\right\}. \tag{2.54}$$

ξ can be rewritten in the form

$$\xi = 1 - \frac{1}{E(X)}\int_0^1 F^{-1}[L(u)]\, du.$$

Compared to the Lorenz curve, the Zenga curve is somewhat more difficult to interpret; it is neither necessarily continuous nor is it convex (or concave). For a classical Pareto distribution with c.d.f. $F(x) = 1 - (x/x_0)^{-\alpha}$, $0 < x_0 \leq x$, and $\alpha > 1$, we obtain $F^{-1}(u) = x_0(1-u)^{-1/\alpha}$ and $F_{(1)}^{-1}(u) = x_0(1-u)^{-1/(\alpha-1)}$ and therefore

$$Z(u) = 1 - (1-u)^{1/[\alpha(\alpha-1)]}, \quad 0 < u < 1.$$

Clearly, the Zenga curve of the Pareto distribution is an increasing function on [0, 1], approaching the u axis with increasing α. Recall from (2.7) that in connection with the Lorenz ordering, an increase in α is associated with a decrease in inequality. It is therefore natural to call a distribution F_2 less concentrated than another distribution F_1 if its Zenga curve is nowhere above the Zenga curve associated with F_1 and thus to define a new ordering via

$$X_1 \geq_Z X_2 :\Longleftrightarrow Z_1(u) \geq Z_2(u) \quad \text{for all } u \in (0, 1). \tag{2.55}$$

In general, the Lorenz and Zenga orderings are unrelated; it is however interesting that the Zenga measure ξ satisfies the Lorenz ordering (Berti and Rigo, 1995). Further research in connection with Zenga and Lorenz orderings may be worthwhile.

2.2 HAZARD RATES, MEAN EXCESS FUNCTIONS, AND TAILWEIGHT

Researchers in the actuarial sciences have addressed the problem of distinguishing among various skewed probability distributions given sparse observations at the right tail. Specifically, it has been suggested to employ what Benktander (1963) called the *mortality of claims*

$$r(x) = \frac{f(x)}{1 - F(x)}, \quad x \geq 0, \tag{2.56}$$

and what Benktander and Segerdahl (1960) called the *average excess claim*

$$e(x) = E(X - x \mid X > x) = \frac{\int_x^\infty (t - x)\, dF(t)}{\int_x^\infty dF(t)}, \quad x \geq 0, \tag{2.57}$$

for distinguishing among potential models. See Benktander (1962, 1963) and Benktander and Segerdahl (1960) for some early work in the actuarial literature.

The mortality of claims and the average excess claim are more widely known under different names. The former is usually called the "hazard rate" or "failure rate" (in reliability theory) and also the "force of mortality" (in life insurance) or "intensity function" (in extreme value theory); the latter is also known as the "mean residual life function" (notably in biometrics and engineering statistics) or the "mean excess function" (in actuarial applications). We shall use the terms *hazard rate* and *mean excess function* in the sequel.

The hazard rate gives the rate at which the risk of large claims is decreasing when x grows. Shpilberg (1977) argued that this function has a direct connection with the physical progress of fire in fire insurance: Most fires are extinguished quickly after they start and the amount of any related claims remains slight. However, if early extinction fails, then the chance of rapidly stopping the fire decreases, which in the case of large risk units results in large claims and consequently long tails of the distribution. In Benktander's (1963) view, the lower the claims' rate of mortality, the skewer and more dangerous is the claim distribution.

Integration by parts shows that the mean excess function can alternatively be expressed in the form

$$e(x) = \frac{1}{\bar{F}(x)} \int_x^\infty \bar{F}(t)\, dt, \quad x_0 \leq x, \tag{2.58}$$

where $\bar{F} = 1 - F$. Conversely, in the continuous case the c.d.f. can also be recovered from the mean excess function via

$$F(x) = 1 - \frac{e(x_0)}{e(x)} \exp\left\{ - \int_{x_0}^x \frac{1}{e(t)}\, dt \right\}, \quad x \geq x_0. \tag{2.59}$$

Often (but not always), we shall encounter the case where $x_0 = 0$. It is a direct consequence of (2.59) that a continuous c.d.f. with $F(0) = 0$ is uniquely determined by its mean excess function (e.g., Kotz and Shanbhag, 1980).

There is a simple relationship between the mean excess function and the hazard rate $r(x)$: Rewriting (2.58) in the form $\bar{F}(x)e(x) = \int_x^\infty \bar{F}(t)\, dt$ and differentiating, we see that $\bar{F}(x)e'(x) + \bar{F}'(x)e(x) = -\bar{F}(x)$, which can be rearranged in the form

$$r(x) = -\frac{\bar{F}'(x)}{\bar{F}(x)} = \frac{1 + e'(x)}{e(x)}. \tag{2.60}$$

In the context of income distributions, the mean excess function arises in connection with the somewhat neglected van der Wijk's (1939) law. It asserts that the average income of everybody above a certain level x is proportional to x itself. Formally,

$$\frac{\int_x^{\infty} tf(t)\,dt}{\int_x^{\infty} f(t)\,dt} = \gamma x, \quad \text{for some } \gamma > 0. \tag{2.61}$$

The l.h.s. is clearly the mean excess or mean residual life function, and since this function characterizes a distribution (within the class of continuous distributions), the r.h.s. defines a specific income distribution. In the case of (2.61), this is the Pareto type I distribution; see Chapter 3 below.

In order to discuss the properties of the functions $r(x)$ and $e(x)$ for large losses—or, in economic applications, large incomes—in a unified manner, it is convenient to introduce a concept from classical analysis.

To set the stage, the classical Pareto distribution possesses the survival function

$$\bar{F}(x) = x^{-\alpha}, \quad x \geq 1, \quad \text{for some } \alpha > 0. \tag{2.62}$$

Its slow decrease implies that the moments $E(X^k)$ exist if and only if $k < \alpha$. This is typical for many size distributions; hence, it is useful to have a general framework for describing distributions of the Pareto type.

This framework is provided by the concept of *regularly varying functions*. A function $g : \mathbb{R}_+ \to \mathbb{R}_+$ is regularly varying at infinity with index $\rho \in \mathbb{R}$, symbolically, $g \in \mathrm{RV}_\infty(\rho)$, if

$$\lim_{x\to\infty} \frac{g(tx)}{g(x)} = t^{\rho} \tag{2.63}$$

for all $t > 0$. If $\rho = 0$, g is called slowly varying at infinity. Clearly, x^ρ, $x^\rho \log x$, $x^\rho \log\log x$, $\rho \neq 0$, are all regularly varying at infinity with index ρ. The functions $\log x$, $\log\log x$ are slowly varying. In general, a regularly varying function $g \in \mathrm{RV}_\infty(\rho)$ has a representation of the form $g(x) = x^\rho \ell(x)$, where ℓ is a slowly varying function to which we refer as the slowly varying part of g. In our context, the function g of (2.63) is the density, distribution function, or survival function of the size distribution. We refer to Bingham, Goldie, and Teugels (1987) for an encyclopedic treatment of regularly varying functions. Here we require only some rather basic properties of these functions, all of which may be found in Chapter 1 of Bingham, Goldie, and Teugels:

RV 1. The concept can be extended to regular variation at points x_0 other than infinity; one then replaces $g(x)$ by $g(x_0 - 1/x)$ in the above definition.

RV 2. Products and ratios of regularly varying functions are also regularly varying. Specifically, if $f \in \mathrm{RV}_\infty(\rho)$ and $g \in \mathrm{RV}_\infty(\tau)$, then $fg \in \mathrm{RV}_\infty(\rho + \tau)$ and $f/g \in \mathrm{RV}_\infty(\rho - \tau)$.

RV 3. Derivatives and integrals of regularly varying functions are also regularly varying, under some regularity conditions (which we omit). Loosely speaking, the index of regular variation increases by 1 upon integration, it decreases by 1 upon differentiation. The precise results are referred to as Karamata's theorem and the monotone density theorem, respectively.

We can now describe the behavior of the hazard rate and the mean excess function of distributions with regularly varying tails. It is clear from (2.62) that the Pareto distribution is the prototypical size distribution with a regularly varying tail.

The hazard rate of a Pareto distribution is given by

$$r(x) = \frac{\alpha}{x}, \quad 1 \leq x, \tag{2.64}$$

which is in $\mathrm{RV}_\infty(-1)$, and for the mean excess function a straightforward calculation yields [compare van der Wijk's law (2.61)]

$$e(x) = \frac{x}{\alpha - 1}, \quad 1 \leq x, \tag{2.65}$$

which is in $\mathrm{RV}_\infty(1)$.

These properties can be generalized to distributions with regularly varying tails: If we are given a distribution with $\bar{F} \in \mathrm{RV}_\infty(-\alpha)$, $\alpha > 0$, it follows from property RV 2 that

$$r \in \mathrm{RV}_\infty(-1). \tag{2.66}$$

Similarly, the mean excess function of such a distribution can be shown to possess the property

$$e \in \mathrm{RV}_\infty(1). \tag{2.67}$$

In view of the r.h.s. of (2.58) and property RV 3, the latter result is quite transparent.

The preceding results imply that empirical distributions possessing slowly decaying hazard rates or approximately linearly increasing mean excess functions can be modeled by distributions with regularly varying tails. This is indeed a popular approach in applied actuarial work. The work of Benktander and Segerdahl (1960) constitutes an early example of the use of mean excess plots. More recently, Hogg and Klugman (1983, 1984); Beirlant, Teugels, and Vynckier (1996); and Embrechts, Klüppelberg, and Mikosch (1997) suggested employing the empirical mean excess function for selecting a preliminary model. In addition, Benktander and Segerdahl (1960) and Benktander (1970) defined two new loss distributions in terms of their mean excess function; see Section 7.4 for further details.

In connection with income distributions, the concept of regular variation is also useful in several respects. Firstly it helps to clarify the meaning of Pareto's coefficient α. Clearly, the usefulness of (regression-type) estimates of the Pareto parameter is questionable in the absence of an underlying exact Pareto distribution. On the other hand, empirical Pareto plots are often approximately linear for large incomes and thus a Pareto-type distribution seems to be an appropriate model. How does one define "Pareto type"?

In the economic literature, Mandelbrot (1960) referred to the relation

$$\frac{1 - F(x)}{x^{-\alpha}} = 1 \text{ (for all } x).$$

as the *strong Pareto law.* This is equivalent to F following an exact Pareto distribution. If this property is to be retained for large incomes, an appropriate condition appears to be

$$\lim_{x \to \infty} \frac{1 - F(x)}{x^{-\alpha}} = 1. \tag{2.68}$$

This is Mandelbrot's *weak Pareto law.* [Further weak Pareto laws were introduced by Kakwani (1980b) and Esteban (1986).] Condition (2.68) is closely related to regular variation, although the two concepts are not equivalent, as claimed by Merkies and Steyn (1993). The problem is that there are regularly varying functions such as $\bar{F}(x) \propto x^{\alpha} \log x$ (for large x) for which the slowly varying part $\log x$ is unbounded. Here $\bar{F}$ is clearly regularly varying, but (2.68) does not hold. It would thus seem that Mandelbrot's weak Pareto law should perhaps be rephrased as requiring that the size distribution be regularly varying in the upper tail.

A second issue concerning the meaning of α in the absence of an underlying Pareto distribution is its relation to income inequality. For an exact Pareto distribution, the situation is clear (see Section 3.5 below): A smaller α is associated with greater inequality in the sense of the Lorenz ordering and several associated inequality measures. What can be said if the income distribution is just of the Pareto type? Authors such as Bowman (1945) argued that a steeper Pareto curve is associated with a more equal distribution of income, but a formal proof of this fact has been lacking until recently.

Kleiber (1999b, 2000a) showed how this can be proven using properties of regularly varying functions within the framework of the Lorenz ordering. Briefly, the argument runs as follows.

The property $X_1 \geq_L X_2$ is equivalent to (Arnold, 1987)

$$\int_0^x \tilde{F}_1(t)\, dt \geq \int_0^x \tilde{F}_2(t)\, dt \quad \text{for all } x \in \mathbb{R}_+ \tag{2.69}$$

and also to

$$\int_x^\infty \bar{\tilde{F}}_1(t)\,dt \geq \int_x^\infty \bar{\tilde{F}}_2(t)\,dt \quad \text{for all } x \in \mathbb{R}_+. \tag{2.70}$$

Here $\tilde{F}_i$ denotes the c.d.f. of the mean-scaled random variable $X_i/E(X_i)$, $i = 1, 2$. If $\bar{F}_i \in \text{RV}_\infty(-\alpha_i)$, $\alpha_i > 0$, $i = 1, 2$, it follows using property RV 3 under appropriate regularity conditions that the integrated upper tails are in $\text{RV}_\infty(-\alpha_i + 1)$. Hence, we obtain from (2.70)

$$g(x) := \frac{\int_x^\infty \bar{\tilde{F}}_1(t)\,dt}{\int_x^\infty \bar{\tilde{F}}_2(t)\,dt} \geq 1, \quad \text{for all } x \in \mathbb{R}_+, \Longrightarrow \alpha_1 \leq \alpha_2.$$

This shows that if size distributions with regularly varying tails are ordered in the Lorenz sense, then the more unequal distribution necessarily exhibits heavier tails (namely, a smaller α). An analogous argument yields that if $F_i \in \text{RV}_0(-\beta_i)$, $\beta_i > 0$, $i = 1, 2$, then $X_1 \geq_L X_2$ implies $\beta_1 \leq \beta_2$. Hence, there exists a similar condition for the lower tails.

This argument provides a useful tool for deriving the necessary conditions for Lorenz dominance in parametric families. Many distributions studied in this book are regularly varying at infinity and/or the origin, and the index of regular variation can usually directly be determined from the density or c.d.f. Consequently, necessary conditions for the Lorenz ordering are often available in a simple manner.

As a by-product, it turns out that Pareto's alpha can be considered an inequality measure even in the absence of an underlying exact Pareto distribution, provided it is interpreted as an index of regular variation. See Kleiber (1999b, 2000a) for further details and implications in the context of income distributions.

The preceding discussion has shown how Pareto tail behavior can be formalized using the concept of regular variation. Although many of the distributions studied in detail in the following chapters are of this type, there are some that cannot be discussed within this framework.

To conclude this section, we therefore introduce a somewhat broader classification of size distributions according to tail behavior. The distributions we shall encounter below comprise three types of models that we may call Pareto-type distributions, lognormal-type distributions and gamma-type distributions, respectively. A preliminary classification is given in Table 2.2 (for $x \to \infty$).

Here we have distributions with polynomially decreasing tails (type I), exponentially decreasing tails (type III) as well as an intermediate case (type II). These three types can be modified to enhance flexibility in the left tail.

Type I. The basic form is clearly the Pareto (I) distribution that is zeromodal. Unimodal generalizations are of the forms (1) $f(x) \propto e^{-1/x}x^{-\alpha}$, leading to

Table 2.2 Three Types of Size Distributions

Type I	Type II	Type III
(Pareto type) $f(x) \sim x^{-\alpha} = e^{-\alpha \log x}$	(Lognormal type) $f(x) \sim e^{-(\log x)^{\alpha}}$	(Gamma type) $f(x) \sim e^{-\alpha x}$

distributions of the inverse gamma (or Vinci) type that exhibit a light (non-Paretian) left tails or (2) $f(x) \propto x^p(1+x)^{-(\alpha+p)}$, leading to distributions of the beta (II) type that exhibit heavy (Paretian) left tails and may therefore be considered "double Pareto" distributions.

Type II. For $\alpha = 2$ this yields a distribution of lognormal type. For a general α the densities will be unimodal due to the difference in behavior of $\log x$ for $0 < x < 1$ and $x \geq 1$.

Type III. The prototypical type III distribution is the exponential distribution with density $f(x) = \alpha e^{-\alpha x}$ that is zeromodal. A more flexible shape is obtained upon introducing a polynomial term, leading to densities of type $f(x) \propto x^p e^{-\alpha x}$, the prime example being the gamma density. "Weibullized" versions also fall under type III.

2.3 SYSTEMS OF DISTRIBUTIONS

The most widely known system of statistical distributions is the celebrated Pearson system, derived by Karl Pearson in the 1890s in connection with his work on evolution. It contains many of the best known continuous univariate distributions. Indeed, we shall encounter several members of the Pearson system in the following chapters, notably Chapters 5 and 6 that comprise models related to the gamma and beta distributions.

The Pearson densities are defined in terms of the differential equation

$$f'(x) = \frac{(x-a)f(x)}{c_0 + c_1 x + c_2 x^2}, \tag{2.71}$$

where a, c_0, c_1, c_2 are constants determining the particular type of solution. (The equation originally arose from a corresponding difference equation satisfied by the hypergeometric distribution by means of a limiting argument.) The most prominent solution of (2.71) is the normal p.d.f. that is obtained for $c_1 = c_2 = 0$. All solutions are unimodal; however, the maxima may be located at the ends of the support.

There are three basic types of solutions of (2.71), referred to as types I, VI, and IV, depending on the type of roots of the quadratic in the denominator (real and opposite signs, real and same sign, and complex, respectively). Ten further types arise as special cases; see Table 2.3. (For the Pearson type XII distributions, the constants g and h are functions of the skewness and kurtosis coefficients.)

Table 2.3 The Pearson Distributions

Type	Density	Support
I	$(1+x)^{m_1}(1-x)^{m_2}$	$-1 \leq x \leq 1$
VI	$x^{m_2}(1+x)^{-m_1}$	$0 \leq x \leq \infty$
IV	$(1+x^2)^{-m}\exp(-v\arctan x)$	$-\infty < x < \infty$
Normal	$\exp\left(-\frac{1}{2}x^2\right)$	$-\infty < x < \infty$
II	$(1-x^2)^m$	$-1 \leq x \leq 1$
VII	$(1+x^2)^{-m}$	$-\infty < x < \infty$
III	$x^m \exp(-x)$	$0 \leq x < \infty$
V	$x^{-m}\exp(-1/x)$	$0 \leq x < \infty$
VIII	$(1+x)^{-m}$	$0 \leq x \leq 1$
IX	$(1+x)^m$	$0 \leq x \leq 1$
X	$\exp(-x)$	$0 \leq x < \infty$
XI	x^{-m}	$1 \leq x < \infty$
XII	$[(g+x)(g-x)]^h$	$-g \leq x \leq g$

In applications the variable x is often replaced by $(z-\mu)/\sigma$ for greater flexibility. The table contains several familiar distributions: Type I is the beta distribution of the first kind, type VI is the beta distribution of the second kind (with the F distribution as a special case), type VII is a generalization of Student's t, types III and V are the gamma and inverse (or inverted, or reciprocal) gamma, respectively, type X is the exponential, and type XI is the Pareto distribution.

A key feature of the Pearson system is that the first four moments (provided they exist) may be expressed in terms of the four parameters a, c_0, c_1, c_2; in turn, the moment ratios

$$\beta_1 = \frac{\mu_3^2}{\mu_2^3} \quad \text{(skewness)}$$

and

$$\beta_2 = \frac{\mu_4}{\mu_2^2} \quad \text{(kurtosis)}$$

provide a complete taxonomy of the Pearson curves. Indeed, Pearson suggested selecting an appropriate density based on estimates of β_1, β_2 that should then be fitted by his method of moments.

The main applications of the Pearson system are therefore in approximating sampling distributions when only low-order moments are available and in providing a family of reasonably typical non-Gaussian shapes that may be used, among other things, in robustness studies. For further information on the Pearson distributions, we refer the reader to Johnson, Kotz, and Balakrishnan (1994, Chapter 12) or Ord (1985) and the references therein.

In this book we mainly require the Pearson system for classifying size distributions, several of which fall into this system (possibly after some simple transformation). Specifically, we shall encounter the Pearson type XI distribution (under the more familiar name of Pareto distribution) in Chapter 3, the types III and V (under the names gamma and inverse gamma distribution, respectively) in Chapter 5, and the type VI distribution (under the name beta distribution of the second kind) in Chapter 6.

Of the many alternative systems of continuous univariate distributions, we will also encounter some members of a system introduced by Irving Burr in 1942. Like the Pearson system of distributions, the Burr family is defined in terms of a differential equation; unlike the Pearson system, this differential equation describes the distribution function and not the density. This has the advantage of closed forms for the c.d.f., sometimes even for the quantile function, which is rarely the case for the members of the Pearson family. The Burr system comprises 12 distributions that are usually referred to by number; see Table 2.4.

Table 2.4 The Burr Distributions

Type	c.d.f.	Support
I	x	$0 < x < 1$
II	$(1+e^{-x})^{-p}$	$-\infty < x < \infty$
III	$(1+x^{-a})^{-p}$	$0 < x < \infty$
IV	$\left[1+\left(\frac{c-x}{x}\right)^{1/c}\right]^{-q}$	$0 < x < c$
V	$[1+c\exp(-\tan x)]^{-q}$	$-\pi/2 < x < \pi/2$
VI	$[1+\exp(-c\sinh x)]^{-q}$	$-\infty < x < \infty$
VII	$2^{-q}(1+\tanh x)^q$	$-\infty < x < \infty$
VIII	$\left[\frac{2}{\pi}\arctan(e^x)\right]^q$	$-\infty < x < \infty$
IX	$1-\frac{2}{2+c[(1+e^x)^q-1]}$	$-\infty < x < \infty$
X	$[1-\exp(-x^2)]^a$	$0 \leq x < \infty$
XI	$\left[x-\frac{1}{2\pi}\sin(2\pi x)\right]^q$	$0 < x < 1$
XII	$1-(1+x^a)^{-q}$	$0 \leq x < \infty$

The c.d.f.'s of all Burr distributions satisfy the differential equation

$$F'(x) = F(x)[1 - F(x)]g(x), \tag{2.72}$$

where g is some nonnegative function.

The uniform distribution is clearly obtained for $g \equiv [F(1-F)]^{-1}$. The most widely known of the (nonuniform) Burr distributions is the Burr XII distribution, frequently just called the *Burr distribution*. In practice, one often introduces location and scale parameters upon setting $x = (z - \mu)/\sigma$ for additional flexibility. See Kleiber (2003a) for a recent survey of the Burr family.

In Chapter 6 below we shall encounter the Burr III and Burr XII distributions, albeit under the names of the Dagum and Singh–Maddala distributions.

Stoppa (1990a) proposed a further system of distributions that is closely related to the Burr system. Rewriting the differential equation defining the Burr distributions in the form

$$\frac{F'(x)}{F(x)} = [1 - F(x)] \cdot g[x, F(x)], \tag{2.73}$$

we see upon setting $g(x, y) =: \tilde{g}(x, y)/x$ that Burr's equation amounts to a specification of the elasticity $\eta(x, F) = F'(x) \cdot x/F(x)$ of a distribution function. Stoppa then proposed a differential equation for the elasticity

$$\eta(x, F) = \frac{1 - [F(x)]^{1/\theta}}{[F(x)]^{1/\theta}} \cdot g[x, F(x)], \quad x \geq x_0 > 0, \tag{2.74}$$

where $\theta > 0$, and $g(x, y)$ is positive in $0 < y < 1$.

For $g(x, y) = g(x)$, $F(x) \neq 0, 1$, and $dF^{1/\theta}/dF = F^{1-1/\theta}/\theta$, the solution of this differential equation is

$$F(x) = \left\{1 - \exp\left(\int_0^{\infty} \ell(\tau)\, d\tau\right)\right\},$$

where $\ell(\tau) = g(\tau)/(\theta\tau)$ is a real function integrable on a subset of $\mathbb{R}_+$. For example, if g is chosen as $b\theta x/(1 - bx)$, b, bx, $1/(b - x)$, $bx \sec^2 x$, respectively, with $b > 0$, we obtain the following c.d.f.'s:

I.	$F(x) = (bx)^{\theta}$	$0 < x < 1/b$
II.	$F(x) = (1 - x^{-b})^{\theta}$	$1 < x < \infty$
III.	$F(x) = (1 - e^{-bx})^{\theta}$	$0 < x < \infty$
IV.	$F(x) = [1 - (bx^{-1} - 1)^{1/(\theta b)}]^{\theta}$	$b/2 < x < b$
V.	$F(x) = (1 - be^{-\tan x})^{\theta}$	$-\pi/2 < x < \pi/2$

For $\theta = 1$ we arrive at various cases of the Burr system, whereas for $\theta \neq 1$ type I defines a power function distribution, type II a generalized Pareto, and type III a generalized exponential distribution, respectively.

In a later paper, Stoppa (1993) presented a classification of distributions inspired by the classical table of chemical elements of Mendeleyev. Stoppa's table comprises 15 so-called *periods* defined by families of distributions for which $\log \eta(x, F)$ depends on a single parameter. Within each period there are subfamilies of distributions characterized by up to five parameters.

As far as income distributions are concerned, special interest is focused on family no. 31 of the system whose c.d.f. is given by

$$\left\{-b_1\left(\frac{x^{b_3}}{b_3}+c\right)\right\}^{-1/b_1}, \tag{2.75}$$

where in general $b_1 \neq 0$ and $b_3 \neq 0$. For $b_1 < 0$, $b_3 < 0$, we get the generalized Pareto type I (the Stoppa distribution), for $b_1 > 0$, $b_3 < 0$, we obtain the inverted Stoppa distribution; $b_1 < 0$, $b_3 > 0$ yields a generalized power function; $b_1 > 0$, $b_3 > 0$ gives us the Burr III (Dagum type I) distribution; and finally for $b_1 = -1$, $b_3 < 0$, we obtain the classical Pareto type I distribution.

Transformation Systems

The Pearson curves were designed in such a manner that for any possible pairs of values $\sqrt{\beta_1}$, β_2 there is just one corresponding member of the Pearson family of distributions. Alternatively, one may be interested in a transformation, to normality, say, such that for any possible pairs of values $\sqrt{\beta_1}$, β_2 there is one corresponding normal distribution. Unfortunately, no such single transformation is available; however, Johnson (1949) has described a set of three transformations which, when combined, do provide one distribution corresponding to each pair of values $\sqrt{\beta_1}$, β_2. These transformations are

$$Z = \mu + \sigma \log(X - \lambda), \quad X > \lambda, \tag{2.76}$$

$$Z = \mu + \sigma \log\left\{\frac{X-\lambda}{\lambda+\xi-X}\right\}, \quad \lambda < X < \lambda + \xi, \tag{2.77}$$

and

$$Z = \mu + \sigma \sinh^{-1}\left\{\frac{X-\lambda}{\xi}\right\}, \quad -\infty < X < \infty. \tag{2.78}$$

Here Z follows a standard normal distribution and μ, σ, λ, ξ represent parameters of which ξ must be positive and σ non-negative.

The distributions defined by the preceding equation are usually denoted as S_L, S_B, and S_U. Below we shall deal only with the S_L distributions, under the more familiar name of *three-parameter lognormal distributions*.

It is natural to extend Johnson's approach to nonnormal random variables Z. For a Z following a standard logistic distribution with c.d.f.

$$F(z) = \frac{1}{1 + e^{-z}}, \quad -\infty < z < \infty,$$

the distributions associated with the corresponding sets of transformations have been described by Tadikamalla and Johnson (1982) and Johnson and Tadikamalla (1992). The resulting distributions are usually denoted as L_L, L_B, and L_U. For a Z following a Laplace (or double exponential) distribution with p.d.f.

$$f(z) = \frac{1}{2} e^{-|z|}, \quad -\infty < z < \infty,$$

the corresponding distributions have been discussed by Johnson (1954); they are denoted as S'_L, S'_B, and S'_U.

We shall encounter distributions of the L_L type—that is, log-logistic distributions—and generalizations thereof in Chapter 6 and (generalizations of) S'_L distributions in Chapter 4, albeit under the name of *generalized lognormal distributions*. In fact, with little exaggeration, this book can be considered a monograph on exponential transformations of some of the more familiar statistical distributions. Specifically, if Z is

- Exponential, then $\exp(Z)$ follows a Pareto distribution—a distribution studied in Chapter 3.
- Normal, then $\exp(Z)$ follows, of course, a lognormal distribution—a distribution studied in Chapter 4.
- Gamma, then $\exp(Z)$ follows a loggamma distribution—a distribution studied in Chapter 5.
- Logistic, then $\exp(Z)$ follows a log-logistic distribution—a distribution studied, along with its generalizations, in Chapter 6.
- Rayleigh, then $\exp(Z)$ follows a Benini distribution—a distribution studied in Chapter 7.

Less prominent choices for Z include the exponential power (Box and Tiao, 1973) and the Perks (1932) distributions, their exponential siblings are called generalized lognormal and Champernowne distributions and are explored in Sections 4.10 and 7.3, respectively. Very few distributions studied here do not originate from an exponential transformation, mainly the gamma-type models of Chapter 5.

2.4 GENERATING SYSTEMS OF INCOME DISTRIBUTIONS

As was already mentioned in the preceding chapter, a huge variety of size distributions, including almost all of the best known continuous univariate distributions supported on the positive halfline, have been introduced during the last hundred years. It is therefore of particular interest to have, apart from classification systems like those surveyed in the preceding section, generating systems that yield, starting from a few basic principles, models that should be useful for the modeling of size phenomena. Not surprisingly, the largest branch of the size distributions literature, the literature on income distributions, has come up with several generating systems for the derivation of suitable models. We present the systems proposed by D'Addario (1949) and Dagum (1980b,c, 1990a, 1996).

D'Addario's System

Following the idea of transformation functions applied earlier by Edgeworth (1898), Kapteyn (1903), van Uven (1917), and Fréchet (1939), D'Addario (1949) specified his system by means of the *generating function*

$$g(y) = A\{b + \exp(y^{1/p})\}^{-1}, \tag{2.79}$$

where $p > 0$ and b is real, and the *transformation function*

$$y^q \frac{dy}{dx} = \frac{\alpha}{x-c}, \quad c \leq x_0 \leq x < \infty, \tag{2.80}$$

where $\alpha \neq 0$ and q is real. Here x is the income variable and A is a normalizing constant. The differential equation (2.80) yields

$$y = h(x) = a(x-c)^\alpha, \quad q = -1, a > 0, \alpha \neq 0, \tag{2.81}$$

$$y = h(x) = [(1+q)\{\alpha \log(x-c) + a\}]^{1/(1+q)}, \quad q \neq -1, \alpha \neq 0, \tag{2.82}$$

where a is a constant of integration. Equations (2.79) and (2.80) imply that the transformed variable $y = h(x)$ is a monotonic function of income, taking values on the interval $[x_0, x_1]$ if $h(x)$ is increasing and on the interval $[x_1, x_0]$ if $h(x)$ is decreasing, where $x_0 = h(y_0)$ and $x_1 = \lim_{x\to\infty} h(x)$.

The general form of solution of D'Addario's system is given by

$$f(x) = A\left|\frac{dh(x)}{dx}\right|\{b + \exp[h(x)^{1/p}]\}^{-1}, \tag{2.83}$$

where $dh(x)/dx$ is obtained from (2.81) or (2.82), depending on the value of q.

Table 2.5 presents the income distributions that can be deduced from D'Addario's system.

Table 2.5 D'Addario's Generating System

Distribution	Generating Function		Transformation Function			Support
	b	p	α	c	q	
Pareto (I)	0	1	>0	0	0	$x_0 \le x < \infty$
Pareto (II)	0	1	>0	$\neq 0$	0	$c < x_0 \le x < \infty$
Lognormal (2 parameters)	0	1/2	>0	0	0	$0 \le x < \infty$
Lognormal (3 parameters)	0	1/2	>0	$\neq 0$	0	$c \le x < \infty$
Generalized gamma	0	>0	$\neq 0$	$\neq 0$	-1	$c < x_0 \le x < \infty$
Davis	-1	>0	$-p$	$\neq 0$	-1	$c < x_0 \le x < \infty$

It is worth noting that the Davis distribution (see Section 7.2), a distribution not easily related to any other system of distributions, is a member of D'Addario's transformation system. Also, the four-parameter generalized gamma distribution, introduced by Amoroso (1924–1925), comprises a host of distributions as special or limiting cases, including the gamma and inverse gamma, Weibull and inverse Weibull, chi and chi square, Rayleigh, exponential, and half-normal distributions. The Amoroso distribution will be discussed in some detail in Chapter 5.

Dagum's Generalized Logistic System

The Pearson system is a general-purpose system not necessarily derived from observed stable regularities in a given area of application. D'Addario's system is a translation system with flexible generating and transformation functions constructed to encompass as many income distributions as possible. In contrast, the system specified by Dagum (1980b,c, 1983, 1990a) starts from the characteristic properties of empirical income and wealth distributions. He observes that the income elasticity

$$\eta(x, F) = \frac{d\log\{F(x)\}}{d\log x}$$

of the c.d.f. of income is a decreasing and bounded function of F starting from a finite and positive value as $F(x) \to 0$ and decreasing toward zero as $F(x) \to 1$, that is, for $x \to \infty$. This pattern leads to the specification of the following generating system for income and wealth distributions:

$$\frac{d\log\{F(x) - \delta\}}{d\log x} = \vartheta(x)\phi(F) \le k, \quad 0 \le x_0 < x < \infty, \tag{2.84}$$

where $k > 0$, $\vartheta(x) > 0$, $\phi(x) > 0$, $\delta < 1$, and $d\{\vartheta(x)\phi(F)\}/dx < 0$. These constraints assure that the income elasticity of the c.d.f. is indeed a positive, decreasing, and bounded function of F and therefore of x. For each specification of ϕ

Table 2.6 Dagum's Generating System

Distribution	$\vartheta(x)$	$\phi(F)$	(δ, β)	Support
Pareto (I)	α	$(1-F)/F$	(0, 0)	$0 < x_0 \leq x < \infty$
Pareto (II)	$\dfrac{\alpha x}{x-c}$	$(1-F)/F$	(0, 0)	$0 < x_0 \leq x < \infty$
Pareto (III)	$\beta x + \dfrac{\alpha x}{x-c}$	$(1-F)/F$	(0, +)	$0 < x_0 \leq x < \infty$
Benini	$2\alpha \log x$	$(1-F)/F$	(0, 0)	$0 < x_0 \leq x < \infty$
Weibull	$\beta x(x-c)^{\alpha-1}$	$(1-F)/F$	(0, +)	$c \leq x < \infty$
log-Gompertz	$-\log \alpha$	$-\log F$	(0, 0)	$0 \leq x < \infty$
Fisk	α	$1-F$	(0, 0)	$0 \leq x < \infty$
Singh–Maddala	α	$\dfrac{1-(1-F)^{\beta}}{F(1-F)^{-1}}$	(0, +)	$0 \leq x < \infty$
Dagum (I)	α	$1-F^{1/\beta}$	(0, +)	$0 \leq x < \infty$
Dagum (II)	α	$1-\left(\dfrac{F-\alpha}{1-\alpha}\right)^{1/\beta}$	(+, +)	$0 \leq x < \infty$
Dagum (III)	α	$1-\left(\dfrac{F-\alpha}{1-\alpha}\right)^{1/\beta}$	(−, +)	$0 < x_0 \leq x < \infty$

and ϑ an income distribution is obtained. Table 2.6 provides a selection of models that can be deduced from Dagum's system.

Among those models, the Dagum (II) distribution is mainly used as a model of wealth distribution.

Since the Fisk distribution is also known as the log-logistic distribution (see Chapter 6) and the Burr III and Burr XII distributions are generalizations of this distribution, Dagum (1983) referred to his system as the *generalized logistic-Burr system*. Needless to say, this collection of distributions can be enlarged further by introducing location and scale parameters or using transformation functions.

CHAPTER THREE

Pareto Distributions

The Pareto distribution is the *prototypical size distribution. In view of the unprecedented information explosion on this distribution during the last two decades, it would be very easy to write a four-volume compendium devoted to this magical model rehashing the wealth of material available in the periodical and monographic literature. This distribution—attributed to Vilfredo Pareto (1895)—in analogy with the Lorenz curve is the pillar of statistical income distributions.*

However, due to space limitations we can in this volume only provide a brief but hopefully succinct account, and we shall concentrate on economic and actuarial applications. For the literature up to the early 1980s, we refer the interested reader to the excellent text by Arnold (1983). We shall emphasize contributions from the last 20 years. These include, among others, the unbiased estimation of various Pareto characteristics. We shall also discuss numerous recent generalizations of the Pareto distribution in which Stoppa's contributions play a prominent role (not sufficiently well represented in the English language literature).

3.1 DEFINITION

The classical Pareto distribution is defined in terms of its c.d.f.

$$F(x) = 1 - \left(\frac{x}{x_0}\right)^{-\alpha}, \quad x \geq x_0 > 0, \tag{3.1}$$

where $\alpha > 0$ is a shape parameter (also measuring the heaviness of the right tail) and x_0 is a scale. The density is

$$f(x) = \frac{\alpha x_0^{\alpha}}{x^{\alpha+1}}, \quad x \geq x_0 > 0, \tag{3.2}$$

Statistical Size Distributions in Economics and Actuarial Sciences, By Christian Kleiber and Samuel Kotz.
ISBN 0-471-15064-9

and the quantile function equals

$$F^{-1}(u) = x_0(1-u)^{-1/\alpha}, \quad 0 < u < 1. \tag{3.3}$$

We shall use the standard notation $X \sim \text{Par}(x_0, \alpha)$.

In his pioneering contributions at the end of the nineteenth century, Pareto (1895, 1896, 1897a) suggested three variants of his distribution. The first variant is the classical Pareto distribution as defined in (3.1). Pareto's second model possesses the c.d.f.

$$F(x) = 1 - \left(1 + \frac{x-\mu}{x_0}\right)^{-\alpha}, \quad x \geq \mu, \tag{3.4}$$

and is occasionally called the three-parameter Pareto distribution. The special case where $\mu = 0$,

$$F(x) = 1 - \left(1 + \frac{x}{x_0}\right)^{-\alpha}, \quad x \geq 0, \tag{3.5}$$

where $x_0, \alpha > 0$, is often referred to as the Pareto type II distribution. This distribution was rediscovered by Lomax (1954) some 50 years later in a different context. In our classification, the Pareto type II distribution falls under "beta-type distributions"; therefore, it will not be discussed in the present chapter but rather in Chapter 6 below, although in more general form. It can be considered a special case of the Singh–Maddala distribution (case $a = 1$, in the notation of Chapter 6); there is also a simple relation with the Pareto type I model, namely,

$$X \sim \text{Par(II)}(x_0, \alpha) \Longleftrightarrow X + x_0 \sim \text{Par}(x_0, \alpha). \tag{3.6}$$

It is worth emphasizing that the term "Pareto distribution" is used in connection with both the Pareto type I and Pareto type II versions. Rytgaard (1990) asserted that "Pareto distribution" usually means Pareto type I in the European and Pareto type II in the American literature, but we have not been able to verify this pattern from the references available to us.

It should also be noted that the Pareto type II distribution belongs to the second period of Stoppa's (1993) classification involving two parameters (see Section 2.3).

The third distribution proposed by Pareto—the Pareto type III distribution (Arnold, 1983, uses a different terminology!)—has the c.d.f.

$$F(x) = 1 - \frac{Ce^{-\beta x}}{(x-\mu)^{\alpha}}, \quad x \geq \mu, \tag{3.7}$$

where $\mu \in \mathbb{R}$, $\beta, \alpha > 0$, and C is a function of the three parameters. It arises from the introduction of a linear term βx in the doubly logarithmic representation

$$\log\{1 - F(x)\} = \log C - \alpha \log(x - \mu) - \beta x.$$

The exponential term in (3.7) assures the finiteness of all moments. For income data the values of β are usually very small—Pareto (1896) obtained a value of $\hat{\beta} = 0.0000274$ for data on the Grand Duchy of Oldenburg in 1890—so the Pareto type III distribution does not seem to be attractive in the sense of income distribution theory and applications. On the other hand, Creedy (1977) pointed out that "...the values of β and $[\mu]$ are not invariant with respect to the units of measurement, and β is likely to be very small since the units of x are large and are elsewhere transformed by taking logarithms." Nonetheless, the Pareto type III model has not been used much.

3.2 HISTORY AND GENESIS

The history of the Pareto distribution is lucidly and comprehensively covered in the above-mentioned monograph by Arnold (1983). We shall therefore provide selected highlights of his exposition supplemented by several additional details that have emerged in the last 20 years or so and describe the very few earlier historical sources not covered in Arnold (1983). The history of Pareto distributions is still a vibrant subject of modern research. The Web site sponsored by the University of Lausanne where Vilfredo Pareto spent some 15 productive years devoted to Walras and Pareto (http://www.unil.ch/cwp/) constantly updates the information on this topic, and we encourage the interested reader to consult this valuable source of historical research to enrich his or her perspective on the income distributions.

3.2.1 Early History

As was already mentioned in Chapter 1, Pareto (1895, 1896) observed a decreasing linear relationship between the logarithm of income and the logarithm of N_x, the number of income receivers with income greater than x, $x \geq x_0$, when analyzing income reported for income tax purposes. Hence, he specified

$$\log N_x = A - \alpha \log x, \tag{3.8}$$

that is,

$$N_x = e^A x^{-\alpha}, \tag{3.9}$$

where $A, \alpha > 0$. Normalizing by the number of income receivers $N := N_{x_0}$, one obtains

$$\frac{N_x}{N} = 1 - F(x) = \left(\frac{x}{x_0}\right)^{-\alpha}, \quad x \geq x_0 > 0. \tag{3.10}$$

Almost immediately, public interest was aroused and other economists began to criticize the idea of a universal form with a single shape parameter permitting inappropriate comparisons between societies. It became clear fairly soon that the

Pareto distribution is only a good approximation of high incomes above a certain threshold and also that the mysterious and some claim notorious α is not always close to 1.5 as Pareto initially believed. Nonetheless, as late as 1941 H. T. Davis considered the value $\alpha = 1.5$ to be a dividing line between egalitarian societies ($\alpha > 1.5$) and inegalitarian ones ($\alpha < 1.5$). [Kakwani (1980b), reminiscent of Mandelbrot (1960), referred to a Pareto distribution with $\alpha = 1.5$ as the *strongest* Pareto law.]

It should be noted that Pareto's discovery was initially met with some resentment by the English and American school (see our biography of Pareto in Appendix A for further details), with notable exceptions such as Stamp (1914) and Bowley (1926). And as late as 1935 Shirras (p. 680) asserted that for Indian income tax and super tax data from the 1910s and 1920s (notably for the year 1929–1930)

> There is indeed no Pareto law. It is time that it should be entirely discarded in studies on the distribution of income.

From the graphical evidence Shirras provided, one is inclined to conclude that if anything, the data are very much in agreement with a Pareto distribution (Adarkar and Sen Gupta, 1936).

In his defense of the Pareto distribution as an appropriate model for personal incomes and wealth, MacGregor (1936) opened with the statement

> Economics has not so many inductive laws that it can afford to lose any.

and asserted that "the law and the name mark a stage in investigation, like Boyle's Law or Darwin's Law, and, although amended, they remain authoritative as first approximations not to be lightly gone back on." One year later Johnson (1937) was able to confirm, for U.S. income tax data for each year for the period 1914–1933, that α does, in fact, not vary substantially, obtaining estimates $\hat{\alpha} \in [1.34, 1.90]$.

In the actuarial literature an early contribution was made by a Norwegian actuary Birger Meidell, who in 1912 employed the Pareto distribution when trying to determine the maximum risk in life insurance. His working hypothesis was that the sums insured are proportional to the incomes of the policy holders, thus incorporating Pareto's pioneering work. The same idea was later expressed by Hagstrœm (1925). In subsequent actuarial investigations the Pareto distribution was more often used in connection with nonlife insurance, notably automobile and fire insurance, and we shall mention several relevant contributions in Section 3.7.

3.2.2 Pareto Income Distribution Derived from the Distribution of Aptitudes

Many (particularly early) writers alluded to a relationship between the distribution of income and the distribution of talents or aptitudes (e.g., Ammon, 1895; Pareto,

1897a; Boissevain, 1939). Rhodes (1944), in a neglected paper, assumed that "talent" is a continuous variable Z, with density $h(z)$, say, with an average income accruing to those with talent z being given by $m(z) = E(X|Z = z)$. His crucial assumption is that the conditional coefficient of variation $\lambda = CV(X|Z = z)$ is constant for all talent groups.

If we write the conditional survival function of income (for those with talent z) in standardized form

$$\bar{F}\left[\frac{x - m(z)}{\lambda m(z)}\right] \tag{3.11}$$

where, by assumption, $\lambda m(z)$ is the conditional standard deviation of income, the unconditional distribution is given by

$$\bar{F}(x) = \int_0^\infty h(z)\bar{F}\left[\frac{x - m(z)}{\lambda m(z)}\right]dz. \tag{3.12}$$

Setting $x - m(z) =: \lambda m(z)v$, we can write

$$m(z) = \frac{x}{1 + \lambda v}. \tag{3.13}$$

This allows us to obtain z as a function of x and v. However, we require the distribution of X. In order to obtain an expression for the c.d.f. of this random variable, we use

$$\frac{\partial m(z)}{\partial v}\frac{dz}{dv} = -\frac{\lambda x}{(1 + \lambda v)^2}. \tag{3.14}$$

Setting $M(v, x) := \partial m(z)/\partial v$ and $P(v, x) := h(z)$, we can now write

$$\int_{v_1}^{v_2} \frac{P(v, x)\bar{F}(v)\lambda x}{M(v, x)(1 + \lambda v)^2}\, dv. \tag{3.15}$$

The new limits of integration are given from (3.13), yielding $m_0 := m(0) = x/(1 + \lambda v_2)$ for the lowest value of z corresponding to no talent. Thus, $v_2 = (x/m_0 - 1)/\lambda$. For simplicity, assume that $m(z) \to \infty$ for $z \to \infty$ ("infinite talent implies infinite income"), which further yields $v_1 = -1/\lambda$. Hence, (3.15) can be rewritten as

$$\int_{-1/\lambda}^{(x/m_0 - 1)/\lambda} \frac{P(v, x)\bar{F}(v)\lambda x}{M(v, x)(1 + \lambda v)^2}\, dv. \tag{3.16}$$

Note that the upper limit is a function of x. This expression is not very tractable without introducing further assumptions. However, if $P(v, x)/M(v, x)$ is separable in the form $\phi(x)\psi(v)$, this integral simplifies to

$$\lambda x\phi(x)\int_{-1/\lambda}^{(x/m_0-1)/\lambda}\frac{\psi(v)\bar{F}(v)}{(1+\lambda v)^2}\,dv =: \lambda x\phi(x)\cdot\chi(x). \tag{3.17}$$

Rhodes then considered the special case where $\chi(x)$ is approximately constant for large x, which implies $\bar{F}(x) \propto c\cdot x\phi(x)$, for some $c > 0$. It remains to determine the form of $\phi(x)$. To this end we introduce $w(z) := M(v, x)/P(v, x)$, yielding $w\cdot\phi\psi = 1$ under the separability assumption. Taking logarithms and differentiating after x, we obtain

$$\frac{w'(z)}{w(z)}\frac{dz}{dx} = -\frac{\phi'(x)}{\phi(x)}.$$

On the other hand, $M(v, x)(dz/dx)(1+\lambda v) = 1$, whereby

$$\frac{w'(z)}{w(z)}\frac{1}{M(v, x)(1+\lambda v)} = -\frac{\phi'(x)}{\phi(x)},$$

or, using $M(v, x) = h(z)\cdot w(z)$, we get

$$\frac{w'(z)}{w^2(z)h(z)}\frac{1}{1+\lambda v} = -\frac{\phi'(x)}{\phi(x)}.$$

Since $-\phi'(x)/\phi(x)$ does not depend on v, a further differentiation with respect to v yields

$$\left[\frac{w'(z)}{w^2(z)h(z)}\right]'\frac{dz}{dv}\frac{1}{1+\lambda v} - \frac{w'(z)}{w^2(z)h(z)}\frac{\lambda}{(1+\lambda v)^2} = 0.$$

In view of (3.13) and (3.14), this can be rearranged in the form

$$\left[\frac{w'(z)}{w^2(z)h(z)}\right]' \bigg/ \left[\frac{w'(z)}{w^2(z)h(z)}\right] = -\frac{M(v, x)}{m(z)}.$$

All the functions in this equation are functions of z (talent) alone. An integration yields

$$\frac{w'(z)}{w^2(z)h(z)} = \frac{a}{m(z)}$$

for some constant a. If we use $M(v, x) = w(z)h(z)$ again, this equals

$$\frac{w'(z)}{w(z)} = \frac{aM(v, x)}{m(z)}.$$

Hence, $w(z) = c \cdot m(z)^a$, for some constant c. Also,

$$\frac{w'(z)}{w^2(z)h(z)} \frac{1}{1 + \lambda v} = -\frac{\phi'(x)}{\phi(x)}$$

yields

$$\frac{\phi'(x)}{\phi(x)} = -\frac{a}{m(z)(1 + \lambda v)} = -\frac{a}{x}.$$

Hence, $\phi(x) = k \cdot x^{-a}$, for some constant k. From the definition $w\phi\psi = 1$ we further obtain $\psi(v) \cdot c \cdot m(z)^a \cdot k \cdot x^{-a} = 1$; therefore,

$$\psi(v) = \frac{1}{ck}\left[\frac{x}{m(z)}\right]^a = \frac{1}{ck}(1 + \lambda v)^a,$$

a function of v alone. Thus, (3.17) is reduced to

$$\chi(x) = \frac{1}{ck}\int_{-1/\lambda}^{(x/m_0-1)/\lambda} \bar{F}(v)(1 + \lambda v)^{-a-2}\, dv,$$

and if this expression is approximately constant for large x, we finally get

$$\bar{F}(x) \approx x\phi(x) = k \cdot x^{-a-1}. \tag{3.18}$$

Thus, the distribution of income is (approximately) a Pareto distribution.

More recently, the idea of explaining size by (unobserved) aptitudes has also been used in the literature on the size distribution of firms. Lucas (1978) presented a model postulating that the observed size distribution is a solution to the problem of how to allocate productive factors among managers of differing abilities so as to maximize output. If "managerial talent" follows a Pareto distribution, the implied size distribution is also of this form in his model.

3.2.3 Markov Processes Leading to the Pareto Distribution

Champernowne (1953) demonstrated that under certain assumptions the stationary income distribution of an appropriately defined Markov process will approximate the Pareto distribution irrespectively of the initial distribution.

Champernowne viewed income determination as a discrete-time Markov chain: Income for the current period—the state of the Markov chain—depends only on one's income for the last period and a random influence. He assumed that there is some minimum income x_0 and that the income intervals defining each state form a geometric (not arithmetic) progression (the limits of class j are higher than limits of class $j-1$ by a certain factor, c, say, rather than a certain absolute amount of income). Thus, a person is in class j if his or her income is between x_0c^{j-1} and x_0c^j. The transition probabilities p_{ij} are defined as the probability of being in class j at time $t+1$ given that one was in class i at time t.

Champernowne required the assumption that the probability of a jump from one income class to another depends only on the width of the jump, but not on the position from which one starts (a form of the law of proportionate effect). In other words, the (time invariant) transition probability p_{ij} is a function of $j-i=k$ only, which is independent of i. If $x_j(t)$ is the number of income earners in the income class j in period t, the process evolves according to

$$x_j(t+1) = \sum_{k=-\infty}^{j} x_{j-k}(t)p_k. \tag{3.19}$$

Champernowne further assumed (his "basic assumption") that transitions are possible only in the range between $-n$ and 1. If the process continues for a long time, the income distribution reaches an equilibrium in which the action of the transition matrix leaves the distribution unchanged. The equilibrium state is thus described by

$$x_j = \sum_{k=-n}^{1} x_{j-k}p_k, \quad j > 0. \tag{3.20}$$

The solution of this difference equation is obtained upon setting $x_j = z^j$, yielding the characteristic equation

$$g(z) := \sum_{k=-n}^{1} z^{1-k}p_k - z = 0. \tag{3.21}$$

This equation has two positive real roots, one of which is clearly unity; in order to ensure that the other will be between zero and 1, Champernowne introduced the *stability assumption*

$$g'(1) = -\sum_{k=-n}^{1} kp_k > 0. \tag{3.22}$$

Since $g(0) = p_1 > 0$, and from the stability condition $g'(1) > 0$, the other root must satisfy

$$0 < b < 1,$$

yielding the required equilibrium distribution of the form

$$x_j = b^j.$$

The total number of incomes is therefore $1/(1 - b)$ and for any other given number of incomes N the equilibrium distribution becomes

$$x_j = N(1 - b)b^j.$$

The number of incomes greater than or equal to $\tilde{x}_j := x_0 c^j$ is therefore

$$N_{\tilde{x}_j} = Nb^j$$

or

$$\log N_{\tilde{x}_j} = \log N + j \log b.$$

Setting $\alpha = -\log b / \log c$ and $\gamma = \log N + \alpha \log x_0$, we finally obtain

$$\log N_{\tilde{x}_j} = \gamma - \alpha \log \tilde{x}_j. \qquad (3.23)$$

This means that the logarithm of the number of incomes exceeding $\tilde{x}_j$ is a linear function of $\log \tilde{x}_j$, thus giving the Pareto distribution in its original form.

An essential feature of the model is the stability condition (3.22), which means that the expectation of possible transitions is always a reduction in income, from whatever amount income one starts with. Steindl (1965) argued that the economic justification of this assumption is implicit in another feature of Champernowne's model: He considered a constant number of incomes and accounted for deaths by assuming that for any income earner who drops out, there is an heir to his or her income. But this means that on changing from an old to a young income earner, there will usually be a considerable drop in income, especially in the case of high incomes. In fact, the proper economic justification for the stability assumption is that the growing dispersion of incomes of a given set of people is counteracted by the limited span of their lives and the predominantly low and relatively uniform income of new entrants.

Champernowne discussed several generalizations of his basic model. If transitions are possible in an extended range between $-n$ and m, $m > 1$, only a distribution asymptotic to a Pareto distribution can be derived. This is still possible if people are allowed to fall into groups (by age or occupation) and movements from one group to another are allowed. However, a Markov process would not yield a stationary distribution unless the transition matrix is constant. It is hard to imagine a society whose institutional framework is so static. Also, a crucial assumption is that the probabilities of advancing or declining are independent of the size of income.

Mandelbrot (1961) constructed a Markov model that approximates a Pareto distribution similarly to Champernowne but does not require a law of proportionate effect (random shocks additive in logarithms). He emphasized weak Pareto laws whose frequency distributions are asymptotic to the Pareto. Total income is a sum of many i.i.d. components (e.g., income in different occupations, incomes from different sources), at least one of which is nonnegligible in size. If the overall income also follows this probability law, we have "stable laws," that is, either normal distributions or a family of Pareto-type laws. To get a normal distribution, one requires that the largest component is negligible in size. Mandelbrot argued that in common economic applications the largest component is not negligible; hence, the sum as well as the limit of properly normalized partial sums can be expected to follow a nonnormal stable law (with a Pareto exponent $\alpha < 2$).

Wold and Whittle (1957) offered a further model (in continuous time) that generates the Pareto distribution, in their case as the distribution of wealth. They assumed that stocks of wealth grow at a compound interest rate during the lifetime of a wealth holder and then divide equally among his or her heirs. Death occurs randomly with the known mortality rate per unit time. Applying the model to wealth above a certain minimum, Wold and Whittle arrived at the Pareto distribution and expressed α as a function of (1) the number of heirs and (2) the ratio of the growth rate of wealth to the mortality rate. [Some 25 years later Walter (1981) derived all solutions of the Wold–Whittle differential equation, showing that not all of them are of the desired Paretian form.]

It is remarkable that the Pareto coefficient is here determined as the ratio of certain growth rates—namely, the ratio of the growth of wealth to the mortality rate of wealth owners—that apparently represent the dissipative and stabilizing tendencies in the process. As noted by Steindl (1965, p. 44), this confirms the intuition of Zipf (1949), who considered the Pareto coefficient the expression of an equilibrium between counteracting forces.

3.2.4 Lydall's Model of Hierarchical Earnings

Lydall (1959) assumed that the people working in an organization or firm are arranged hierarchically and that their salaries reflect the organization of the firm.

Suppose that the levels l_i, $i = 1, 2, \ldots, k$, are numbered from the lowest upward and let x_i be the salary at l_i and y_i the number of employees at that level. Two assumptions are made:

$$\frac{y_i}{y_{i+1}} = n, \quad \text{where } n > 1 \text{ is constant for every } i, \tag{3.24}$$

and

$$\frac{x_{i+1}}{nx_i} = p, \quad \text{where } p < 1 \text{ is constant for all } i. \tag{3.25}$$

These assumptions reflect the fact that the managers on every level supervise a constant number of people on the level below them (3.24), and that the salary of a certain manager is a constant proportion of the aggregate salary of the people whom he or she supervises directly (3.25). It is then natural to assume that $x_{i+1}/x_i = np > 1$. In the highest level there will be one person, in the next level there will be n persons, then n^2, etc. Hence,

$$y_i = n^{k-i}.$$

The total number N_i of persons on levels l_i or above is therefore

$$N_i = 1 + n + n^2 + \cdots + n^{k-i} = \frac{n^{k-i+1} - 1}{n - 1}$$

and the proportion of all employees, Q_i, in the firm who are working on level l_i or above is

$$Q_i = \frac{N_i}{N_1} = \frac{n^{k-i+1} - 1}{n^k - 1} \approx n^{1-i}.$$

From (3.25) we see that

$$x_i = (np)^{i-1} \cdot x_1.$$

Hence,

$$\log Q_i = \frac{\log n}{\log np} \log x_1 - \frac{\log n}{\log np} \log x_i =: \log c - \alpha \log x_i$$

and therefore,

$$Q_i = c \cdot x_i^{-\alpha},$$

which is the Pareto distribution. Here the levels l_i are discrete, but it is easy to cover the continuous case as well (Lydall, 1968, Appendix 4).

3.2.5 Further Approaches

Mandelbrot (1964) derived a Pareto distribution of the amount of fire damage from the assumption that the probability of the fire increasing its intensity at any instant of time is constant. He assumed that the intensity is described by an integer-valued random variable N—there is no fire when $N = 0$, the fire starts when N becomes equal to 1, and it ends when either N becomes equal to zero again or when all that possibly can be destroyed has already been burned. Assume further that, at any instant in time, there is a probability $p = 1/2$ that the fire encounters new material, increasing its intensity by 1, and a probability $q = 1/2$ that the absence of new materials or the actions of dedicated firefighters decreases the fire's intensity by 1. In the absence of a maximum extent of damage and of a lower bound on recorded damages, the duration of the fire will be an even number given by a well-known result (familiar from the context of coin tossing),

$$P(D = x) = 2\binom{1/2}{x/2}(-1)^{x/2-1},$$

which is proportional to $x^{-3/2}$ for reasonably large x. Under the assumption that small damages, below a threshold x_0, say, are not even properly recorded, the extent of damage is then given by

$$P(D > x) = (x/x_0)^{-1/2}, \quad x_0 \leq x,$$

a Pareto distribution with shape parameter $\alpha = 1/2$. [The value of the parameter $\alpha = 1/2$ may seem somewhat extreme here; however, values in the vicinity of 0.5 were found to describe the distribution of fire damage in post-war Sweden (Benckert and Sternberg, 1957).]

Shpilberg (1977) presented a further argument leading to a Pareto distribution as the distribution of fire loss amount. Suppose that the "mortality rate" $\lambda(t)$ of the fire is constant, equal to α, say, so that the duration of the fire T is exponentially distributed. Under the assumption that the resulting damage X is exponentially related to the duration of the fire,

$$X = x_0 \exp(kT),$$

for some x_0, $k > 0$, the c.d.f. of X is given by

$$F(x) = 1 - \left(\frac{x_0}{x}\right)^{\alpha/k}, \quad x \geq x_0,$$

and so it follows a Pareto distribution. (Clearly, different specifications of the hazard (mortality) rate can be used to motivate other loss models, and indeed we shall encounter the Weibull and Benini distributions that can be derived along similar lines.)

3.3 MOMENTS AND OTHER BASIC PROPERTIES

The Pareto density has a polynomial right tail; specifically, it is regularly varying at infinity with index $-\alpha - 1$. Thus, the right tail is heavier as α is smaller, implying that only low-order moments exist. In particular, the kth moment of the Pareto distribution exists only if $k < \alpha$, in that case, it equals

$$E(X^k) = \frac{\alpha x_0^k}{\alpha - k}. \tag{3.26}$$

Specifically, the mean is

$$E(X) = \frac{\alpha x_0}{\alpha - 1} \tag{3.27}$$

and the variance equals

$$\text{var}(X) = \frac{\alpha x_0^2}{\alpha(\alpha - 1)^2(\alpha - 2)}. \tag{3.28}$$

Hence, the coefficient of variation is given by

$$\text{CV} = \frac{1}{\sqrt{\alpha(\alpha - 2)}}, \tag{3.29}$$

and the shape factors are

$$\sqrt{\beta_1} = 2\frac{\alpha + 1}{\alpha - 3}\sqrt{1 - \frac{2}{\alpha}}, \quad \alpha > 3, \tag{3.30}$$

$$\beta_2 = \frac{3(\alpha - 2)(3\alpha^2 + \alpha + 2)}{\alpha(\alpha - 3)(\alpha - 4)}, \quad \alpha > 4. \tag{3.31}$$

It follows that $\sqrt{\beta_1} \to 2$ and $\beta_2 \to 9$, as $\alpha \to \infty$.

For the extremely heavy-tailed members (with $\alpha < 1$) of this class of distributions, other measures of location than the mean must be used. Options include the geometric mean $x_g = \exp\{E(\log X)\}$, here given by

$$x_g = x_0 \exp\left(\frac{1}{\alpha}\right), \tag{3.32}$$

and the harmonic mean $x_h = \{E(X^{-1})\}^{-1}$, which equals

$$x_h = x_0(1 + \alpha^{-1}). \tag{3.33}$$

A comparison with (3.3) shows that the geometric mean is equal to the $(1 - e^{-1})$th quantile. Also, formulas (3.27), (3.32), and (3.33) provide an illustration of the familiar inequalities $E(X) \geq x_g \geq x_h$.

It is easy to see that for $\alpha_1 \leq \alpha_2$ and $x_{01} \geq x_{02}$, the c.d.f.s' of two Par(x_{0i}, α_i) distributions, $i = 1, 2$, do not intersect, so under these conditions Pareto distributions are stochastically ordered, with $X_1 \geq_{\text{FSD}} X_2$. This implies, among other things, that the moments are also ordered (provided they exist), in particular $E(X_1) \geq E(X_2)$.

The density is decreasing; thus, the mode of this distribution is at x_0. Figure 3.1 provides some examples of Pareto densities.

From (3.3), the median is $F^{-1}(0.5) = 2^{1/\alpha}x_0$.

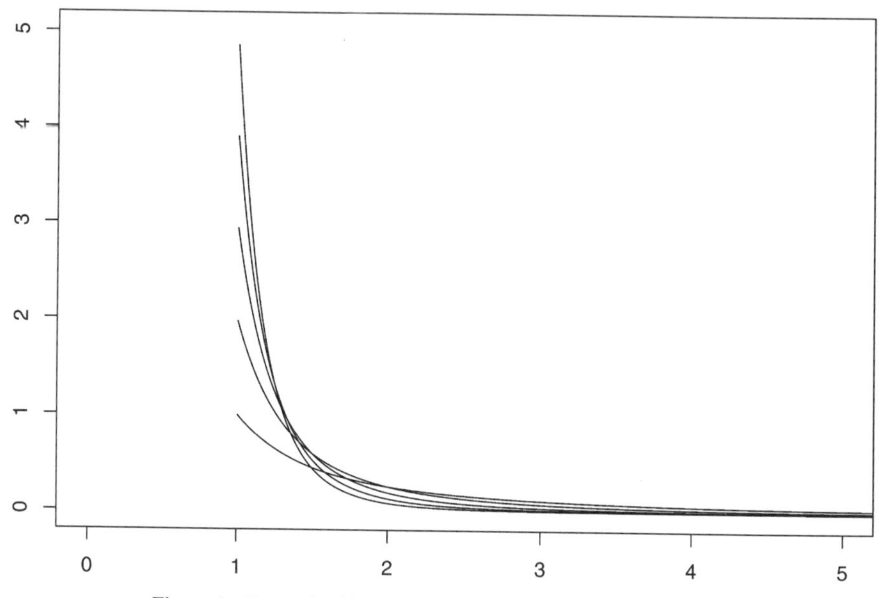

Figure 1 Pareto densities: $x_0 = 1$ and $\alpha = 1(1)5$ (from bottom left).

The Pareto distribution is closed with respect to minimization, in the sense that

$$X \sim \mathrm{Par}(x_0, \alpha) \Longrightarrow X_{1:n} \sim \mathrm{Par}(x_0, n\alpha). \tag{3.34}$$

However, other order statistics do not possess Pareto distributions. In particular, the distribution of the sample maximum is given by

$$F_{n:n}(x) = \left[1 - \left(\frac{x}{x_0}\right)^{-\alpha}\right]^n, \quad 0 < x_0 \leq x. \tag{3.35}$$

Fairly recently, Stoppa (1990a,b) proposed this distribution, for a general $n \in \mathbb{R}$, as a model for the size distribution of personal income. It will be discussed in Section 3.8 below in some detail.

The Pareto distribution is also closed with respect to the formation of moment distributions and with respect to power transformations: if

$$F \sim \mathrm{Par}(x_0, \alpha) \Longrightarrow F_{(k)} \sim \mathrm{Par}(x_0, \alpha - k), \tag{3.36}$$

provided $\alpha < k$, and, for $a > 0$,

$$X \sim \mathrm{Par}(x_0, \alpha) \Longrightarrow X^a \sim \mathrm{Par}\left(x_0^a, \frac{\alpha}{a}\right). \tag{3.37}$$

Also, if X follows a standard Pareto distribution, then $W = X^{-1}$ has the density

$$f(w) = \alpha x_0^{\alpha} w^{\alpha-1}, \quad 0 < w < x_0^{-1}. \tag{3.38}$$

This is a power function distribution and thus a special case of the Pearson type I distribution that will be encountered in Chapter 6.

The hazard rate is given by

$$r(x) = \frac{\alpha}{x}, \quad x > x_0, \tag{3.39}$$

which is monotonically decreasing. Its slow decrease reflects the heavy-tailed nature of the Pareto distribution. The mean excess function is

$$e(x) = \frac{x}{\alpha - 1}, \quad x \geq x_0. \tag{3.40}$$

Thus, the Pareto distribution obeys van der Wijk's law (1.13). (Since the mean excess function characterizes a distribution, the Pareto distribution is, in fact, characterized by van der Wijk's law within the class of continuous distributions.) From (3.39) and (3.40) it follows that the product of the hazard rate and the mean excess function is constant

$$r(x) \cdot e(x) = \frac{\alpha}{\alpha - 1}. \tag{3.41}$$

The distribution theory associated with samples from Pareto distributions is generally somewhat complicated. From general results in, for example, Feller (1971, p. 279), it follows that convolutions of Pareto distributions also exhibit Paretian tail behavior; however, expressions for the resulting distributions are quite involved. Nonetheless, asymptotically the situation is under control, an important probabilistic property of the Pareto distribution being associated with the central limit theorem: properly normalized partial sums of i.i.d. Par(x_0, α) random variables are asymptotically normally distributed only if $\alpha \geq 2$, for $\alpha < 2$ nonnormal stable distributions arise (see, e.g., Zolotarev, 1986). The Pareto distribution is perhaps the simplest distribution with this property.

A basic distributional property of the standard Par(x_0, α) distribution is its close relationship with the exponential distribution

$$X \stackrel{d}{=} x_0 \exp\left(\frac{Y}{\alpha}\right), \tag{3.42}$$

where Y is a standard exponential variable, that is, $f_Y(y) = \exp(-y)$, $y > 0$. The classical Pareto distribution may therefore be viewed as the "log-exponential" distribution. This yields, for example, the following relationship between Pareto and gamma distributions: If X_i, $i = 1, \ldots, n$, are i.i.d. with $X_i \sim \text{Par}(x_0, \alpha)$, then

$$\alpha \log \prod_{i=1}^{n} \left(\frac{X_i}{x_0}\right) = \alpha \sum_{i=1}^{n} \log\left(\frac{X_i}{x_0}\right) = Ga(n, 1). \tag{3.43}$$

More general results on distributions of products and ratios of Pareto variables with possibly different parameters were derived by Pederzoli and Rathie (1980) using Mellin transforms.

In general, the intimate relationship between the Pareto and exponential distributions implies that one can obtain many properties of the former from properties of the latter, in particular characterizations to which we now turn.

3.4 CHARACTERIZATIONS

A large number of characterizations of Pareto distributions are based on the behavior of functions of order statistics, which is a consequence of the vast literature on characterizations of the exponential distribution in terms of these functions. Since we are unaware of meaningful interpretations in connection with size phenomena, we only present the most prominent example to illustrate the flavor of the results.

The example under consideration is

$$X_{i:n}, \ \frac{X_{i+1:n}}{X_{i:n}} \ \text{independent} \Longrightarrow X \sim \text{Pareto}. \tag{3.44}$$

This is essentially an "exponentiated" version of a classical exponential characterization in terms of spacings due to Fisz (1958). (The quantities $X_{i+1:n}/X_{i:n}$ are often referred to as geometric spacings.) Refinements and related results may be found in Galambos and Kotz (1978) and Arnold (1983).

Several characterizations of the Pareto distributions are based on the linearity of the mean excess (or mean residual life) function, although sometimes in heavily disguised form. This essentially goes back to Hagstrœm (1925) and D'Addario (1939), however rigorous proofs are of a more recent date (Arnold, 1971; Huang, 1974).

A closely related class of characterizations can be subsumed under the heading "truncation equivalent to rescaling," a line of research initiated by Bhattacharya (1963). The basic result is that

$$P(X > y \,|\, X > z) = P\left(\frac{z}{x_0} X > y\right), \quad \text{for all } y > z \geq x_0, \tag{3.45}$$

characterizes the $\text{Par}(x_0, \alpha)$ distribution. If one supposes that moments exist, this characterization is reduced to the linearity of the mean excess function as a function of z (e.g., Kotz and Shanbhag, 1980). Note that income distributions arising from income tax statistics are generally of the truncated form, and Lorenz curves and inequality measures are quite often calculated for such data. The preceding characterization implies that only if the entire distribution is of the Pareto form, the truncated distribution can be safely used for inferences about the inequality associated with the entire distribution. Otherwise, reported inequality statistics will often be too low. See Ord, Patil, and Taillie (1983) for an additional discussion of the effect. Further variations on this theme are the truncation invariance of the Gini and generalized entropy indices (Ord, Patil, and Taillie, 1983) and the truncation invariance, both from above and from below, of the Lorenz curve of a two-component Pareto-power function mixture (Moothathu, 1993). (Recall that the power function distribution is the inverse Pareto distribution.)

Several characterizations of the Pareto distribution have been cast in the framework of income underreporting, although it is difficult to believe that the proposed mechanisms describe the actual underreporting process.

Krishnaji (1970) provided widely quoted characterization assuming that reported income Y is related to true (but unobservable) income X through a multiplicative error

$$Y \stackrel{d}{=} RX,$$

where R, X are independent. In order to be meaningful as an underreporting factor, it is clearly necessary that R take values in $[0, 1]$. (In an actuarial context, the appropriate framework would be the overreporting of insurance claims. If we set $Y \stackrel{d}{=} X/R$, the characterizations given below apply with minor modifications.) Assuming that R possesses a power function distribution (a special case of the beta distribution) with the density

$$f(r) = pr^{p-1}, \quad 0 \leq r \leq 1, \tag{3.46}$$

where $p > 0$, Krishnaji obtained the following characterization: If $P(X > x_0) = 1$, for some $x_0 > 0$, and $P(RX > x_0) > 0$ with R, X independent, then

$$P(RX > y \,|\, RX > x_0) = P(X > y), \quad y > x_0, \tag{3.47}$$

if and only if $X \sim \text{Par}(x_0, \alpha)$. However, there is nothing special about the distribution (3.46) of R. Fosam and Sapatinas (1995) have shown that the result holds true for any R supported on (a subset of) $(0, 1)$ such that the distribution of $\log R$ is non-arithmetic. Indeed, (3.47) may be rewritten as

$$P\left\{R\left(\frac{X}{x_0}\right) > y^* \middle| R\left(\frac{X}{x_0}\right) > 1\right\} = P\left\{\frac{X}{x_0} > y^*\right\}, \quad y^* = \frac{y}{x_0} > 1,$$

which is seen to be equivalent to

$$P\left\{\log\left(\frac{X}{x_0}\right) > -\log R + x \,\middle|\, \log\left(\frac{X}{x_0}\right) > -\log R\right\} = P\left\{\log\left(\frac{X}{x_0}\right) > x\right\},$$

$$x = \log y^* > 0.$$

The conclusion follows therefore directly from the (strong) lack of memory property of the exponential distribution; hence, this characterization is a consequence of (3.42).

The result remains unaffected if condition (3.47) is replaced by the condition

$$E(Y - x \mid Y > x) = E(X - x \mid X > x), \quad x > x_0 > 0, \tag{3.48}$$

with $E(X^+) < \infty$, as follows from Kotz and Shanbhag (1980).

An interesting alternative characterization in the framework described above is as follows. If there exists a random variable Z such that the regression $E(Z \mid X = y)$ is linear, then, under some smoothness conditions, $E(Z \mid RX = y)$ is also linear if and only if X follows a Pareto distribution (Krishnaji, 1970). The result was extended by Dimaki and Xekalaki (1990) and further generalized by Fosam and Sapatinas (1995) who weakened (3.46) to $R \sim \text{beta}(p, q)$, $q \in \mathbb{N}$, requiring only $E(Z \mid X = y) = \delta + \beta x^\alpha$, for some positive α.

Revankar, Hartley, and Pagano (1974) provided a second characterization in terms of underreported incomes. However, in contrast to Krishnaji's approach utilizing a multiplicative reporting error, they postulated an additive relation between true and reported income. Let the random variables X, Y, and U denote the actual (unobserved) income, reported income, and the underreporting error, respectively, and define

$$Y = X - U, \quad \text{where } 0 < U < \max\{0, X - m\},$$

where m is the tax-exempt level. Under the assumptions that (1) the average amount of underreported income from a given $X = x > m$ is proportional to $x - m$, that is,

$$E(U \mid X = x) = b(x - m) = a + bx, \tag{3.49}$$

$0 < b < 1$ and $a = -bm$, and (2) $E(X) < \infty$, it follows that

$$E(U \mid X > y) = \alpha + \beta y, \tag{3.50}$$

with $\beta > b > 0$, if and only if X follows a Pareto distribution with a finite first moment. In particular, for $a = \alpha$ the $\text{Par}(x_0, \alpha)$ distribution is obtained. [For $a < \alpha$ we obtain a Pareto (II) distribution, while $\beta = b$ yields a characterization of the exponential distribution. The case $\beta < b$ was investigated by Stoppa (1989).]

The classical Pareto distribution is also the maximum entropy density on $[x_0, \infty)$ subject to the constraint of a fixed geometric mean (e.g., Näslund, 1977; Kapur, 1989, p. 56). This may also be considered a restatement of the corresponding exponential characterization: The exponential distribution is the maximum entropy distribution on $[0, \infty)$ when the first moment is prescribed (Kapur, 1989, p. 56). Thus, the Pareto result follows from (3.42) together with (3.32).

Nair and Hitha (1990) provided a characterization in terms of the "equilibrium distribution" defined via the p.d.f. $f_Z(x) = 1/E(X)\{1 - F_X(x)\}$, a concept of special significance in renewal theory. They showed that the condition $E(X) = kE(Z)$, for some $k > 1$, characterizes the Pareto distribution within a subclass of the one-parameter exponential family specified by $f(x) = u(\theta)v(x)\exp\{-\theta \log x\}$, where θ lies in an open interval on the positive halfline. They also showed that, for a distribution with a finite mean, the condition $F_{(1)}(x) = F(x)$ for all $x > x_0 > 0$ [where $F_{(1)}$ denotes the c.d.f. of the first-moment distribution] characterizes the Pareto distribution.

Moothathu (1990b) obtained the following characterization in terms of the independence of certain random variables. For a fixed integer $k \geq 2$, consider the i.i.d. random variables $Z_1, \ldots, Z_k$, with $P(Z_1 > 1) = 1$ and let $Y = (Y_1, \ldots, Y_{k-1})$ be a further vector of random variables supported on $\mathbb{R}_+^{k-1}$, independent of the Z's. Set $T = 1 + Y_1 + \cdots + Y_{k-1}$ and define

$$V = \min\left\{Z_1^T, Z_2^{T/Y_1}, \ldots, Z_k^{T/Y_{k-1}}\right\}.$$

Now the random variable Z_1 follows a Pareto distribution if and only if V and Y are independent. (The result follows from a characterization of the Weibull distribution obtained in the same paper.) This result has potential applications in that it enables one to treat a problem of testing k-sample homogeneity as a problem of testing independence.

3.5 LORENZ CURVE AND INEQUALITY MEASURES

The Lorenz curve, which exists whenever $\alpha > 1$, is given by

$$L(u) = 1 - (1 - u)^{1-1/\alpha}, \quad 0 < u < 1. \tag{3.51}$$

As was already mentioned in the preceding chapter, it follows that Pareto Lorenz curves never intersect and that, for $X_i \sim \text{Par}(x_0, \alpha)$,

$$X_1 \geq_L X_2 \iff \alpha_1 \leq \alpha_2, \tag{3.52}$$

provided $\alpha_i > 1$, $i = 1, 2$. There is an interesting alternative but less direct argument leading to this result. Arnold et al. (1987) observed that every distribution F corresponding to an unbounded random variable and possessing a strongly unimodal density generates an ordered family of Lorenz curves via $L_\tau(u) = F(F^{-1}(u) - \tau)$, $\tau \geq 0$. The Pareto distribution admits such a representation, the generating distribution being the (Gumbel-type) extreme value distribution.

The Gini coefficient of the Pareto distribution is

$$G = \frac{1}{2\alpha - 1}, \tag{3.53}$$

and the generalized Gini coefficients are given by (Kleiber and Kotz, 2002)

$$G_\nu = \frac{\nu - 1}{\nu\alpha - 1}, \quad \nu \geq 2. \tag{3.54}$$

Hence for $\alpha = 1.5$, the value originally obtained by Pareto for most of his data, we have $G = 0.5$.

The Pietra index equals

$$P = \frac{(\alpha - 1)^{-1}}{\alpha^\alpha}. \tag{3.55}$$

and the Theil coefficient is

$$T_1 = \frac{1}{\alpha - 1} - \log\left(\frac{\alpha}{\alpha - 1}\right). \tag{3.56}$$

All of these expressions are decreasing with increasing α, showing that the parameter α, or rather its inverse, may be considered a measure of inequality.

In the Italian literature the Zenga curve and inequality measures derived from it have also been considered. As was mentioned in the preceding chapter, the Zenga curve is given by (Zenga, 1984)

$$Z(u) = 1 - (1 - u)^{1/[(\alpha-1)\alpha]}, \quad 0 < u < 1, \tag{3.57}$$

and the two Zenga coefficients are (Zenga, 1984)

$$\xi = \int_0^1 Z(u)\, du = \frac{1}{1 + \alpha(\alpha - 1)} \tag{3.58}$$

and (Pollastri, 1987a)

$$\xi_2 = 1 - \exp\left\{\frac{-1}{\alpha(\alpha - 1)}\right\}. \tag{3.59}$$

This shows that the Zenga curve of the Pareto distribution is an increasing function on [0, 1], approaching the x axis with increasing α, and that the two coefficients are decreasing as α increases.

3.6 ESTIMATION

The estimation of Pareto characteristics is covered in depth in Arnold (1983) and Johnson, Kotz, and Balakrishnan (1994). Here we shall only include the classical regression-type estimators, ML estimation, and several recent developments, notably in connection with UMVU estimation. For the method of moments, quantile and Bayes estimators, as well as methods based on order statistics, we refer the interested reader to the above-mentioned works.

3.6.1 Regression Estimators

Since the Pareto distribution was originally discovered as the distribution whose survival function is linear in a double-logarithmic plot, the Pareto diagram, it is not surprising that the regression estimators of its parameters have been used from the very beginning. The least-squares estimators

$$\hat{\alpha}_{\mathrm{LS}} = \frac{-n\sum_{i=1}^{n}\log X_i \cdot \log \bar{F}(X_i) + \sum_{i=1}^{n}\log X_i \cdot \sum_{i=1}^{n}\log \bar{F}(X_i)}{n\sum_{i=1}^{n}(\log X_i)^2 - \left(\sum_{i=1}^{n}\log X_i\right)^2} \tag{3.60}$$

and $\hat{x}_0$, defined by

$$\overline{\log \bar{F}(X)} = \hat{\alpha}_{\mathrm{LS}} \cdot \log \hat{x}_0 - \hat{\alpha}_{\mathrm{LS}} \cdot \overline{\log (X)}$$

(where a bar denotes averaging), are still quite popular in applied work; Quandt (1966b) has shown that they are consistent.

Early writers such as Pareto (1896, 1897a,b), Benini (1897), or Gini (1909a) employed Cauchy regression (e.g., Linnik, 1961), a method that is seldom used nowadays. Interestingly, in a fairly recent small Monte Carlo study Pollastri (1990) found that Cauchy's method is often slightly better, in terms of MSE, than least-squares regression. More recently, Hossain and Zimmer (2000) recommended that the least-squares estimators be generally preferred over the maximum likelihood and related estimators for estimating x_0, and also for estimating α for small values of the parameter ($\alpha \leq 4$).

3.6.2 Maximum Likelihood Estimation

The likelihood for a sample from a Pareto distribution is

$$L = \prod_{j=1}^{n} \frac{\alpha x_0^{\alpha}}{x_j^{\alpha+1}}. \tag{3.61}$$

This yields the MLE of α

$$\hat{\alpha} = n\left[\sum_{j=1}^{n} \log\left(\frac{X_j}{\hat{x}_0}\right)\right]^{-1}, \tag{3.62}$$

whereas for the threshold x_0 the estimator is given by

$$x_0 = X_{1:n}. \tag{3.63}$$

It should be noted that, since $x_0 \le x$ and $\hat{x}_0$, $\hat{x}_0$ overestimates x_0. If this estimator is used to solve for $\hat{\alpha}$ in (3.62), it is seen that α is also overestimated.

Some direct calculations show that

$$E(\hat{\alpha}) = \frac{n\alpha}{n-2}, \quad n > 2,$$

$$\text{var}(\hat{\alpha}) = \frac{n^2\alpha^2}{(n-2)^2(n-3)}, \quad n > 3,$$

yielding

$$\text{MSE}(\hat{\alpha}) = \frac{\alpha^2(n^2+4n-12)}{(n-2)^2(n-3)}, \quad n > 3.$$

The corresponding results for $\hat{x}_0$ are

$$E(\hat{x}_0) = \frac{nx_0\alpha}{n\alpha - 1}, \quad n > \frac{1}{\alpha},$$

$$\text{var}(\hat{x}_0) = \frac{nx_0\alpha^2}{(n\alpha-1)^2(n\alpha-2)}, \quad n > \frac{2}{\alpha},$$

yielding

$$\text{MSE}(\hat{x}_0) = \frac{2x_0^2}{(n\alpha-1)(n\alpha-2)}, \quad n > \frac{2}{\alpha}.$$

Both estimators are consistent (Quandt, 1966b). Since $1/\hat{\alpha}$ is asymptotically efficient in the exponential case, the same is true of $\hat{\alpha}$. On the other hand, $n(X_{1:n} - x_0)$ follows asymptotically an Exp$(0, x_0/\alpha)$ distribution and hence is biased. Saksena and Johnson (1984) showed that the maximum likelihood estimators are jointly complete.

If the parameter α is known, then $T = X_{1:n}$ will be the complete sufficient statistic with p.d.f.

$$f(t) = \frac{n\alpha}{t^{n\alpha+1}} x_0^{n\alpha}, \quad x_0 \le t, \tag{3.64}$$

a Par$(n\alpha, x_0)$ distribution [cf. (3.34)].

The MLE for the shape parameter α, for known scale x_0, equals

$$\hat{\alpha} = \frac{n}{\sum_{i=1}^{n} \log(X_j/x_0)}.$$

This estimator is also complete and sufficient. Incidentally, its distribution has the p.d.f.

$$f(x) = \frac{(\alpha n)^n}{\Gamma(n)} x^{-(n+1)} e^{-(\alpha n)/x}, \quad x > 0, \tag{3.65}$$

which can be recognized as the density of a Vinci (1921) distribution that is discussed in greater detail in Chapter 5. It should be noted that the situation where x_0 is known is not uncommon in actuarial applications, where, for example, in reinsurance the reinsurer is only concerned with losses above a certain predetermined level, the retention level.

If the parameters α, x_0 are both unknown, then the complete sufficient statistic is (T, S), where

$$T = X_{1:n}, \quad S = \sum_{i=1}^{n} \log\left(\frac{X_i}{X_{1:n}}\right), \tag{3.66}$$

which has the density function

$$f(s, t) = \frac{n\alpha^n x_0^{n\alpha}}{\Gamma(n-1)} \cdot \frac{s^{n-2} e^{-\alpha s}}{t^{n\alpha+1}}, \qquad x_0 \le t, 0 \le s, \tag{3.67}$$

showing that $\hat{\alpha}$ and $\hat{x}_0$ are mutually independent. [This is also a direct consequence of the independence of exponential spacings, the property underlying characterization (3.44).] Hence, the marginal density of $\hat{\alpha}$ is now given by

$$f(x) = \frac{(\alpha n)^{n-1}}{\Gamma(n-1)} x^{-n} e^{-\alpha n/x}, \quad x > 0, \tag{3.68}$$

again an inverse gamma (Vinci) distribution.

In view of the importance of grouped data in connection with size distributions, we briefly mention relevant work pertaining to the classical Pareto distribution. Maximum likelihood estimation from grouped data when x_0 is known was studied by Fisk (1961b) and Aigner and Goldberger (1970). Let $x_0 = 1 < x_1 < x_2 < \cdots < x_{k+1} = \infty$ be the boundaries of the $k+1$ groups and denote the number of observations, in a random sample of size n, falling into the interval $[x_i, x_{i+1})$ by n_i ($\sum_{i=0}^{k} n_i = n$). The MLE of α is then the solution of

$$\sum_{i=0}^{k-1} n_i \left(\frac{x_{i+1}^{-\alpha} \log x_{i+1} - x_i^{-\alpha} \log x_i}{x_{i+1}^{-\alpha} - x_i^{-\alpha}} \right) + n_k \log x_k = 0. \tag{3.69}$$

It is worth noting that if the group boundaries form a geometric progression, with $x_{i+1} = cx_i$, $i = 1, \ldots, k$, then the MLE can be expressed in closed form, namely,

$$\hat{\alpha} = \frac{1}{\log c} \log\left(1 + \frac{n}{\sum_{i=0}^{k} in_i}\right). \tag{3.70}$$

3.6.3 Optimal Grouping

The prevalence of grouped data in connection with the distribution of income or wealth leads naturally to the question of optimal groupings. Only fairly recently, Schader and Schmid (1986) have addressed this problem in a likelihood framework.

Suppose that a sample $X = (X_1, \ldots, X_n)^\top$ of size n is available and the parameter of interest is α. (Clearly, this is the relevant parameter in connection with inequality measurement as the formulas in Section 3.5 indicate.) By independence, the Fisher information of the sample on α is

$$I(\alpha) = nI_1(\alpha) = \frac{n}{\alpha^2},$$

where I_1 denotes the information in a single observation. For a given number of groups k with group boundaries $X_0 = a_0 < a_1 < \cdots < a_{k-1} < a_k = \infty$, define the class frequencies N_j as the number of X_i in $[a_{j-1}, a_j)$, $j = 1, \ldots, k$. Thus, the joint distribution of $N = (N_1, \ldots, N_k)^\top$ is multinomial with parameters n and $p_j = p_j(\alpha)$, where

$$p_j(\alpha) = \int_{a_{j-1}}^{a_j} f(x \mid \alpha)\, dx, \quad j = 1, \ldots, k.$$

Now the Fisher information in N is

$$I_N(\alpha) = n \sum_{i=1}^{k} \frac{[\partial p_j(\alpha)/\partial\alpha]^2}{p_j(\alpha)} = \frac{n}{\alpha^2} \sum_{i=1}^{k} \frac{[\log(z_j)z_j - \log(z_{j-1})z_{j-1}]^2}{z_{j-1} - z_j},$$

where $z_0 = 1$, $z_k = 0$, $z_j = (a_j/x_0)^{-\alpha}$, $j = 1, \ldots, k-1$. This expression is, for fixed α, a function of k and the $k-1$ class boundaries $a_1, \ldots, a_{k-1}$.

Passing from the complete data X to the class frequencies N implies a loss of information that may be expressed in terms of the decomposition

$$I_X(\alpha) = I_N(\alpha) + I_{X|N}(\alpha).$$

The relative loss of information is then given by

$$L = 1 - \alpha^2 I_N(\alpha).$$

L is a function of k, α, and $a_1, \ldots, a_{k-1}$, but not a function of n. Given k and α, the loss of information is now minimized if and only if the class boundaries $a_1^*, \ldots, a_{k-1}^*$ are defined by

$$I_N(\alpha; a_1^*, \ldots, a_{k-1}^*) = \sup_{a_1^* < \cdots < a_{k-1}^*} I_N(\alpha; a_1, \ldots, a_{k-1}).$$

The boundaries $a_1^*, \ldots, a_{k-1}^*$ are now called optimal class boundaries and the corresponding intervals $[a_{j-1}^*, a_j^*), j = 1, \ldots, k$, are called an optimal grouping. By using these boundaries, the number of classes k may now be determined in such a way that the loss of information due to grouping does not exceed a given bound, γ, say. Thus, one requires, for given $\gamma \in (0, 1)$ and α, the smallest integer k^* such that

$$L(\alpha; k^*; a_1^*, \ldots, a_{k-1}^*) \leq \gamma.$$

The relevant k can now be found by determining the boundaries and the information loss for $k = 1, 2, \ldots$ until for the first time L is less than or equal to γ. In practice, the parameter α will remain unknown and have to be replaced by an estimate. Schader and Schmid (1986) argued that, although the corresponding class boundaries will not be optimal in that case, the loss of information will often be less than using ad hoc determined class boundaries.

Table 3.1 provides optimal class boundaries z_j^* based on the least number of classes k^* for which the loss of information is less than or equal to a given value of γ, for $\gamma = 0.1$, 0.05, 0.025, and 0.01. From the table, optimal class boundaries a_j^* for a Pareto distribution with parameters α and x_0 can be obtained by setting $a_j^* = x_0 z_j^{*-1/\alpha}$.

3.6.4 Unbiased Estimation of Pareto Characteristics

Over the last 10–15 years there has been considerable interest in the UMVU estimation of various Pareto characteristics, notably the parameters, the density, the c.d.f., the moments, and several inequality indices.

An improvement over the MLEs is obtained by removing their biases. In the case where both parameters are unknown, this amounts to replacing $\hat{x}_0$ by the estimator

$$x_0^* = X_{1:n}\left\{1 - \frac{1}{(n-1)\hat{\alpha}}\right\}, \tag{3.71}$$

Table 3.1 Optimal Class Boundaries for Pareto Data

γ	k^*	$z_1^*, \ldots, z_{k^*-1}^*$
0.1	5	0.5486 0.2581 0.0933 0.0190
0.05	7	0.6521 0.3958 0.2171 0.1021 0.0369 0.0075
0.025	9	0.7173 0.4935 0.3218 0.1953 0.1071 0.0504 0.0182
0.01	15	0.8192 0.6616 0.5256 0.4097 0.3122 0.2315 0.1660
		0.1142 0.0745 0.0452 0.0248 0.0117 0.0042 0.0009

Source: Schader and Schmid (1986).

and replacing $\hat{\alpha}$ by

$$\alpha^* = \left(1 - \frac{2}{n}\right)\hat{\alpha}. \tag{3.72}$$

These are, in fact, the UMVU estimators of the parameters. This is a direct consequence of the fact that $(\hat{\alpha}, \hat{x}_0)$ is a sufficient statistic and (α^*, x_0^*) is a function of $(\hat{\alpha}, \hat{x}_0)$. These estimators were obtained by Likeš (1969), with simplified derivations later given by Baxter (1980).

The variances of the UMVUEs are found to be

$$\operatorname{var}(\alpha^*) = \frac{\alpha^2}{n-3}, \quad n > 3, \tag{3.73}$$

and

$$\operatorname{var}(x_0^*) = \frac{x_0^2}{\alpha(n-1)(\alpha n - 2)}, \quad n > \frac{2}{\alpha}. \tag{3.74}$$

Equation 3.73 shows that the UMVUE almost attains the Cramér–Rao bound for α, which is α^2/n.

Being a rescaled ML estimator, the UMVUE of α evidently also follows an inverse gamma distribution. In addition, note that, although x_0 is necessarily positive, x_0^* is negative if $\hat{\alpha} < 1/(n-1)$, that is, if $2n\alpha/\hat{\alpha} > 2n(n-1)\alpha$. Since $2n\alpha/\hat{\alpha}$ follows an inverse gamma distribution, this occurs with nonzero probability. However, this probability appears to be quite small in practice.

If x_0 is known, the UMVUE of α is

$$\alpha^* = \left(1 - \frac{1}{n}\right)\hat{\alpha},$$

with variance

$$\operatorname{var}(\alpha^*) = \frac{\alpha^2}{n-2}, \quad n > 2.$$

Similarly, for known α the UMVUE of x_0 is

$$x_0^* = X_{1:n}\left\{1 - \frac{1}{n\alpha}\right\},$$

with variance

$$\operatorname{var}(x_0^*) = \frac{x_0^2}{\alpha n(\alpha n - 2)}, \quad n > \frac{2}{\alpha}.$$

Moothathu (1986) has studied the UMVU estimation of the quantiles, the mean, as well as the geometric and harmonic means, for both known and unknown x_0. [For known x_0, the estimators of the moments, the geometric mean, and the median were derived by Kern (1983) somewhat earlier.] The estimators are conveniently expressed in terms of various special functions.

The estimators of the mean are given by

$$x_0 \, {}_1F_1(1; n; nS_1) \quad \text{if } x_0 \text{ is known} \tag{3.75}$$

$$X_{1:n} \, {}_1F_1(1; n-1; nS_2) - \frac{S_2}{n-1} \, {}_1F_1(1; n-1; nS_2) \quad \text{if } x_0 \text{ is unknown,} \tag{3.76}$$

where ${}_1F_1$ is Kummer's confluent hypergeometric function.

Here

$$S_1 = \frac{1}{n} \sum_{i=1}^{n} \log\left(\frac{X_i}{x_0}\right), \tag{3.77}$$

$$S_2 = \frac{1}{n} \sum_{i=1}^{n} \log\left(\frac{X_i}{X_{1:n}}\right). \tag{3.78}$$

The estimators for the uth quantile are

$$x_0 \, {}_0F_1[-; n; -nS_1 \log(1-u)] \quad \text{if } x_0 \text{ is known,} \tag{3.79}$$

$$X_{1:n} \Big\{ {}_0F_1[-; n-1; -nS_2 \log(1-u)] - \frac{S_2}{n-1} \, {}_0F_1[-; n; -nS_2 \log(1-u)] \Big\} \quad \text{if } x_0 \text{ is unknown,} \tag{3.80}$$

where ${}_0F_1$ is a Bessel function. (See p. 288 for a definition of ${}_pF_q$.)

Being equal to the $(1 - 1/e)$th quantile, the geometric mean is therefore estimated by

$$x_0 \, {}_0F_1(-; n; nS_1) \quad \text{if } x_0 \text{ is known,} \tag{3.81}$$

$$X_{1:n} \Big\{ {}_0F_1(-; n-1; nS_2) - \frac{S_2}{n-1} \, {}_0F_1(-; n; nS_2) \Big\} \quad \text{if } x_0 \text{ is unknown.} \tag{3.82}$$

In the case of the harmonic mean, the corresponding estimators are

$$x_0(1 + S_1) \quad \text{if } x_0 \text{ is known,} \tag{3.83}$$

$$X_{1:n} \, {}_1F_1(-1; n-1; -nS_2) - \frac{S_2}{n-1} \, {}_1F_1(-1; n; -nS_2) \quad \text{if } x_0 \text{ is unknown.} \tag{3.84}$$

All of the previously given estimators are moreover strongly consistent.

Further results are available for the coefficient of variation, for which the UMVUEs are (Moothathu, 1988)

$$S_1 \, {}_1F_1\left(\frac{1}{2}; n+1; 2nS_1\right) \quad \text{if } x_0 \text{ is known,} \tag{3.85}$$

$$S_2 \, {}_1F_1\left(\frac{1}{2}; n; 2nS_2\right) \quad \text{if } x_0 \text{ is unknown,} \tag{3.86}$$

the mode (Moothathu, 1986), the skewness and kurtosis coefficients (Moothathu, 1988), the p.d.f. and c.d.f. (Asrabadi, 1990), and the mean excess function (Rytgaard, 1990).

Independently, Woo and Kang (1990) obtained the UMVUEs for a whole class of functions of the two Pareto parameters, namely, $x_0^p \alpha^q (\alpha - r)^s$, such that $n - s - q - 1 > 0$ and $n\alpha > p$, along with their variances. This clearly includes the parameters themselves as special cases, as well as the moments, the harmonic mean, the mode, the coefficient of variation, the mean excess function, and the Gini index (see Section 3.6.7 below), but not the quantiles, the geometric mean, and the skewness and kurtosis coefficients. The resulting family of estimators can be expressed in terms of a difference in the confluent hypergeometric function and the variances in terms of a bivariate hypergeometric function.

3.6.5 Robust Estimation

A well-known problem with ML estimators (and indeed many classical estimators) is that they are very sensitive to extreme observations and model deviations such as gross errors in the data. Victoria-Feser (1993) and Victoria-Feser and Ronchetti (1994) proposed robust alternatives to ML estimation in the context of income distribution models. Following Hampel et al. (1986), they assessed the robustness of a statistic $T_n = T_n(x_1, \ldots, x_n)$ in terms of the influence function. In order to define this function, it is convenient to consider T_n as a functional of the empirical distribution function

$$F_n(x) = \frac{1}{n} \sum_{i=1}^{n} \delta_{x_i}(x),$$

where δ_x denotes a point mass in x. If we write $T(F_n) := T_n(x_1, \ldots, x_n)$, the influence function (IF) at the (parametric) model F_θ, $\theta \in \Theta \subseteq \mathbb{R}^k$, is defined by the population counterpart of $T(F_n)$, namely, $T(F_\theta)$, as

$$\text{IF}(x; T; F_\theta) = \lim_{\epsilon \to 0} \frac{T[(1-\epsilon)F_\theta + \epsilon\delta_x] - T(F_\theta)}{\epsilon}, \tag{3.87}$$

that is, as the directional derivative of T at F_θ in the direction of δ_x.

The IF describes the effect of a small contamination (namely, $\epsilon\delta_x$) at a point x on the functional/estimate, standardized by the mass ϵ of the contamination. Hence, the linear approximation $\epsilon\text{IF}(x; T; F_\theta)$ measures the asymptotic bias of the estimator caused by the contamination.

In the case of the ML estimator, the IF is proportional to the score function $s(x; \theta) = (\partial/\partial\theta) \log f(x; \theta)$. In the Pareto case we have

$$s(x; \alpha) = \frac{1}{\alpha} - \log x + \log x_0,$$

which is seen to be unbounded in x. Thus, a single point can carry the MLE arbitrarily far. (This is also the case for most other size distributions.) Clearly, a desirable robustness property for an estimator is a bounded IF. Estimators possessing this property are referred to as bias-robust (or, more concisely, B-robust) estimators. An optimal B-robust estimator (OBRE) as defined by Hampel et al. (1986) belongs to the class of M estimators, that is, it is a solution T_n of the system of equations

$$\sum_{i=1}^{n} \psi(x_i; T_n) = 0$$

for some function ψ.

The OBRE is optimal in the sense that it is the M estimator that minimizes the trace of the asymptotic covariance matrix under the constraint that it has a bounded influence function. There are several variants of this estimator depending on the way one chooses to bound the IF. Victoria-Feser and Ronchetti (1994) employed the so-called standardized OBRE. For a given bound, c, say, on the IF, it is defined implicitly by

$$\sum_{i=1}^{n} \psi(x_i; T_n) = \sum_{i=1}^{n} \{s(x_i; \theta) - a(\theta)\} W_c(x_i; \theta) = 0,$$

where

$$W_c(x_i; \theta) = \min\left\{1; \frac{c}{\|A(\theta)\{s(x; \theta) - a(\theta)\}\|}\right\}.$$

Here the $k \times k$ matrix $A(\theta)$ and $k \times 1$ vector $a(\theta)$ are defined implicitly by

$$E\{\psi(x; \theta)\psi(x; \theta)^\top\} = \{A(\theta)A(\theta)^\top\}^{-1}$$

and

$$E\psi(x; \theta) = 0.$$

The idea behind the OBRE is to have an estimator that is as similar as possible to the ML estimator for the bulk of the data (for efficiency reasons) and therefore to use the score as its ψ function for those values and to truncate the score if a certain bound c is exceeded (for robustness reasons). The constant c can be considered the regulator between robustness and efficiency: for small c the estimator is quite robust but loses efficiency relative to the MLE, and vice versa for large c. The matrix $A(\theta)$ and vector $\psi(x; \theta)$ can be considered the Lagrange multipliers for the constraints resulting from a bounded IF and the condition of Fisher consistency, $T(F_\theta) = \theta$.

Victoria-Feser and Ronchetti suggested using a c so that 95% efficiency at the model is achieved. In the Pareto case this means that $c = 3$ should be employed. For computational purposes they recommend an algorithm based on the Newton–Raphson method using the MLE, a trimmed moment estimate, or a less robust OBRE (large c) as the initial value. We refer the interested reader to Victoria-Feser and Ronchetti (1994) for further algorithmic details.

Also aiming at a favorable tradeoff between efficiency and robustness, Brazauskas and Serfling (2001a) proposed a new class of estimators called the generalized median (GM) estimators. These are defined by taking the median of the evaluations $h(X_1, \ldots, X_n)$ of a given kernel $h(x_1, \ldots, x_n)$ over all subsets of observations taken k at a time [of which there are $\binom{n}{k}$]. Specifically, in the case of the Pareto parameter α,

$$\hat{\alpha}_{\mathrm{GM}} = \mathrm{med}\{h(X_{i_1}, \ldots, X_{i_k})\}, \tag{3.88}$$

where $\{i_1, \ldots, i_k\}$ is a set of distinct indices from $\{1, \ldots, n\}$, with two particular choices of kernel $h(x_1, \ldots, x_n)$:

$$h^{(1)}(x_1, \ldots, x_n) = \frac{1}{C_k} \frac{1}{k^{-1} \sum_{j=1}^{k} \log x_j - \log X_{1:k}} \tag{3.89}$$

and

$$h^{(2)}(x_1, \ldots, x_n; X_{1:n}) = \frac{1}{C_{n,k}} \frac{1}{k^{-1} \sum_{j=1}^{k} \log x_j - \log X_{1:n}}. \tag{3.90}$$

Here C_k and $C_{n,k}$ are median unbiasing factors, chosen in order to assure that in each case the distribution of $h^{(j)}(X_{i_1}, \ldots, X_{i_k}), j = 1, 2$, has median α. For $n = 50$, 100, and 200, Brazauskas and Serfling provided the approximation $C_{n,k} \approx k/[k(1 - 1/n) - 1/3]$. Note that the kernel $h^{(1)}$ can be viewed as providing the MLE based on a particular subsample and thus inherits the efficiency properties of the MLE in extracting the information about α pertaining to that sample. $h^{(2)}$ is a modification that always employs the minimum of the whole sample instead of the minimum of the particular subsample.

In Brazauskas and Serfling (2001b), these estimators are compared to several estimators of trimmed mean and quantile type with respect to efficiency-robustness tradeoffs. Efficiency criteria are exact and asymptotic relative MSEs with respect to the MLE, robustness criterion is the breakdown point, with upper outliers receiving special attention. It turns out that the GM types dominate all trimmed mean and quantile-type estimators for all sample sizes under consideration. For instance, if $n \geq 25$ and protection against 10% upper outliers is considered sufficient, $\hat{\alpha}_{\mathrm{GM}}^{(2)}$ with $k = 4$ yields a relative efficiency of more than 0.89 and a breakdown point ≥ 0.12.

3.6.6 Miscellaneous Estimators

Kang and Cho (1997) derived the MSE-optimal estimator of the shape parameter α, for unknown scale x_0, within the class of estimators of the form $c/\log(X_j/\hat{x}_0)$ [note that both the MLE (3.62) and UMVUE (3.72) are of this form] obtaining

$$\tilde{\alpha} = \frac{n-3}{\sum_{i=1}^{n} \log(X_i/\hat{x}_0)} \tag{3.91}$$

with an MSE of

$$\mathrm{MSE}(\tilde{\alpha}) = \frac{\alpha^2}{n-2}, \quad n > 2. \tag{3.92}$$

They referred to this estimator as the minimum risk estimator (MRE). Compared to the MLE (3.62), it not only has smaller MSE but a smaller bias as well

$$\mathrm{bias}(\tilde{\alpha}) = -\frac{\alpha}{n-2} \tag{3.93}$$

(half of the bias of the MLE, in absolute value). Kang and Cho also studied jackknifed and bootstrapped versions of the MLEs and MREs when one of the parameters is known.

In Section 3.4 we saw that the Pareto distribution can be characterized in terms of maximum entropy, and it is also possible to obtain estimators of its parameters based on this principle. This leads to the estimating equations

$$\check{x}_0 = \sqrt{\frac{1}{\mathrm{var}(\log X)}} \tag{3.94}$$

and

$$\check{\alpha} = \exp\left\{E(\log X) - \frac{1}{\check{x}_0}\right\}. \tag{3.95}$$

Singh and Guo (1995) have studied the performance of these estimators in relation to the MLEs and two further estimators by Monte Carlo. It turns out that the maximum entropy estimators perform about as well as the MLEs in terms of bias and root mean square error, and often much better than the method of moments estimators, especially in small samples.

3.6.7 Inequality Measures

The maximum likelihood estimators of the Lorenz curve (3.51), Gini index (3.53), and Theil index (3.56) are obtained by replacing the shape parameter α by its estimator $\hat{\alpha}$ depending on whether x_0 is known or unknown and then introducing the condition $L(u) \geq 0$.

Moothathu (1985) has studied the sampling distributions of the ML estimators of $L(u)$ and the Gini index, which are of the mixed type, and shown that the estimators are strongly consistent. In the case of the Gini coefficient, the sampling distribution of the MLE possesses the density $f_{\hat{G}}(w) = p + (1-p)f(w)$, where $p = P(\hat{G} \geq 1) = P(\hat{\alpha} \leq 1)$ and

$$f(w) = \frac{(2n\alpha)^{n-s+1}}{\Gamma(n-s+1)}\left(\frac{w}{1+w}\right)^{n-s}(1+w)^{-2}\exp\left(\frac{-2n\alpha w}{1+w}\right), \quad 0 < w < 1, \quad (3.96)$$

with $s = 1(2)$ if x_0 is known (unknown).

UMVUEs of the Lorenz curve, Gini index, and Theil coefficient were derived by Moothathu (1990c). [The UMVUE of the Gini coefficient for both parameters unknown may already be found in Arnold (1983, p. 200).] In what follows, S_i, $i = 1, 2$, are defined as in (3.77) and (3.78). For known x_0, an unbiased estimator of $L(u)$ is

$$\begin{aligned}\hat{L}^* &= 1 - (1-u)\,{}_0F_1[-; n; -nS_1\log(1-u)] \\ &= 1 - (1-u)\sum_{j=0}^{\infty}\frac{\{-nS_1\log(1-u)\}^j}{(n)_j\, j!},\end{aligned} \quad (3.97)$$

where ${}_0F_1$ is a Bessel function and $(n)_j$ is Pochhammer's symbol for the forward factorial function (see p. 287 for a definition), and in the case where x_0 is unknown, the UMVUE is

$$\hat{L}^* = 1 - (1-u)\,{}_0F_1[-; n-1; -nS_2\log(1-u)]. \quad (3.98)$$

The UMVUEs of the Gini coefficient are

$${}_1F_1\left(1; n; \frac{nS_1}{2}\right) - 1 \quad \text{if } x_0 \text{ is known} \quad (3.99)$$

$${}_1F_1\left(1; n-1; \frac{nS_2}{2}\right) - 1 \quad \text{if } x_0 \text{ is unknown}, \quad (3.100)$$

where ${}_1F_1$ is Kummer's confluent hypergeometric function.

The UMVUEs of the coefficient of variation were given in Section 3.6.4 above. For the Theil coefficient, the UMVUEs are

$$\frac{nS_1^2}{2(n+1)}\,{}_2F_2(2, 2; 3, n+2; nS_1) \quad \text{if } x_0 \text{ is known} \tag{3.101}$$

$$\frac{nS_2^2}{2(n-1)}\,{}_2F_2(2, 2; 3, n+1; nS_2) \quad \text{if } x_0 \text{ is unknown,} \tag{3.102}$$

where ${}_2F_2$ is a generalized hypergeometric function (see p. 288 for a definition).

Ali et al. (2001) showed that the UMVUE of the Gini coefficient has a variance in the vicinity of the Cramér–Rao bound and is considerably more efficient than the MLE in terms of MSE, in small samples. The same holds true for the Lorenz curve for large quantiles.

Latorre (1987) derived the asymptotic distributions of the ML estimators of the Gini coefficient and of Zenga's measures ξ and ξ_2, and the sampling distributions of the latter indices were obtained by Stoppa (1994). Since $P(\hat{\alpha} \leq 1) > 0$ although the coefficients only exist for $\alpha > 1$, these distributions are all of the mixed type.

3.7 EMPIRICAL RESULTS

Having already sketched the early history of income distributions in Chapter 1 we shall here confine ourselves to the comparatively few more recent studies employing Pareto distributions in that area and add some material on the distribution of wealth and on the size distributions of firms and insurance losses.

Income Data

In a reexamination of Pareto's (1896, 1897b) results for the Pareto type II distribution, Creedy (1977) noted some inconsistencies in Pareto's statements. He obtained parameter estimates of the location parameter μ that are highly significant and moreover have opposite signs (compared to Pareto's work), contradicting Pareto's remark that the shift parameter μ is positive for earnings from employment.

In a study aimed at a reassessment of the "conventional wisdom" that earnings are approximately lognormally distributed but with an upper tail that is better modeled by a Pareto distribution, Harrison (1981) considered the gross weekly earnings of 91,968 full-time male workers aged 21 and over from the 1972 British New Earnings Survey, disaggregated by occupational groups and divided into 34 earnings ranges. Although for the aggregated data he found that the main body of the distribution comprising 85% of the total number of employees is "tolerably well described" by the lognormal distribution for the (extreme) upper tail, the fit provided by the Pareto distribution is "distinctly superior." Specifically, there is evidence for a fairly stable Pareto tail with a coefficient in the vicinity of 3.85. However, he pointed out that "...a strict interpretation...suggests that...[the Pareto distribution] applies to only a small part of the distribution rather than to the top 20% of all employees" [as implied by Lydall's (1959) model of hierarchical earnings]. When disaggregated data divided into 16 occupational groups are considered, the stability of the Pareto tail often disappears, with estimates for the Pareto coefficient varying quite markedly for different lower thresholds.

Ratz and van Scherrenberg (1981) studied the distribution of incomes of registered professional engineers, in all disciplines, in the province of Ontario, Canada, annually for the period 1955–1978. Modeling the relationship between the distribution's parameters and years of experience by regression techniques, they found a negative relation between that variable and the shape parameter α. Thus, evidence exists that the incomes of more experienced engineers are more spread out than those of engineers at the beginning of their careers, a fairly intuitive result.

Ransom and Cramer (1983) argued that in most studies income distribution functions are put forth as approximate descriptive devices that are not meant to hold exactly. They therefore suggested employing a measurement error model, viewing observed income y as the sum of two independent variates, $y = x + u$, where x is the systematic component and u is a $N(0, \sigma^2)$ error term. Fitting a model of this type with a systematic component following a Pareto distribution to U.S. family incomes for 1960 and 1969, they discovered that the error accounts for about a third of the total variation, rendering the underlying distribution almost meaningless. The results are also inferior to those of other three-parameter models, notably the Singh–Maddala distribution.

Cowell, Ferreira, and Litchfield (1998) studied income distribution in Brazil over the 1980s (the decade of the international debt crisis). Brazil is a rather interesting country for scholars researching the distribution of income because it exhibits one of the most unequal distributions in the world, with 51% of total income going to the richest 10% and only 2.1% going to the poorest 20% (in 1995). Applying nonparametric density estimates, the authors found that the conventional nonparametric approach employing a normal kernel and a fixed window width does not seem to work well with data as heavily skewed as these. When fitting a Pareto distribution through incomes above \$1,000, it turns out that inequality among the very rich was not too extreme in 1981, with an α in the vicinity of 3. However, the situation considerably worsened over the 1980s and α decreased to about 2 in 1990.

More recently, analyzing the extreme right tail of Japanese incomes and income tax payments for the fiscal year 1998, Aoyama et al. (2000) obtained estimates of α in the vicinity of 2.

From these studies (and the older ones cited in Chapter 1), it emerges that the Pareto distribution is usually unsuitable to approximate the full distribution of income (as has long been known). However, it should be noted that the distribution has been successfully used for interpolation purposes in connection with grouped income data where is often desirable to introduce a distributional assumption for the open-ended category (e.g., "U.S. \$100,000 and over"). See Cowell and Mehta (1982) or Parker and Fenwick (1983) for further discussion of this topic.

Wealth Data

Steindl (1972) obtained an estimate of 1.7 for α from Swedish wealth data of 1955 and 1968. For Dutch data, he found that the Pareto coefficient increased slightly from 1.45 in 1959 to 1.52 in 1967.

Chesher (1979) estimated a Pareto type I model for Irish wealth data (grouped into 26 classes) of 1966, obtaining an estimate of α as low as 0.45. However, it emerges that the lognormal distribution performs much better on these data, with χ^2 and likelihood improvements of about 93%. The fit of the Pareto distribution is very poor in the upper tail, 274 times the observed number of individuals being predicted in the highest wealth class. Chesher attributed this to the fact that the sparsely populated upper classes carry little weight in the multinomial ML procedure he employed. When attempting to incorporate the 65% of individuals whose estate size is unrecorded, the Pareto distribution outperforms the lognormal distribution only for individuals whose wealth exceeds £40,000 (comprising only 0.4% of the population). Thus, the Pareto distribution does not seem to be appropriate for these data.

Analyzing data from the 1996 *Forbes* 400 list of the richest people in the United States, Levy and Solomon (1997) obtained an estimate of the Pareto coefficient of 1.36. Since the data comprise only extremely large incomes, their agreement with a Pareto tail is quite adequate.

Firm Sizes

In a classical study on the size distribution of firms, Steindl (1965) obtained Pareto coefficients in the range between 1.0 and 1.5. For all corporations in the United States in 1931 and 1955 (by assets), the parameter α is approximately equal to 1.1; for German firms in 1950 and 1959 (by turnover), it is about 1.1 in manufacturing and about 1.3 in retail trade, whereas for German firms in 1954 (by employment) it is about 1.2 in manufacturing.

Quandt (1966a) investigated the distribution of firm sizes (size being measured in terms of assets) in the United States. Using the *Fortune* lists of the 500 largest firms in 1955 and 1960 and 30 samples representing industries according to four-digit S.I.C. classes, he concluded that the Pareto types I–III (the Pareto type III is erroneously referred to as the Champernowne distribution in his paper) are appropriate models for only about half of the samples. The best of the three appears to be the type III variant, while the classical Pareto type I provides only six adequate fits. Pareto type I and II distributions seem to be appropriate for the two *Fortune* samples, however. Overall, the lognormal distribution does considerably better than Pareto distributions for these data.

Engwall (1968) studied the largest firms (according to sales) in 1965 within five areas: the United States, all countries outside the United States, Europe, Scandinavia, and Sweden, obtaining a shape parameter α between 1 and 2 in all cases.

More recently, Okuyama, Takayasu, and Takayasu (1999) obtained Pareto coefficients in the range (0.7, 1.4). They considered annual company incomes utilizing Moody's Company Data and Moody's International Company Data as well as Japanese data on companies having incomes above 40 million yen (the former databases comprise about 10,000 U.S. companies and 11,000 non-U.S. companies, respectively; the latter comprises 85,375 Japanese firms). It turns out that there are not only differences between countries but also between industry sectors, thus confirming earlier work by Takayasu and Okuyama (1998).

It becomes clear that for firm sizes the Pareto coefficient is somewhat smaller than for incomes and bounded by 2 from above, implying firm size distributions in the domain of attraction of a nonnormal stable law.

Insurance Losses

A number of researchers have suggested the use of the Pareto distribution as a plausible model for fire loss amount.

Benckert and Sternberg (1957) postulated the Pareto law for the distribution of fire losses in Swedish homes during the period 1948–1952, obtaining estimates in the vicinity of 0.5 for the damages to four types of houses.

Andersson (1971) used the Pareto distribution to model fire losses in the Northern countries (Denmark, Finland, Norway, and Sweden) for the periods 1951–1958 and 1959–1966, obtaining Pareto coefficients in the range from 1.25 to 1.76 and confirming an international trend toward an increase in the number of large claims from the first to the second period, as measured by a decrease in the parameter α. However, this trend appeared to be less pronounced for the Northern countries.

As was the case with firm sizes, it is noteworthy that fire insurance data seem to imply an α less than 2, thus pointing toward distributions in the domain of attraction of a nonnormal stable law. However, for automobile insurance data Benktander (1962) obtained a considerably larger estimate of $\hat{\alpha} = 2.7$. The paper by Seal (1980) contains a more extensive list of estimates of α compiled from the early actuarial literature up to the 1970s.

We must reitereate that a number of studies using "the Pareto distribution," notably in the actuarial literature, actually employ not the classical Pareto distribution (3.1) but the Pareto type II distribution that is a special case of the beta type II (Pearson type VI) distribution. They will therefore be mentioned in Chapter 6.

3.8 STOPPA DISTRIBUTIONS

Stoppa (1990a,b) proposed a generalization of the classical Pareto distribution by introducing a power transformation of the Pareto c.d.f. Thus, the c.d.f. of the Stoppa distribution is given by

$$F(x) = \left[1 - \left(\frac{x}{x_0}\right)^{-\alpha}\right]^{\theta}, \quad 0 < x_0 \leq x, \tag{3.103}$$

where $\alpha, \theta > 0$. The classical Pareto distribution is obtained for $\theta = 1$. The p.d.f. is

$$f(x) = \theta\alpha x_0^{\alpha} x^{-\alpha-1}\left[1 - \left(\frac{x}{x_0}\right)^{-\alpha}\right]^{\theta-1}, \quad 0 < x_0 \leq x, \tag{3.104}$$

and the quantile function is given by

$$F^{-1}(u) = x_0(1 - u^{1/\theta})^{-1/\alpha}, \quad 0 < u < 1. \tag{3.105}$$

Stoppa's generalized Pareto distribution can be derived from a differential equation for the elasticity $\eta(x, F)$ of the distribution function. If one supposes that (1) the resulting income density is either unimodal or decreasing, (2) the support of the distribution is $[x_0, \infty)$, for some $x_0 > 0$, (3) $\eta(x, F)$ is a decreasing function of $F(x)$, with $\lim_{x\to x_0} \eta(x, F) = \infty$ and $\lim_{x\to\infty} \eta(x, F) = 0$, the differential equation

$$\eta(x, F)\left[= \frac{F'(x)}{F(x)} \cdot x\right] = \alpha\theta \frac{1 - [F(x)]^{1/\theta}}{[F(x)]^{1/\theta}}, \quad \alpha, \theta > 0, \tag{3.106}$$

leads to an income distribution with the p.d.f. (3.104).

For integer values n of θ, the distribution can be viewed as the distribution of $X_{n:n}$ from a Pareto parent distribution; cf. (3.35). Thus, the distribution itself is closed under maximization, namely,

$$X \sim \text{Stoppa}(x_0, \alpha, \theta) \Longrightarrow X_{n:n} \sim \text{Stoppa}(x_0, \alpha, n\theta). \tag{3.107}$$

In contrast, the Pareto distribution is closed under minimization; cf. (3.34).

Compared to the classical Pareto distribution, the Stoppa distribution is more flexible since it has an additional shape parameter θ that allows for unimodal (for $\theta > 1$) and zeromodal (for $\theta \le 1$) densities. The mode is at

$$x_{\text{mode}} = x_0\left(\frac{1 + \theta\alpha}{1 + \alpha}\right)^{1/\alpha}, \quad \theta > 1, \tag{3.108}$$

and at x_0 otherwise. As θ increases, the mode shifts to the right. Figure 3.2 illustrates the effect of the new parameter θ.

The kth moment exists for $k < \alpha$ and equals

$$E(X^k) = \theta x_0^k B\left(1 - \frac{k}{\alpha}, \theta\right). \tag{3.109}$$

The Lorenz curve is of the form

$$L(u) = \frac{B_z(\theta, 1 - 1/\alpha)}{B(\theta, 1 - 1/\alpha)}, \quad 0 < u < 1, \tag{3.110}$$

where $z = u^{1/\theta}$ and B_z denotes the incomplete beta function, and the Gini coefficient is given by

$$G = \frac{2B(2\theta, 1 - 1/\alpha)}{B(\theta, 1 - 1/\alpha)} - 1. \tag{3.111}$$

It follows that for $X_i \sim \text{Stoppa}(x_0, \alpha_i, \theta)$, $i = 1, 2$, with $1 < \alpha_1 \le \alpha_2$, we have $X_1 \ge_L X_2$, and for $X_i \sim \text{Stoppa}(x_0, \alpha, \theta_i)$, $i = 1, 2$, with $\theta_1 \le \theta_2$, we have $X_1 \le_L X_2$. Analogous implications hold true for the Zenga ordering (Polisicchio, 1990). Consequently, the Gini coefficient is an increasing function of θ, for fixed α, and a

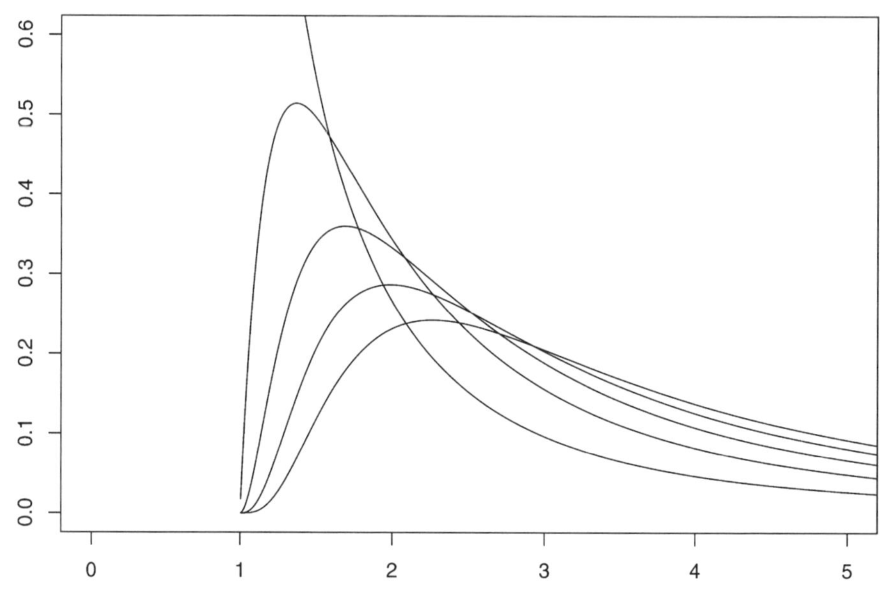

Figure 2 Stoppa densities: $x_0 = 1$, $\alpha = 1.5$, and $\theta = 1(1)5$ (from top left).

decreasing function of α, for fixed θ. Also, for fixed θ and $\alpha \to 1$ ($\alpha \to \infty$) the Lorenz curve tends to the Lorenz curve associated with maximal (minimal) concentration, whereas for fixed α and $\theta \to 0$ ($\theta \to \infty$) the Lorenz curve tends to the Lorenz curve associated with minimal (maximal) concentration (Domma, 1994).

Zenga's inequality measure ξ_2 is given by (Stoppa, 1990c)

$$\xi_2 = 1 - \exp\left\{\frac{1}{\alpha}\left[\psi(\theta + 1) - \psi\left(\theta - \frac{1}{\alpha} + 1\right) + \psi\left(1 - \frac{1}{\alpha}\right) - \gamma\right]\right\}, \tag{3.112}$$

where γ is Euler's constant.

The parameters of (3.103) can be estimated in several ways. Stoppa (1990b,c) considered nonlinear least-squares estimation in the Pareto diagram as well as ML estimation. The corresponding estimators are not available in closed form and must be derived numerically. In the ML case the parameter x_0, defining the endpoint of the support of F, poses the usual problem arising in connection with threshold parameters, the likelihood being unbounded in the x_0 direction. Stoppa suggested using modified ML estimators as discussed by Cohen and Whitten (1988). Starting values for ML estimation may be obtained from, for example, regression estimators as proposed by Stoppa (1995). He pointed out that the c.d.f. can be rewritten as

$$\log\{[1 - F^{1/\theta}(x)]\} = \alpha \log x_0 - \alpha \log x,$$

yielding for the elasticity

$$\log \eta(F, x) = \log(\theta \alpha x_0^\alpha) - \alpha \log x - \frac{1}{\theta} \log F(x).$$

Thus, $\log(\theta \alpha x_0^\alpha)$, α and $1/\theta$ can be estimated by least squares, say, and estimates for the original parameters are then obtained by solving the defining equations of the new parameters.

For known x_0 the Fisher information matrix is given by

$$I(\theta, \alpha) = \begin{pmatrix} \frac{n}{\theta^2} & \frac{n[\psi(2) - \psi(\theta+2)]}{\alpha(\theta+1)} \\ \cdot & \frac{n}{\alpha^2} + \frac{\theta}{\alpha(\theta+1)(\theta+2)}\{2 \log x_0 [\psi(2) - \psi(\theta+4)] \\ & - \frac{1}{\alpha^3}[\psi'(1) + \psi'(\theta+3)]\} \end{pmatrix}, \tag{3.113}$$

from which an estimate of the asymptotic covariance matrix of $(\hat{\theta}, \hat{\alpha})$ may be obtained by numerical inversion of $I(\hat{\theta}, \hat{\alpha})$. Properties of estimators of functions of α and/or θ can be derived by means of the delta method. It should be noted that most of the functions arising in connection with inequality measurement are scale-free and are therefore functions of only α and θ. In particular, Stoppa (1990c) derived the asymptotic distributions of the ML estimates of the Gini index (3.111) and of Zenga's inequality measure (3.112).

Stoppa (1990b) proposed a second generalized Pareto distribution—which we shall call the Stoppa type II distribution—with the c.d.f.

$$F(x) = \left\{1 - \left[\frac{x - c}{x_0}\right]^{-\alpha}\right\}^\theta, \quad x > c, \tag{3.114}$$

which is seen to be a two-parameter Stoppa type I distribution amended with location and scale parameters c and x_0. It can be derived along similar lines as the type I model.

The procedure leading to the Stoppa distribution (power transformation of the c.d.f.) was recently applied to more general distributions by Zandonatti (2001); see Chapter 6 for generalizations of the Pareto (II) and Singh–Maddala distributions, among others.

3.9 CONIC DISTRIBUTION

In spite of being an unpublished discussion paper, Houthakker (1992) has received substantial attention among the admittedly relatively narrow circle of researchers and users in the area of statistical income distributions.

Resurrecting the geometrical approach to income distributions that takes the Pareto diagram as the starting point, Houthakker introduced a flexible family of generalized Pareto distributions defined in terms of conic sections. A conic section in the Pareto diagram is given by

$$c_0U^2 + 2c_1UV + c_2V^2 + 2c_3U + 2c_4V + c_5 = 0, \tag{3.115}$$

where $U = \log\{\bar{F}(x)\}$, $V = \log x$, and $c_0, c_1, \ldots, c_5$ are parameters. In order to retain the weak Pareto law, a conic section with a linear asymptote is required. Consequently, circles and ellipses have to be excluded and the only admissible conic section in our context is the hyperbola. Since $\exp U$ must define a survival function, further constraints have to be imposed. It is not difficult but somewhat tedious to derive the resulting c.d.f. and p.d.f. We present the basic properties of the conic distributions using a reparameterization suggested by Kleiber (1994). He utilizes the earlier work of Barndorff-Nielsen (1977, 1978) on so-called hyperbolic distributions that leads to a more transparent functional form.

The c.d.f. of a general conic distribution is given by

$$F(x) = 1 - \exp\left\{-\eta\sqrt{1 + (\log x - \lambda)^2} + \xi(\log x - \lambda) + \mu\right\}, \quad x \geq x_0 > 0, \tag{3.116}$$

where $\eta > 0$, $-\infty < \xi \leq \eta$, $\mu = \eta\sqrt{1 + (\log x_0 - \lambda)^2} - \xi(\log x_0 - \lambda)$. It is also required that

$$\frac{\eta(\log x_0 - \lambda)}{\sqrt{1 + (\log x_0 - \lambda)^2}} - \xi \geq 0. \tag{3.117}$$

The density is therefore

$$f(x) = x^{-\xi-1}\exp\left\{-\eta\sqrt{1 + (\log x - \lambda)^2} - \xi\lambda + \mu\right\} \cdot \left[\frac{\eta(\log x - \lambda)}{\sqrt{1 + (\log x - \lambda)^2}} - \xi\right]. \tag{3.118}$$

Houthakker discussed two subclasses of the conic distributions in detail. These are the conic-linear distributions defined via the condition $\eta - \xi = 0$ and the conic-quadratic models defined by $f(x_0) = 0$. The latter condition is equivalent to the l.h.s. of (3.117) being equal to zero. Consequently, for non-"quadratic" conic distributions, we always have $f(x_0) > 0$.

It is not difficult to see that the c.d.f. is asymptotically of the form $x^{-(\eta-\xi)}$; hence, the kth moment of this distribution exists only for $k < \eta - \xi$, and $\eta - \xi$ plays the role of Pareto's α. For the moments we obtain a rather formidable expression involving a complete as well as an incomplete modified Bessel function of the third kind (also known as a MacDonald function), which will not be given here. The probably easiest way to the Gini coefficient is via the representation in terms of moments of order statistics (2.22); the resulting expression also involves incomplete Bessel functions.

Kleiber (1994) pointed out that a simple regression estimator may be used to obtain parameter estimates for the conic distribution. In a doubly logarithmic representation of (3.116), only the parameter λ enters in a nonlinear fashion, so a quick method is to estimate μ, η, ξ via linear regression while performing a grid search over λ. However, it is evident from (3.116) that the regressors $\sqrt{1+(\log x-\lambda)^2}$ and $\log x-\lambda$ are highly collinear, suggesting that the model is essentially overparameterized. Indeed, Brachmann, Stich, and Trede (1996) reported on numerical problems when trying to fit linear and quadratic conic distributions to German household incomes.

3.10 A "LOG-ADJUSTED" PARETO DISTRIBUTION

Ziebach (2000) proposed a generalization of the Pareto distribution by introducing a logarithmic adjustment term that allows for a more flexible shape. Like Houthakker (1992), he started from the Pareto diagram, specifying a c.d.f. with the representation

$$\log\{1-F(x)\} = -\alpha\log x - \beta\log(\log x) + \gamma.$$

Clearly, the new term does not affect the asymptotic linearity in the doubly logarithmic representation, so that the resulting distribution obeys the weak Pareto law. The condition $F(x_0)=0$, for some $x_0>0$, yields $e^\gamma = x_0^\alpha(\log x_0)^\beta$, leading to the c.d.f.

$$F(x) = 1-\left(\frac{x_0}{x}\right)^\alpha\left(\frac{\log x_0}{\log x}\right)^\beta, \quad 1<x_0\le x, \tag{3.119}$$

where either $\alpha>0$ and $\beta\ge-\alpha\log x_0$ or $\alpha=0$ and $\beta>0$. The density is given by a more complex expression

$$f(x) = \frac{x_0^\alpha(\log x_0)^\beta(\alpha\log x+\beta)}{x^{\alpha+1}(\log x)^{\beta+1}}, \quad 1<x_0\le x, \tag{3.120}$$

which is decreasing for $\beta\ge0$ and unimodal for

$$\beta\in\left[-\alpha\log x_0,\ -\alpha\log x_0+\frac{1}{2}\left\{\sqrt{(\log x_0+1)^2+4\alpha\log x_0}-(\log x_0+1)\right\}\right].$$

Only if $\beta=-\alpha\log x_0$, do we have $f(x_0)=0$; otherwise, $f(x_0)>0$. Figure 3.3 illustrates the effect of the new parameter β.

The moments of the distribution are

$$E(X^k) = \frac{\alpha x_0^k}{\alpha-k} - k\beta(\alpha-k)^{\beta-1}x_0^\alpha(\log x_0)^\beta\Gamma[-\beta;(\alpha-k)\log x_0], \tag{3.121}$$

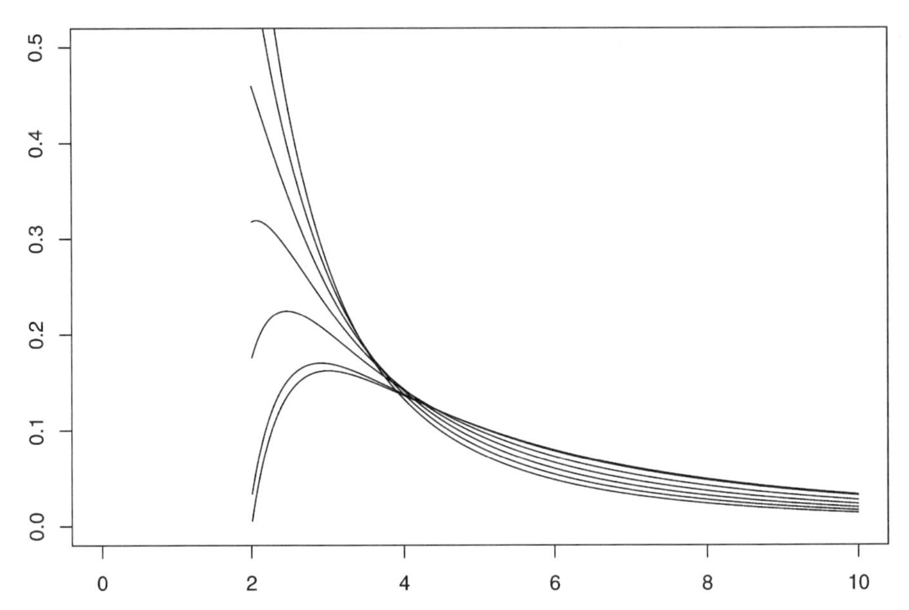

Figure 3 "Log-adjusted" Pareto densities: $x_0 = 2$, $\alpha = 1.5$, $\beta = -1.03, -1(.2)0$ (from bottom left). (Here $\beta = 1.03$ represents the boundary of the admissible parameter space.)

for $k < \alpha$, and

$$E(X^k) = x_0^k + kx_0^k \cdot \frac{\log x_0}{\beta - 1},$$

for $k = \alpha$ and $\beta > 1$; otherwise, they do not exist. Equation (3.121) illustrates the effect of the new parameter β: The first term is simply the kth moment of a classical Pareto distribution, the second terms describes the adjustment due to β.

The Lorenz curve is available via the first-moment distribution and the Gini coefficient via the representation (2.22) in terms of expectations of order statistics, but the resulting expressions are again somewhat involved.

The "log-Pareto" distribution is quite similar to the log-gamma distribution

$$f(x) = \frac{\alpha^\beta x_0^\alpha}{\Gamma(\beta)} x^{-\alpha-1} \left\{ \log\left(\frac{x}{x_0}\right) \right\}^{\beta-1}, \quad 0 < x_0 \leq x, \tag{3.122}$$

where $\beta \geq 1$, $\alpha > 1$, a model to be discussed in Section 5.4. It appears that the latter distribution is perhaps more tractable, especially since there is a well-developed theory of estimation for the gamma distribution.

3.11 STABLE DISTRIBUTIONS

In a series of pioneering contributions around 1960, Mandelbrot (1959, 1960, 1961) argued that incomes follow what he calls a *Pareto–Lévy distribution*, that is, a maximally skewed stable distribution with a characteristic exponent α between 1 and 2. Here stability means that if X_1 and X_2 are independent copies of X and b_i, $i = 1, 2$, are positive constants, then

$$b_1X_1 + b_2X_2 \stackrel{d}{=} bX + a$$

for some positive b and real a. In other words, the shape of X is preserved under addition, up to location and scale.

There is by now a very substantial literature on stable distributions, partially furnished by successful applications, notably in finance, and partially due to recent advances in statistical computing. We refer the interested reader to the classical text by Zolotarev (1986) for probabilistic properties of stable laws. Their main attraction is that they represent the only limits of properly normalized sums of i.i.d. random variables and therefore generalize the normal distribution. Indeed, nonnormal stable laws arise in the (generalized) central limit problem for i.i.d. random variables if these variables do not possess finite variances. In that case, the limit distribution does not have finite variance either, its right tail being regularly varying and therefore of the Pareto type (without following an exact Pareto distribution), thus justifying the inclusion of stable distributions here.

Mandelbrot argued that the total income of any income recipient is obtained by adding incomes from different sources (e.g., family income is obtained as the sum of the incomes of all family members). If all these types of income follow the same type of distribution and one is willing to make an independence assumption, one expects that total income will be adequately approximated by a stable distribution.

Although that sounds very attractive, the drawback is that stable laws must generally be defined in terms of their characteristic functions, as only in three exceptional cases—the normal and Cauchy distributions and a special case of the inverse gamma distribution (see Section 5.4)—are the densities and/or distribution functions are available in terms of elementary functions. The characteristic function of the stable laws is

$$\log \phi(t) = \log E(e^{itx}) = i\mu t - \lambda|t|^{\alpha} \exp\left\{\frac{\pi}{2} i\gamma \operatorname{sign}(t)\right\}. \tag{3.123}$$

Here $0 < \alpha < 1$ or $1 < \alpha \leq 2$. For $\alpha = 1$, comprising a small family of Cauchy-type distributions, a slightly different form must be used. The parameter α is usually referred to as the characteristic exponent or index of stability; in a certain sense, it is the most important of the four parameters since it governs the tail behavior of the distribution and therefore the existence of moments. The kth absolute moment $E|X|^k$ exists iff $k < \alpha$. The parameter $\mu \in \mathbb{R}$ is a location parameter, $\lambda > 0$ is a scale, and γ is a skewness parameter, with $|\gamma| \leq 1 - |1 - \alpha|$. Clearly, $\alpha = 2$ yields the normal distribution. [We warn our readers that there are a multitude of alternative parameterizations of the stable laws! The form (3.123) of the characteristic function is a variant of Zolotarev's (1986, p. 12) parameterization (B).]

The connection with income distributions is as follows: If $1 < \alpha \leq 2$, which is broadly consistent with Pareto's original $\alpha = 1.5$, we have $|\gamma| \leq 2 - \alpha$. Thus, the distribution is maximally skewed to the right if $\gamma = 2 - \alpha$. Mandelbrot suggested that this is the relevant case for applications to income data because otherwise the probability of negative incomes might become too large. [Stable densities are positive on the whole real line unless $\alpha < 1$ and the distribution is maximally positively (negatively) skewed, in which case the support is the positive (negative) halfline.] Thus, the maximally skewed stable laws define a three-parameter subfamily of all stable distributions. In view of $|\gamma| \leq 1 - |1 - \alpha|$ for all stable laws, the parameter γ becomes less and less meaningful (as well as harder to estimate) as α approaches 2. This means that in the maximally skewed stable case α must stay well below 2 in order to retain a reasonably skewed distribution.

It took almost 20 years after Mandelbrot's discovery for an attempt to be made to fit a maximally skewed stable distribution to income data. Van Dijk and Kloek (1978, 1980) addressed the problem of estimation from grouped data. They employed multinomial maximum likelihood and minimum χ^2 estimates that are based on numerical inversion of the characteristic function, followed by numerical integration of the resulting densities. (This was a somewhat burdensome computational procedure in 1980.) Van Dijk and Kloek's estimates of α, for Australian family disposable income in 1966–1968 and for the Dutch gross incomes in 1973, ranged from 1.17 to 1.72.

These authors also considered log-stable distributions, assuming that not income itself but rather its logarithm (income power) follows a stable distribution. This is quite natural since the normal distribution is a particular stable distribution and its offspring, the lognormal, is one of the classical size distributions. The appearance of the stable distributions in connection with income data could therefore be justified by appealing to a generalized form of Gibrat's law of proportionate effect (leading to the lognormal distribution; see Section 4.2). Although this approach is arguably best discussed within the framework of the following chapter, dealing with lognormal distributions, we mention it here since the estimation of stable and log-stable models is completely analogous. In a log-stable framework it is no longer necessary to confine ourselves to the three-parameter subclass of Pareto-Lévy distributions, and consequently the unrestricted four-parameter family of stable distributions can be used. Van Dijk and Kloek obtained estimates of α between 1.53 and 1.96 for the logarithmic incomes.

Van Dijk and Klock preferred log-stable distributions over stable ones on the grounds of some pooled tests. We are however somewhat skeptical as far as the usefulness of the log-stable model is concerned. From Kleiber (2000b) the log-stable densities lack moments of any order and moreover they exhibit a pole at the origin, a feature that does not seem to be consistent with the data. (Of course, the latter feature cannot be verified from grouped data.)

Van Dijk and Kloek reported that both the maximally skewed stable and log-stable family perform generally better than the log-t and Champernowne distributions. However, they also noted that "the data considered were not sufficient to settle the dispute about the question what is the correct model to describe the right-hand tail of an income distribution" (van Dijk and Kloek, 1978, p. 19). In fact, no data however abundant can settle this basic dispute.

On the actuarial side, Holcomb (1973; see also Paulson, Holcomb, and Leitch, 1975, p. 169) asserted that the claim experience for a line of nonlife insurance is a mixture of independent random variables, from nonidentical distributions all lying within the domain of attraction of a stable law with a support that is bounded from below ($\alpha < 1$). This would imply that the aggregate claim distribution converges to a stable law.

Holcomb considered three data sets. The first comprised data for the period 1965–1970, detailing 2,326 national losses due to burglaries experienced by a U.S. chain of retail stores. The second contained 2,483 records of vandalism for the same chain and the same period, whereas the third comprised 1,142 records constituting a different chain's combined history of insured losses due to all property and liability perils. Holcomb obtained estimates of 0.82, 0.88, and 0.67, respectively, for α.

A problem with this approach is that the employed distributions possess infinite means that are difficult to reconcile with common actuarial premium principles, most of which require a finite first moment. This may explain to some extent why maximally skewed stable distributions have not been used much in actuarial applications.

3.12 FURTHER PARETO-TYPE DISTRIBUTIONS

Krishnan, Ng, and Shihadeh (1990) proposed a generalized Pareto distribution by introducing a more flexible (polynomial) form for the elasticity

$$\frac{\partial \log\{1 - F(x)\}}{\partial \log x} = -\alpha - \sum_{i=1}^{k} \beta_i x^i. \tag{3.124}$$

This approach yields c.d.f.'s of the type

$$1 - F(x) \propto x^{-\alpha} \exp\left\{-\sum_{i=1}^{k} \beta_i x^i\right\}, \tag{3.125}$$

called polynomial Pareto curves by Krishnan, Ng, and Shihadeh and comprising the Pareto type I (for $\beta_1 = \cdots = \beta_k = 0$) distribution as a special case. When we apply their linear specification ($\beta_2 = \cdots = \beta_k = 0$), which is very close to Pareto's third proposal (3.7), to two data sets, it turns out that the estimates of the new parameter β_1 are rather small, confirming Pareto's work.

Perhaps somewhat unexpectedly, the size distribution of prizes in many popular lotteries and prize competitions is also intimately related to the Pareto distribution. Observing that the prizes in German lotteries are often characterized by the fact that the number of prizes n_i of a certain class i is inversely proportional to their values x_i,

Bomsdorf (1977) was led to a study of the discrete distributions with probability mass function

$$f(x) = \frac{p}{d_i}, \quad x_i = d_i \cdot a, \quad i = 1, \ldots, k,$$

where $a > 0$, $d_1 = 1$, $d_i < d_{i+1}$, $p = 1/\sum_{i=1}^{k} d_i^{-1}$, which he called the *prize-competition distribution*. A continuous analog is clearly given by the density

$$f(x) = \frac{c}{x}, \quad a_1 \le x \le a_2. \tag{3.126}$$

If we rewrite the supporting interval in the form $[a^n, a^{n+1}]$, where $n = \log_a a_1 \in \mathbb{R}$ and $a = a_2/a_1$, the normalizing constant is found to be $c = \log_a e$. Thus for the case where $a = e$, matters simplify to the hyperbolic function

$$f(x) = \frac{1}{x}, \quad e^n \le x \le e^{n+1}, \tag{3.127}$$

with the c.d.f.

$$F(x) = \log x - n, \quad e^n \le x \le e^{n+1},$$

an expression that is seen to describe a doubly truncated Pareto-type distribution corresponding to $\alpha = 0$ in (3.2). [Note that in (3.2) $\alpha = 0$ is inadmissible because on an unbounded support the resulting function would not be a distribution function.]

The moment generating function of the prize-competition distribution (3.127) is given by the formula

$$m(t) = 1 + \sum_{i=1}^{\infty} t^i \frac{e^{in+i}}{i \cdot i!} - \sum_{i=1}^{\infty} t^i \frac{e^{in}}{i \cdot i!},$$

from which the first moment and the variance are found to be

$$E(X) = e^{n+1} - e^n$$

and

$$\operatorname{var}(X) = 2 \cdot e^{2n+1} - \frac{e^{2n+2}}{2} - \frac{3e^{2n}}{2}.$$

The Bomsdorf (1977) distribution (the continuous version) has been extended to the distribution with c.d.f.

$$F(x) = k \cdot x \frac{(\log x)^b}{b+1}, \quad b \ne -1, \quad 1 \le x \le e^{1/b},$$

by Stoppa (1993) and constitutes family 6 in the first period of his extensive table of distributions based on a differential equation for the elasticity described in Section 2.3. This is an L-shaped distribution with the mode at zero for $b > 1$ and at $e^{(1-b)/b}$ for $b \leq 1$.

A so-called *generalized Pareto distribution* with c.d.f.

$$F_c(x) = \begin{cases} 1 - \left(1 - \frac{cx}{b}\right)^{1/c}, & \text{if } c \neq 0, \\ 1 - e^{-x}, & \text{if } c = 0, \end{cases} \tag{3.128}$$

where

$$x \geq 0 \quad \text{if } c \geq 0,$$

$$0 \leq x \leq -\frac{1}{c} \quad \text{if } c < 0,$$

is of great importance in the analysis of extremal events (see, e.g., Embrechts, Klüppelberg, and Mikosch, 1997, or Kotz and Nadarajah, 2000). In the context of income distributions, this model was fitted to the 1969 personal incomes in 157 counties of Texas, and to lifetime tournament earnings of 50 professional golfers through 1980 by Dargahi-Noubary (1994). [The data are given in Arnold (1983), Appendix B.]

Colombi (1990) proposed a "Pareto-lognormal" distribution that is defined as the distribution of the product $X := Y \cdot Z$ of two independent random variables, one following a Par$(1, \alpha)$ distribution and the other following a two-parameter lognormal distribution (to be described in detail in the next chapter). The p.d.f. of the product is given by

$$f(x) = \frac{\theta}{x^{\theta+1}} \exp\left\{\theta\left(\mu + \frac{\theta\sigma^2}{2}\right)\right\} \Lambda(x; \mu + \theta\sigma^2, \sigma), \quad x > 0,$$

where $\theta > 0$, $\mu \in \mathbb{R}$, $\sigma > 0$, and

$$\Lambda(x; \mu + \theta\sigma^2, \sigma) = \Phi\left(\frac{\log x - \mu - \theta\sigma^2}{\sigma}\right), \quad x > 0,$$

denotes the c.d.f. of the two-parameter lognormal distribution.

The kth moment of the Pareto-lognormal distribution exists for $k < \theta$ and equals

$$E(X^k) = \frac{\theta}{\theta - k} e^{k\mu + (k\mu)^2/2}.$$

Similarly to the Pareto distribution, the Pareto-lognormal family is closed with respect to the formation of moment distributions.

Colombi showed that the distribution is unimodal and provided an implicit expression for the mode. He also discussed sufficient conditions for Lorenz ordering and derived the Gini coefficient, which is of the form

$$G = 1 - 2\Phi\left(-\frac{\sigma}{\sqrt{2}}\right) + \frac{2}{2\theta - 1} e^{\theta\sigma(\theta\sigma - \sigma)} \Phi\left(-\frac{2\theta\sigma - \sigma}{\sqrt{2}}\right),$$

an expression that is seen to be decreasing in σ.

Moreover, he fit the distribution to Italian family incomes for 1984 and 1986. For the 1984 data the model is outperformed by both the Dagum type I and Singh–Maddala distribution in terms of likelihood and minimum chi-square, but for the 1986 data the situation is reversed.

We should like to add that several further distributions included in this book, notably those of Chapter 6, may also be considered "generalized Pareto distributions," in that they possess a polynomially decreasing upper tail but are more flexible than the original Pareto model in the lower-income range.

CHAPTER FOUR

Lognormal Distributions

Naturally we cannot present all that is known about the lognormal distribution in a chapter of moderate length within the scope of this book. For example, in quantitative economics and finance, the lognormal distribution is ubiquitous and it arises, among other things, in connection with geometric Brownian motion, the standard model for the price dynamics of securities in mathematical finance.

At least two books have been devoted to the lognormal distribution: Aitchison and Brown (1957) presented the early contributions with an emphasis on economic applications, whereas the compendium edited by Crow and Shimizu (1988) contains wider coverage. The lognormal distribution is also systematically covered in a 50-page chapter in Johnson, Kotz, and Balakrishnan (1994). Below we shall concentrate on the "size" aspects of the distribution and emphasize topics that were either omitted or only briefly covered in these three sources. These include a generalized lognormal distribution for which the literature available is mainly in Italian.

4.1 DEFINITION

The p.d.f. of the lognormal distribution is given by

$$f(x) = \frac{1}{x\sqrt{2\pi}\sigma} \exp\left\{ -\frac{1}{2\sigma^2} (\log x - \mu)^2 \right\}, \quad x > 0. \tag{4.1}$$

Thus, the distribution arises as the distribution of $X = \exp Y$, where $Y \sim \mathrm{N}(\mu, \sigma^2)$. Hence, the c.d.f. is given by

$$F(x) = \Phi\left(\frac{\log x - \mu}{\sigma} \right), \quad x > 0, \tag{4.2}$$

where Φ denotes the c.d.f. of the standard normal distribution.

Statistical Size Distributions in Economics and Actuarial Sciences, By Christian Kleiber and Samuel Kotz.
ISBN 0-471-15064-9 © 2003 John Wiley & Sons, Inc.

There is also a three-parameter (shifted) lognormal distribution, to be discussed in Section 4.7 below.

4.2 HISTORY AND GENESIS

As far as economic size distributions are concerned, the pioneering study marking initial use of the lognormal distribution was Gibrat's thesis of 1931. (See Appendix A for a short biography of Robert Gibrat.) Gibrat asserted that the income of an individual (or the size of a firm) may be considered the joint effect of a large number of mutually independent causes that have worked during a long period of time. At a certain time t it is assumed that the change in some variate X is a random proportion of a function $g(X_{t-1})$ of the value X_{t-1} already attained. Thus, the underlying law of motion is

$$X_t - X_{t-1} = \epsilon_t g(X_{t-1}),$$

where the ϵ_t's are mutually independent and also independent of X_{t-1}. In the special case where $g(X) \equiv X$, the so-called *law of proportionate effect* results, meaning that the change in the variate at any step of the process is a random proportion of the previous value of the variate. Thus,

$$X_t - X_{t-1} = \epsilon_t X_{t-1},$$

which may be rewritten as

$$\frac{X_t - X_{t-1}}{X_{t-1}} = \epsilon_t,$$

whereby through summation over t

$$\sum_{t=1}^{n} \frac{X_t - X_{t-1}}{X_{t-1}} = \sum_{t=1}^{n} \epsilon_t.$$

Assuming that the effect at each step is small, we get

$$\sum_{t=1}^{n} \frac{X_t - X_{t-1}}{X_{t-1}} \approx \int_{X_0}^{X_n} \frac{dX}{X} = \log X_n - \log X_0,$$

which yields

$$\log X_n = \log X_0 + \sum_{t=1}^{n} \epsilon_t.$$

If we use a suitable central limit theorem (CLT) it follows that X_n are asymptotically lognormally distributed. In the original derivation the ϵ_t were assumed to be i.i.d. so

Table 4.1 Gibrat's Fittings

Variable	Location	Date
Food expenditures	England	1918
Income	England	1893–1894, 1911–1912
	Oldenburg	1890
	France	1919–1927
	Prussia	1852, 1876, 1881, 1886, 1892, 1893, 1894, 1901, 1911
	Saxony	1878, 1880, 1886, 1888, 1904, 1910
	Austria	1898, 1899, 1903, 1910
	Japan	1904
Wealth tax	Netherlands	1893–1910
Wealth	Basel	1454, 1887
Old-age state pensions	France	1922–1927
Postal check accounts	France	1921–1922
Legacies	France	1902–1913, 1925–1927
	England	1913–1914
	Italy	1910–1912
	9 French départements	1912, 1926
Rents	Paris	1878, 1888, 1913
Real estate (tilled)	France	1899–1900
Real estate	Belgium	1850
Real estate sales	France	1914
Security holdings	France	1896
Wages	United Kingdom	1906–1907
	Italy	1906
	Denmark	1906
	Bavaria	1909
Dividends	Germany	1886, 1896
Firm profits	France	1927
Firm sizes	France	1896, 1901, 1906, 1921, 1926
	Alsace-Lorraine	1907, 1921, 1926
	European countries and the United States	around 1907
	Turkey	1925
	Germany	1907, 1925
City sizes	France	1866, 1906, 1911, 1921, 1926
	Europe	1850, 1910, 1920, 1927
	United States	1900, 1920, 1927
Family sizes	France	1926

Source: Gibrat (1931).

that the Lindeberg–Lévy CLT is sufficient to handle the problem. The same limiting result may, of course, be obtained using more general CLTs if the ϵ_t are heterogeneous; see Hart (1973, Appendix B) for a discussion.

It should be noted that Gibrat's approach is very close to the one exhibited in Edgeworth's (1898) *method of translations*, by means of which a exponential transform of a normal variable is carried out. Gibrat followed Kapteyn's (1903) generic approach (the so-called *Kapteyn's engine*) to construct a binomial process and provided a modification of the independence assumption inherent in this process, trying to justify its application in economics. {Much earlier Galton (1879) and McAlister (1879) suggested the exponential of a normal variable, inspired by the Weber–Fechner law in psychophysics [details are presented in a later publication of Fechner (1897): *Kollektivmasslehre*].} Being a French engineer trained to think geometrically, Gibrat did not use "sophisticated" mathematical tools (such as least-squares or Cauchy methods). Gibrat argued in favor of his model as compared with its classical earlier competitor, the Pareto distribution, for a number of cases in various economic areas. See Table 4.1 for a listing of the data to which Gibrat successfully fit lognormal distributions.

An innovative feature of Gibrat's model is its potential application in all economic domains. The Pareto law seemed to be restricted solely to income distributions, whereas the application of the Gibrat law of proportional effect appeared to be more extensive. As indicated in Table 4.1, in his 1931 dissertation Gibrat applied it to other areas of economics such as fortunes and estates, firm profits, firm sizes (by number of workers), number of city inhabitants, and family sizes.

Gibrat's law of proportionate effect was subsequently modified, for example, to prevent the variance of $\log X$ to grow without bound (see Section 1.4). However, for more than 70 years it has proved to be a useful first approximation of firm size dynamics against which alternatives have to be tested; see, for example, Mansfield (1962), Evans (1987a,b), Hall (1987), or Dunne and Hughes (1994). Sutton (1997) provided a recent survey of Gibrat's law in connection with the size distribution of firms.

4.3 MOMENTS AND OTHER BASIC PROPERTIES

In view of the close relationship with the normal distribution, many properties of the lognormal distribution follow directly from corresponding results for the normal distribution. For example, random samples from a lognormal population are commonly generated from normal random numbers via exponentiation. The main difference is the new role of the parameters: $\exp(\mu)$ becomes a scale parameter, whereas σ is now a shape parameter. Figure 4.1 illustrates the effect of σ, showing the well-known fact that for small σ, a $\text{LN}(\mu,\sigma^2)$ distribution can be approximated by a $\text{N}(\exp\mu, \sigma^2)$ distribution.

One of the main attractions of the normal distribution is its stability properties under summation, namely, that sums of independent normal random variables are

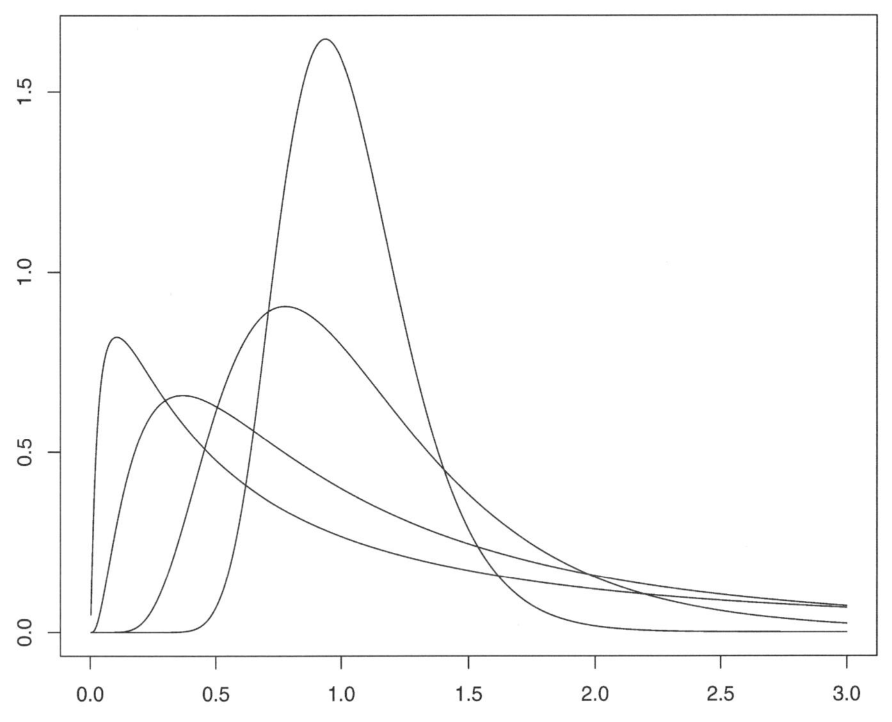

Figure 1 Lognormal densities: $\mu = 0$ and $\sigma = 0.25, 0.5, 1.0, 1.5$ (modes from right to left).

also normally distributed. This translates into the following multiplicative stability property of the lognormal distribution: For independent $X_i \sim \mathrm{LN}(\mu_i, \sigma_i^2)$, $i = 1, 2$,

$$X_1 X_2 \sim \mathrm{LN}(\mu_1 + \mu_2, \sigma_1^2 + \sigma_2^2), \tag{4.3}$$

in particular for the sample geometric mean of n i.i.d. lognormal random variables

$$\left(\prod_{i=1}^{n} X_i\right)^{1/n} \sim \mathrm{LN}\left(\mu, \frac{\sigma^2}{n}\right). \tag{4.4}$$

However, sums of lognormal random variables are not very tractable. Unfortunately, as noted by Mandelbrot (1997), "dollars and firm sizes do not multiply, they add and subtract... [hence] the lognormal has invariance properties but not very useful ones." In his view this constitutes a strong case against the lognormal distribution.

The structure of the lognormal distribution makes it convenient to express its moments. They are obtained in terms of the moment-generating function $m_Y(\cdot)$ of the normal distribution

$$E(X^k) = E(e^{kY}) = \exp\left(k\mu + \frac{1}{2}k^2\sigma^2\right). \tag{4.5}$$

Hence, the lognormal distribution is one of the few well-known distributions that possesses moments of all orders, positive, negative, and fractional. However, the sequence of integer moments does not characterize the lognormal distribution; see Heyde (1963) for a continuous distribution having the same moments as the lognormal and Leipnik (1991) for a discrete counterexample. This somewhat unexpected property is of interest in probability theory and has attracted the attention of numerous researchers.

The low-order moments and other basic characteristics of the lognormal distribution are as follows. [For compactness the notation $\omega = \exp(\sigma^2)$ is used in several formulas.] From (4.5), the mean of the lognormal distribution is

$$E(X) = \exp\left(\frac{\mu + \sigma^2}{2}\right) \tag{4.6}$$

and the variance is given by

$$\text{var}(X) = e^{2\mu+\sigma^2}(e^{\sigma^2} - 1) = e^{2\mu}\omega(\omega - 1). \tag{4.7}$$

Hence, the coefficient of variation equals

$$\text{CV}(X) = \sqrt{\exp\sigma^2 - 1} = \sqrt{\omega - 1}. \tag{4.8}$$

The coefficient of skewness is

$$\alpha_3 = (\omega + 2)\sqrt{\omega - 1}$$

and the coefficient of kurtosis equals

$$\alpha_4 = \omega^4 + 2\omega^3 + 3\omega^2 - 3.$$

Note that $\alpha_3 > 0$ and $\alpha_4 > 3$, that is, the distribution is positively skewed and leptokurtic.

The geometric mean is given by

$$x_{\text{geo}} = \exp\{E(\log X)\} = e^{\mu}, \tag{4.9}$$

which coincides with the median (see below).

The distribution is unimodal, the mode being at

$$x_{\text{mode}} = \exp(\mu - \sigma^2). \tag{4.10}$$

A comparison of (4.6) and (4.10) illustrates the effect of the shape parameter σ: For a fixed scale $\exp\mu$, an increase in σ moves the mode toward zero while at the same time the mean increases, and both movements are exponentially fast. This means that the p.d.f. becomes fairly skewed for moderate increases in σ (compare with Figure 4.1).

By construction, the lognormal quantiles are given by

$$F^{-1}(u) = \exp\{\mu + \sigma\Phi^{-1}(u)\}, \quad 0 < u < 1, \tag{4.11}$$

where Φ is the c.d.f. of the standard normal distribution. In particular, the median is

$$x_{\text{med}} = F^{-1}(0.5) = \exp(\mu), \tag{4.12}$$

which is a direct consequence of the symmetry of the normal distribution.

Thus, the mean–median–mode inequality

$$E(X) > x_{\text{med}} > x_{\text{mode}}$$

is satisfied by the lognormal distribution.

The lognormal distribution is closed under power transformations, in the sense that

$$X \sim \text{LN}(\mu, \sigma^2) \Longrightarrow X^r \sim \text{LN}(r\mu, r^2\sigma^2), \text{ for all } r \in \mathbb{R}. \tag{4.13}$$

Note that power transformation is a popular device in applications of statistical distribution theory.

The entropy is

$$E\{-\log f(X)\} = \frac{1}{2} + \mu - \log\left(\frac{1}{\sigma\sqrt{2\pi}}\right). \tag{4.14}$$

The form of the characteristic function has been a long-standing challenging problem; fairly recently Leipnik (1991) provided a series expansion in terms of Hermite functions in a logarithmic variable. It is of special interest because the formula for the characteristic function of its generator—the normal distribution—is a basic fact in probability theory discovered possibly 150 years ago. It follows from more general results of Bondesson (1979) that the distribution is infinitely divisible.

The mean excess (or mean residual life) function has an asymptotic representation of the form (e.g., Embrechts, Klüppelberg, and Mikosch 1997)

$$e(x) = \frac{\sigma^2 x}{\log x - \mu}[1 + o(1)]. \tag{4.15}$$

Its asymptotically linear increase reflects the heavy-tailed nature of the lognormal distribution.

Sweet (1990) has studied the hazard rate of lognormal distributions. Figure 4.2 exhibits the typical shape, which is unimodal with $r(0) = 0$ [in fact, all derivatives of $r(x)$ are zero at $x = 0$], and a slow decrease to zero as $x \to \infty$.

The value of x that maximizes $r(x)$ is

$$x_M = \exp(\mu + z_M \sigma),$$

where z_M is given by $(z_M + \sigma) = \varphi(z_M)/[1 - \Phi(z_M)]$. Thus, $-\sigma < z_M < -\sigma + \sigma^{-1}$, and therefore

$$\exp(\mu - \sigma^2) < x_M < \exp(\mu - \sigma^2 + 1).$$

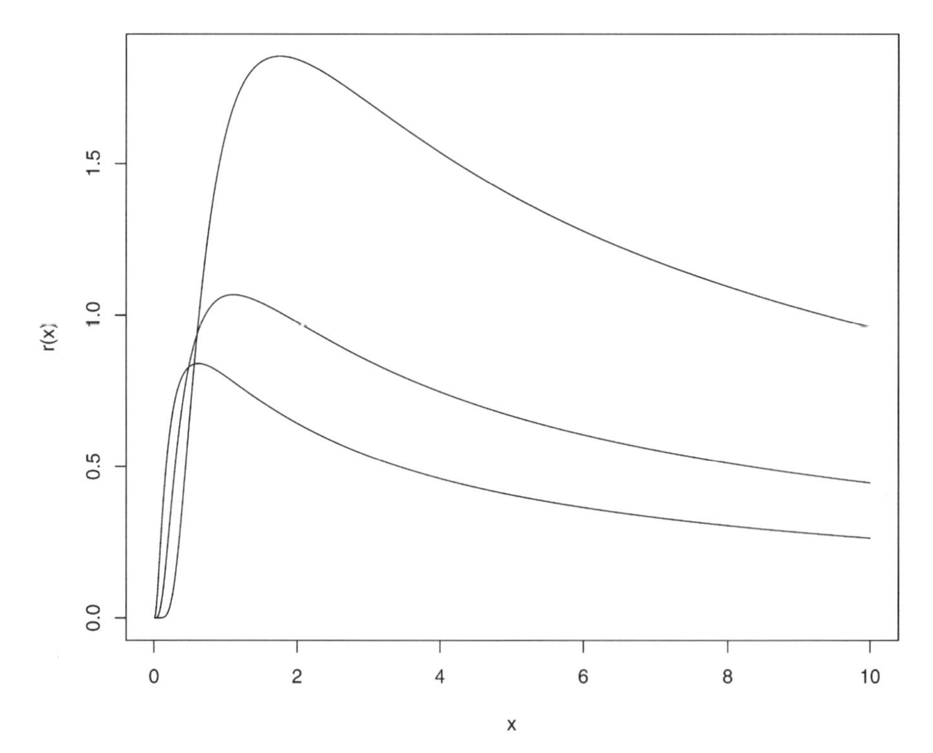

Figure 2 Lognormal hazard rates: $\mu = 0$ and $\sigma = 0.5, 0.75, 1.0$ (from top).

As $\sigma \to \infty$, $x_M \to \exp(\mu - \sigma^2)$, and so for large σ,

$$\max_x r(x) \approx \frac{\exp(\mu - \sigma^2/2)}{\sigma\sqrt{2\pi}}.$$

Similarly, as $\sigma \to 0$, $x_M \to \exp(\mu - \sigma^2 + 1)$, and so for small σ,

$$\max_x r(x) \approx \{\sigma^2 \exp(\mu - \sigma^2 + 1)\}^{-1}.$$

The properties of the order statistics from lognormal parent distributions can usually only be derived numerically. Recently, Balakrishnan and Chen (1999) provided comprehensive tables on moments, variances, and covariances of order statistics for all sample sizes up to 25 and for several choices of the shape parameter σ.

4.4 CHARACTERIZATIONS

The lognormal distribution can conveniently be characterized by the maximum entropy property (similar to the Pareto, exponential, and normal distributions). In this particular case, we have the following result: If $E(\log X)$ and $E(\log^2 X)$ are prescribed, the lognormal distribution is the maximum entropy distribution on $[0, \infty)$ (Kapur, 1989, p. 68). (The result may, of course, also be stated as prescribing the geometric mean and variance of logarithms.)

From (4.3) we know that products of independent lognormal random variables are also lognormally distributed. This can be extended to a characterization parallelling Cramér's (1936) celebrated characterization of the normal distribution: If X_i, $i = 1, 2$, are independent and positive, then their product $X_1 X_2$ is lognormally distributed if and only if each X_i follows a lognormal distribution.

4.5 LORENZ CURVE AND INEQUALITY MEASURES

Unfortunately, the lognormal Lorenz curve cannot be expressed in a simple closed form; it is given implicitly by

$$L(u) = \Phi[\Phi^{-1}(u) - \sigma^2], \quad 0 < u < 1. \tag{4.16}$$

It follows directly from the monotonicity of Φ that the Lorenz order is linear within the family of two-parameter lognormal distributions, specifically

$$X_1 \geq_L X_2 \Longleftrightarrow \sigma_1^2 \geq \sigma_2^2. \tag{4.17}$$

This basic result can be derived in various other ways: First, as was noted above, the parameter $\exp(\mu)$ is a scale parameter and hence plays no role in connection to the Lorenz ordering. Thus, for $X_i \sim \text{LN}(0, \sigma_i^2)$, $i = 1, 2$, we have $X_1 \stackrel{d}{=} \phi(X_2)$, where $\phi(x) = x^{\sigma_1/\sigma_2}$; cf. (4.13). Therefore, $F_1^{-1}F_2 = \phi(F_2^{-1}F_2) = \phi$. This function is clearly convex iff $\sigma_1 \geq \sigma_2$. Hence, X_1 is more spread out than X_2 in the sense of the convex (transform) order that is known to imply the Lorenz ordering (see Chapter 2). Second, a slightly different argument due to Fellman (1976) uses the fact that the function ϕ is star-shaped, that is, that $\phi(x)/x$ is increasing on $[0, \infty)$, which implies that X_1 is more spread out than X_2 in the sense of the star-shaped ordering, an ordering that is also known to imply the Lorenz ordering (see Chapter 2). (In fact, the star-shaped ordering is intermediate between the convex and Lorenz orderings.) For a third approach, see Arnold et al. (1987), who showed that the result (4.17) follows also from the strong unimodality of the normal distribution.

The lognormal Lorenz curve has an interesting geometric property, namely, it is symmetric about the alternate diagonal of the unit square; see Figure 4.3. This can be verified analytically using Kendall's condition (2.8). At the point of intersection with

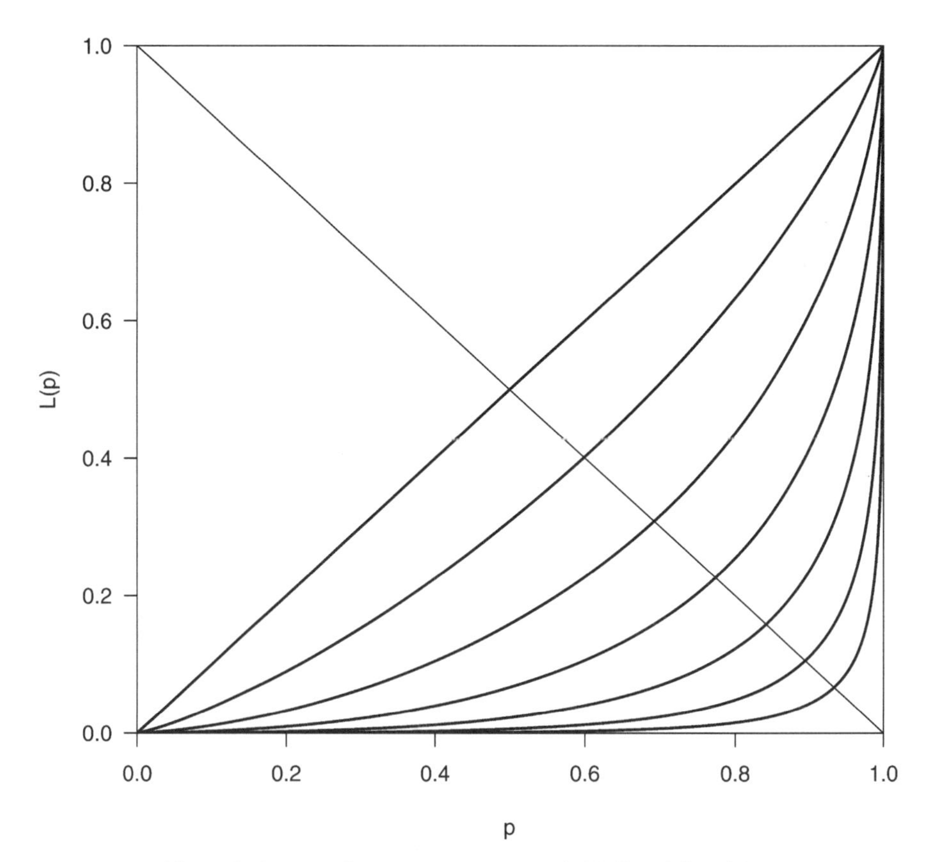

Figure 3 Lognormal Lorenz curves: $\sigma = 0.5(0.5)3$ (from left to right).

the alternate diagonal, that is, the point $[\varphi(\sigma/2), \varphi(-\sigma/2)]$, the slope of the Lorenz curve is unity (e.g., Moothathu, 1981).

The Lorenz curve may alternatively be represented in terms of the first-moment distribution. The c.d.f.'s of the higher-order moment distributions of lognormal distributions can be expressed as the c.d.f.'s of lognormal distributions with a different set of parameters (see, e.g., Aitchison and Brown, 1957). Specifically, for $X \sim \text{LN}(\mu, \sigma^2)$ we have

$$F_{(k);\mu,\sigma^2}(x) = F_{(0)}(x; \mu + k\sigma^2, \sigma^2), \tag{4.18}$$

where $F_{(0)}$ stands for the c.d.f. of the lognormal distribution.

This closure property is most useful in connection with size phenomena. In particular, the lognormal Lorenz curve can now be expressed as

$$\{[u, L(u)] | u \in [0, 1]\} = \{[F_{(0)}(x), F_{(1)}(x)] | x \in [0, \infty)\}.$$

It can also be exploited for the derivation of various inequality measures. The Pietra and Gini coefficients are remarkably simple; the former is given by

$$P = 2\Phi\left(\frac{\sigma}{2}\right) - 1 \tag{4.19}$$

and the Gini coefficient equals

$$G = 2\Phi\left(\frac{\sigma}{\sqrt{2}}\right) - 1. \tag{4.20}$$

The Theil measure is

$$T_1 = E\left[\frac{X}{E(X)} \log\left(\frac{X}{E(X)}\right)\right] = \frac{1}{2}\sigma^2, \tag{4.21}$$

which coincides with the expression for Theil's second measure T_2. The variance of logarithms is, of course,

$$\text{VL}(X) = \text{var}(\log X) = \sigma^2. \tag{4.22}$$

It is interesting that the preceding three coefficients are closely related in the case of the lognormal distribution, namely,

$$T_1 = T_2 = \frac{1}{2}\text{VL}(X)$$

[Theil (1967), see also Maasoumi and Theil (1979)]. It should also be noted that (4.22) is increasing in σ, in agreement with (4.17); thus, in the lognormal case the variance of logarithms satisfies the Lorenz ordering (Hart, 1975). (As mentioned in Chapter 2, this is not true in general.)

Two modifications of the first Theil coefficient with the median and the mode replacing the mean were provided by Shimizu and Crow (1988, p. 11). They are

$$T_{\text{med}}(X) = E\left[\frac{X}{x_{\text{med}}}\log\left(\frac{X}{x_{\text{med}}}\right)\right] = \sigma^2 \exp\left(\frac{\sigma^2}{2}\right)$$

and

$$T_{\text{mode}}(X) = E\left[\frac{X}{x_{\text{mode}}}\log\left(\frac{X}{x_{\text{mode}}}\right)\right] = 2\sigma^2 \exp\left(\frac{3\sigma^2}{2}\right).$$

The Zenga curve is of the form (Zenga, 1984)

$$Z(u) = 1 - e^{-\sigma^2}, \quad 0 \le u \le 1, \tag{4.23}$$

and hence constant. Note that it is an increasing function of the shape parameter σ; hence, inequality is increasing in σ according to both Lorenz and Zenga curves (Polisicchio, 1990). Zenga's first index is given by (Zenga, 1984)

$$\xi = 1 - \exp(-\sigma^2), \tag{4.24}$$

which is also the expression for his second measure ξ_2 (Pollastri, 1987a). Thus, like the two Theil coefficients, the two Zenga coefficients coincide in the lognormal case.

4.6 ESTIMATION

Due to its close relationship with the normal distribution, the estimation of lognormal parameters presents few difficulties. The maximum likelihood estimators are

$$\hat{\mu} = \overline{\log x} = \frac{1}{n}\sum_{i=1}^{n} \log x_j \tag{4.25}$$

and

$$\hat{\sigma}^2 = \frac{1}{n}\sum_{i=1}^{n} (\log x_j - \overline{\log x}). \tag{4.26}$$

The Fisher information on $(\mu, \sigma^2)^\top$ in one observation is

$$I(\mu, \sigma^2) = \begin{bmatrix} \frac{1}{\sigma^2} & 0 \\ 0 & \frac{1}{2\sigma^4} \end{bmatrix}. \tag{4.27}$$

Thus, the lognormal parameterization (4.1) enjoys the attractive property that the parameters are orthogonal.

From these expressions, parametric estimators of, for example, the mean, median, or mode, are easily available via the invariance of the ML estimators; asymptotic standard errors follow via the delta method. For example, Iyengar (1960) and Latorre (1987) derived the asymptotic distributions of the ML estimators of the Gini coefficient; the latter paper also presents the asymptotic distribution of Zenga's inequality measure ξ. For the Gini coefficient,

$$\hat{G} = 2\Phi\left(\frac{\hat{\sigma}}{\sqrt{2}}\right) - 1 \approx N\left[G, \frac{\sigma^2 e^{\sigma^2/2}}{2\pi n}\right]. \tag{4.28}$$

It is also worth noting that the UMVUE of σ^2, namely, $\tilde{\sigma}^2 = \sum_{j=1}^n (\log x_j - \overline{\log x})/(n-1)$, may be interpreted, in our context, as the UMVUE of the variance of logarithms, VL(X).

The unbiased estimation of some classical inequality measures from lognormal populations was studied by Moothathu (1989). He observes that for functionals

$$\tau(b, \lambda) = \text{erf}(b\lambda), \tag{4.29}$$

where $\text{erf}(x) = 2/\sqrt{\pi} \int_0^x e^{-t^2}$ is the error function and b is some known constant, UMVU estimators are given by

$$U_2(b) = h\left(b, \frac{n-1}{2}, -V_2^2\right),$$

when both parameters of the distribution are unknown, and

$$U_1(b) = h\left(b, \frac{n-1}{2}, -V_1^2\right),$$

when only σ is unknown. Here

$$V_1^2 = \frac{1}{2} \sum_{j=1}^n (\log X_j - \mu)^2$$

and

$$V_2^2 = \frac{1}{2}\sum_{j=1}^{n}(\log X_j - \overline{\log X})^2$$

and

$$h(b, m, t) = \frac{2b\sqrt{t}\Gamma(m)}{\sqrt{\pi}\Gamma(m+1/2)}\,{}_1F_2\left(\frac{1}{2}; \frac{3}{2}, m+\frac{1}{2}; -b^2 t\right),$$

where ${}_1F_2$ is a generalized hypergeometric function (see p. 288 for a definition). The Gini and Pietra measures are clearly of the form (4.29) (see Section 4.5 above), namely, $G = \text{erf}(\sigma/2)$ and $P = \text{erf}(\sigma/2^{3/2})$. Moothathu also provided unbiased estimators of their variances, as well as strongly consistent and asymptotically normally distributed estimators of G and P.

The optimal grouping of samples from lognormal populations was studied by Schader and Schmid (1986). Suppose that a sample $X = (X_1, \ldots, X_n)^\top$ of size n is available and the parameter of interest is σ. (This is the relevant parameter in connection with inequality measurement, as the formulas in Section 4.5 indicate.) By independence, the Fisher information of the sample for σ is

$$I(\sigma) = nI_1(\sigma) = \frac{2n}{\sigma^2},$$

where I_1 denotes the information in one observation. For a given number of groups k with group boundaries $0 = a_0 < a_1 < \cdots < a_{k-1} < a_k = \infty$, define the class frequencies N_j as the number of X_i in $[a_{j-1}, a_j), j = 1, \ldots, k$. Thus, the joint distribution of $N = (N_1, \ldots, N_k)^\top$ is multinomial with parameters n and $p_j = p_j(\sigma)$, where

$$p_j(\sigma) = \int_{a_{j-1}}^{a_j} f(x|\sigma)\,dx, \quad j = 1, \ldots, k.$$

Now the Fisher information in N is

$$I_N(\sigma) = n\sum_{j=1}^{k}\frac{[\partial p_j(\sigma)/\partial\sigma]^2}{p_j(\sigma)} = \frac{n}{\sigma^2}\sum_{j=1}^{k}\frac{[\phi(z_j)z_j - \phi(z_{j-1})z_{j-1}]^2}{\Phi(z_j) - \Phi(z_{j-1})},$$

where $z_0 = -\infty$, $z_k = \infty$, $z_j = (\log a_j - \mu)/\sigma, j = 1, \ldots, k-1$, and ϕ and Φ again denote the p.d.f. and c.d.f. of the standard normal distribution. This expression is, for fixed σ, a function of k and the $k-1$ class boundaries $a_1, \ldots, a_{k-1}$.

As was described in greater detail in Section 3.6 in the Pareto case, passing from the complete data X to the class frequencies N implies a loss of information, which

Table 4.2 Optimal Class Boundaries for Lognormal Data

γ	k^*	$z_1^*, \ldots, z_{k^*-1}^*$
0.1	9	−2.5408 −1.9003 −1.3715 −0.8355 0.8355...
0.05	13	−2.8602 −2.2974 −1.8646 −1.4793 −1.1011 −0.6833 0.6833...
0.025	17	−3.0834 −2.5636 −2.1764 −1.8460 −1.5424 −1.2466 −0.9398
		−0.5833 0.5833...
0.01	29	−3.5038 −3.0486 −2.7227 −2.4569 −2.2258 −2.0166 −1.8216
		−1.6355 −1.4541 −1.2736 −1.0898 −0.8971 −0.6861 −0.4341 0.4341...

Source: Schader and Schmid (1986).

may be expressed in terms of the decomposition

$$I_X(\sigma) = I_N(\sigma) + I_{X|N}(\sigma).$$

The relative loss of information is now given by

$$L = \frac{I_X(\sigma) - I_N(\sigma)}{I_X(\sigma)} = 1 - \frac{[\partial p_j(\sigma)/\partial\sigma]^2/p_j(\sigma)}{E_\theta[\partial \log f(X|\sigma)/\partial\sigma]^2} \; 1 - \frac{\sigma^2}{n} I_n(\sigma).$$

Thus, one requires, for given $\gamma \in (0, 1)$ and σ, the smallest integer k^* such that

$$L(\sigma; k^*; a_1^*, \ldots, a_{k-1}^*) \leq \gamma.$$

Schader and Schmid (1986) found that only for k odd there is a unique global optimum of this optimization problem and provided optimal class boundaries z_j^* based on the least odd number of classes k^* for which the loss of information is less than or equal to a given value of γ, for $\gamma = 0.1$, 0.05, 0.025, and 0.01. See Table 4.2. From the table, optimal class boundaries a_j^* for a lognormal distribution with parameters μ and σ can be obtained upon setting $a_j^* := \exp(\sigma z_j^* + \mu)$.

It is interesting that compared to the Pareto distribution (cf. Section 3.6), a considerably larger number of classes is required for a given loss of information.

4.7 THREE- AND FOUR-PARAMETER LOGNORMAL DISTRIBUTIONS

If there exists a $\lambda \in \mathbb{R}$ such that $Z = \log(X - \lambda)$ follows a normal distribution, then X is said to follow a three-parameter lognormal distribution. For this to be the case, it is clearly necessary that X take any value exceeding λ but have zero probability of taking any value below λ. Thus, the p.d.f. of X is

$$f(x) = \frac{1}{(x-\lambda)\sqrt{2\pi}\sigma} \exp\left\{ -\frac{1}{2\sigma^2} [\log(x-\lambda) - \mu]^2 \right\}, \quad x > \lambda. \tag{4.30}$$

As a size distribution, this distribution was already considered by Gibrat (1931). The distribution is obtained through one of the three transformations in Johnsons's (1949) translation system; see (2.76) in Chapter 2.

The location characteristics of the three-parameter form are greater by λ than those of the $\mathrm{LN}(\mu, \sigma^2)$ distribution. The mean is at $\lambda + \exp(\mu + \sigma^2/2)$, the median at $\lambda + \exp(\mu)$, and the mode at $\lambda + \exp(\mu - \sigma^2)$. The quantiles are displaced from $F^{-1}(u)$ to $\lambda + F^{-1}(u)$, $0 < u < 1$. The kth moment about λ is

$$E[(X-\lambda)^k] = \exp\left(k\mu + \frac{k^2\sigma^2}{2}\right) \tag{4.31}$$

so that the moments about the mean and hence the shape factors remain unchanged.

If the threshold parameter λ is known, as is, for example, the case in actuarial applications when it represents a deductible for claim amounts, estimation can, of course, proceed as described in Section 4.6 after adjusting the data by λ. If this parameter is unknown, there are considerable additional difficulties. Since the books by Cohen and Whitten (1988) and Crow and Shimizu (1988) contain several chapters studying estimation problems associated with the three-parameter lognormal distribution, we shall be rather brief here. The main difficulty appears to be that likelihood methods lead to an estimation problem with an unbounded likelihood. Specifically, Hill (1963) demonstrated that there exists a path along which the likelihood function

$$L(x_1, \ldots, x_n; \mu, \sigma^2, \lambda) = \prod_{i=1}^{n} f(x_i; \mu, \sigma^2, \lambda) \tag{4.32}$$

tends to $+\infty$ as (μ, σ^2, λ) approaches $(-\infty, \infty, x_{1:n})$.

One way of circumventing this difficulty consists of considering the observations as measured with error (being recorded only to the nearest unit of measurement), leading to a multinomial model as suggested by Giesbrecht and Kempthorne (1976).

With individual data considered error-free, Hill (1963) and Griffiths (1980) justified using estimates corresponding to the largest *finite* local maximum of the likelihood function. Smith (1985, p. 88) noted the existence of a local maximum that defines an asymptotically normal and efficient estimator. On differentiating the logarithm of the likelihood (4.32), we obtain the estimating equations

$$\frac{\partial \log L}{\partial \mu} = \frac{1}{\sigma^2}\sum_{i=1}^{n}[\log(x_i - \lambda) - \mu] = 0, \tag{4.33}$$

$$\frac{\partial \log L}{\partial \sigma} = -\frac{n}{\sigma} + \frac{1}{\sigma^3}\sum_{i=1}^{n}[\log(x_i - \lambda) - \mu]^2 = 0, \tag{4.34}$$

and

$$\frac{\partial \log L}{\partial \lambda} = \frac{1}{\sigma^2}\sum_{i=1}^{n}\frac{\log(x_i - \lambda) - \mu}{x_i - \lambda} + \sum_{i=1}^{n}\frac{1}{x_i - \lambda} = 0. \tag{4.35}$$

Eliminating μ and σ from these equations, we get (Cohen, 1951)

$$\gamma(\hat{\lambda}) := \sum_{i=1}^{n} \frac{1}{x_i - \hat{\lambda}} \left[\sum_{i=1}^{n} \log(x_i - \hat{\lambda}) - \sum_{i=1}^{n} \log^2(x_i - \hat{\lambda}) + \frac{1}{n} \left\{ \sum_{i=1}^{n} \log(x_i - \hat{\lambda}) \right\}^2 \right] - n \sum_{i=1}^{n} \frac{\log(x_i - \hat{\lambda})}{x_i - \hat{\lambda}} = 0, \quad (4.36)$$

a highly nonlinear equation in one variable. The probably most commonly used approach to ML estimation seems to be solving (4.36) by, for example, Newton–Raphson and subsequently obtaining estimates of μ and σ^2 from this solution (namely, the mean and variance of the logarithms of the data adjusted by $\hat{\lambda}$).

However, a practical difficulty appears to be that the search for the *local MLEs* (LMLEs) must be conducted with great care. Wingo (1984) argued that convergence problems in early attempts (prior to the mid-1970s) resulted from trying to find stationary values of the likelihood function using Newton's or related methods without proper safeguards for avoiding the region of attraction of the infinite maximum. He proposed avoiding the solution of $\gamma(\hat{\lambda}) = 0$ altogether and instead numerically maximizing a reparameterized conditional log-likelihood function with univariate global optimization methods. This involves a parameter transformation $\lambda(\theta) := x_{1:n} - \exp(-\theta)$, $\theta \in (-\infty, \infty)$, discussed by Griffiths (1980), that renders the reparameterized log-likelihood approximately quadratic and symmetric in the neighborhood of its finite (local) maximum.

The Fisher information on $\theta = (\mu, \sigma^2, \lambda)^\top$ in one observation is

$$I(\theta) = \begin{bmatrix} \frac{1}{\sigma^2} & 0 & \frac{e^{-\mu+\sigma^2/2}}{\sigma^2} \\ 0 & \frac{1}{2\sigma^4} & \frac{-e^{-\mu+\sigma^2/2}}{\sigma^2} \\ \frac{e^{-\mu+\sigma^2/2}}{\sigma^2} & \frac{-e^{-\mu+\sigma^2/2}}{\sigma^2} & \frac{e^{-2\mu+2\sigma^2}(1+\sigma^2)}{\sigma^2} \end{bmatrix}, \quad (4.37)$$

from which approximate variances of the MLEs can be obtained by inversion.

It is noteworthy that all ML estimators converge to their limiting values at the usual $\sqrt{n}$ rate. This is even the case for λ despite it being a threshold parameter. The reason is the high contact (exponential decrease) of the density for $x \to \lambda$.

A further estimation technique that is suitable in this context is maximum product-of-spacings (MPS) estimation, proposed by Cheng and Amin (1983). Here the idea is to choose $\theta = (\mu, \sigma, \lambda)^\top$ to maximize the geometric mean of the spacings

$$\text{GM} = \left\{ \prod_{i=1}^{n+1} D_i \right\}^{1/(n+1)},$$

where $D_i = F(x_i) - F(x_{i-1})$, or equivalently its logarithm. For the three-parameter lognormal distribution it can be shown that

$$\sqrt{n}(\tilde{\theta} - \theta) \xrightarrow{d} N[0, I(\theta)^{-1}].$$

Thus, the MPS estimators are asymptotically efficient and also converge at the usual $\sqrt{n}$ rate.

A modified method of moments estimator (MMME) has seen suggested by Cohen and Whitten (1980); see also Cohen and Whitten (1988). Here the idea is to replace the third sample moment by a function of the first order statistic (which contains more information about the shift parameter λ than any other observation). As with the local MLE, this leads to a nonlinear equation in one variable. Cohen and Whitten reported that it can be satisfactorily solved by the Newton–Raphson technique.

We note that a so-called four-parameter lognormal distribution has been defined by

$$Z = \mu^* + \sigma^* \log\left\{\frac{X - \lambda}{\delta}\right\}, \tag{4.38}$$

where Z denotes a standard normal random variable. Since this can be rewritten as

$$Z = \mu^{**} + \sigma^* \log(X - \lambda),$$

with $\mu^{**} = \mu^* - \sigma^* \log \delta$, it is really just the three-parameter lognormal distribution that is defined by (4.30).

4.8 MULTIVARIATE LOGNORMAL DISTRIBUTION

The most natural definition of a multivariate lognormal distribution is perhaps in terms of a multivariate normal distribution as the joint distribution of $\log X_i$, $i = 1, \ldots, k$. This approach leads to the p.d.f.

$$f(x_1, \ldots, x_k) = \frac{1}{(2\pi)^{n/2}\sqrt{|\Sigma|}x_1 \cdots x_k} \exp\left\{-\frac{1}{2}(\log x - \mu)^\top \Sigma^{-1} (\log x - \mu)\right\},$$

$$x_i > 0,\ i = 1, \ldots, k, \tag{4.39}$$

where $x = (x_1, \ldots, x_k)^\top$, $\log x = (\log x_1, \ldots, \log x_k)^\top$, $\mu = (\mu_1, \ldots, \mu_k)^\top$, and $\Sigma = (\sigma_{ij})$. If $X = (X_1, \ldots, X_k)^\top$ is a random vector following this distribution, this is denoted as $X \sim \mathrm{LN}_k\ (\mu, \Sigma)$. From the form of the moment-generating function of the multivariate normal distribution, we get

$$E(X_1^{r_1} \cdots X_k^{r_k}) = \exp\left(r^\top \mu + \frac{1}{2} r^\top \Sigma \mu\right),$$

where $r = (r_1, \ldots, r_k)^\top$. It follows that for any $i = 1, \ldots, k$

$$E(X_i^s) = \exp\left(s\mu_i + \frac{1}{2}s^2\sigma_{ii}^2\right),$$

and for any $i, j = 1, \ldots, k$

$$\operatorname{cov}(X_i, X_j) = \exp\left\{\mu_i + \mu_j + \frac{1}{2}(\sigma_{ii} + \sigma_{jj})\right\}\{\exp(\sigma_{ij}) - 1\}. \tag{4.40}$$

The conditional distributions of, for example, X_1 given $X_2, \ldots, X_k$ may be shown to be also lognormal. However, despite the close relationship with the familiar multivariate normal distribution, some differences arise. For example, although Pearson's measure of (pairwise) correlation ϱ_{Y_i,Y_j} can assume any value between -1 and 1 for the multivariate normal distribution (for which the marginals differ only in location and scale), the range of this coefficient is much narrower in the multivariate lognormal case and depends on the shape parameters σ_{ij}. Specifically, if Y is bivariate normal with unit variances, a calculation based on (4.40) shows that for the corresponding bivariate lognormal distribution the range of ϱ_{X_i,X_j} is

$$\left[\frac{e^{-1} - 1}{e - 1}, 1\right] = [-0.3679, 1],$$

so only a limited amount of negative correlation is possible. For further information on the dependence structure of the multivariate lognormal distribution, see Nalbach-Leniewska (1979).

In the context of income distributions, Singh and Singh (1992) considered a likelihood ratio test for comparing the coefficients of variation in lognormal distributions. Since from (4.8) the coefficients of variation are given by $\sqrt{\exp(\sigma_i^2) - 1}$, $i = 1, \ldots, k$, the problem is equivalent to comparing the variances of $Y_i = \log X_i$, $i = 1, \ldots, k$. Thus, the null hypothesis may be stated as H_0: $\sigma_1^2 = \cdots = \sigma_k^2 =: \sigma_0^2$ (say). More generally, any statistical test for the equality of lognormal characteristics depending only on the shape parameter σ—such as the coefficient of variation, the Gini coefficient, and various other inequality measures—is equivalent to a test for equality of the variances computed for the logarithms of the data, having a normal distribution, as noted by Iyengar (1960) who used this approach when testing for the equality of Gini coefficients. Thus, the problem may be solved, in the case where k independent samples are available, by the classical Bartlett test for the homogeneity of variances. If dependence must be taken into account, Singh and Singh suggested the following LR statistic:

$$Q = n\log|G| + n\operatorname{tr} G - nk,$$

where $G = I + n\hat{\sigma}_0^2 S^{-1} - S^{-1}S_d$, $\hat{\sigma}_0^2 = \operatorname{tr} S_d/nk$, and S is a standard estimator of the covariance matrix, namely, $S = (s_{ij})$, where $s_{ij} = \sum(Y_{il} - \bar{Y}_i)(Y_{jl} - \bar{Y}_j)$ and S_d is a diagonal matrix defined in terms of the diagonal of S. The approximate distribution of Q is a χ^2_{k-1}.

4.9 EMPIRICAL RESULTS

Being one of the two classical size distributions, a large number of empirical studies employing lognormal distributions are available.

Income Data

The lognormal distribution has been fit to various income data for at least the last 50 years. One of the earliest investigations was completed by Kalecki (1945) who considered it for the United Kingdom personal incomes for 1938–1939. He found the two-parameter lognormal fit for the whole range of incomes to be quite poor, but a two-parameter model for incomes above a certain threshold—that is, a three-parameter lognormal distribution—provides a good approximation.

Champernowne (1952) employed the three-parameter lognormal when studying Bohemian data of 1933. He found that a two-parameter log-logistic distribution fits as good as the lognormal.

Steyn (1959, 1966) considered income data for South African white males for 1951 and 1960 that are adequately described by a mixture of a lognormal and a doubly truncated lognormal distribution.

Employing a three-parameter lognormal distribution, Metcalf (1969) studied the changes between three distributional characteristics—the mean as well as income levels at certain bottom and top quantiles divided by median income—and aggregate economic activity by means of regression techniques for the period 1949–1965. He established separate patterns of movement in these measures for each of three family groups: families with a male head and a wife in the paid labor force, families with a male head and a wife not in the labor force, and families with a female head (these groups received about 88% of all personal income and almost 98% of all personal income going to families for the period under study). In particular, increases in real wages and employment rates appear to improve the relative position of low-income families that are labor-force-oriented and to lower the relative—but not absolute—position of high-income families. Also, families with a female head responded less elastically to employment and real wage changes than did families with a male head.

Using nonparametric bounds on the Gini coefficient developed by Gastwirth (1972), Gastwirth and Smith (1972) found that the implied Gini indices derived from two- and three-parameter lognormal distributions fall outside these bounds for U.S. individual gross adjusted incomes for 1955–1969 and concluded that lognormal distributions are inappropriate for modeling these data.

In a very thorough and exhaustive study, Kmietowicz and Webley (1975) fit the lognormal distribution to data for rural households from the 1963–1964 Income and Expenditure Survey of the Central Province of Kenya. They employed various fitting procedures in order to cope with some peculiarities of the data and found that the fit is better for the entire province than for any of its five districts. Also, they used lognormal distributions to "predict" the size distribution for urban households, for which only the average household income was available.

Kloek and van Dijk (1977) fit the lognormal distribution to Australian family disposable incomes for the period 1966–1968, disaggregated by age of the head of the family, occupation and education of the head of the family, and by family size. For some subsamples, the fit of the distribution is comparable to the log-t (which has one additional parameter); however the Champernowne distribution often performs better.

Kloek and van Dijk (1978) considered 1973 Dutch earnings data, to which they fit several income distributions. They found that a substantially better approximation (compared to the two-parameter lognormal distribution) is obtained by using three- and four-parameter families such as the log-t or Champernowne distributions.

McDonald and Ransom (1979a) considered the distribution of U.S. family income for 1960 and 1969 through 1975. When compared to alternative beta, gamma, and Singh–Maddala approximations using three different estimation techniques, the lognormal always provides the worst approximation in terms of sum of squared errors (SSE) and chi-square criteria.

In a detailed study comparing the performance of the Pareto and lognormal distributions, Harrison (1979, 1981) considered the gross weekly earnings of 91,968 full-time male workers aged 21 and over from the 1972 British New Earnings Survey, disaggregated by occupational groups and divided into 34 earnings ranges. For the aggregated data he found that the main body of the distribution comprising 85% of the total number of employees is "tolerably well described" by the lognormal distribution, whereas for the (extreme) upper tail the Pareto distribution is "distinctly superior." However, he pointed out that "the lognormal performs less well, even in the main body of the distribution, than is usually believed . . . ; and a strict interpretation . . . suggests that it [the Pareto distribution] applies to only a small part of the distribution rather than to the top 20% of all employees" [as implied approximately by Lydall's (1968) model of hierarchical earnings]. When disaggregated data divided into 16 occupational groups is considered, the fit of the lognormal distribution improves considerably, the strongest evidence for lognormality being found for the group "textiles, clothing and footwear." Nonetheless, there are still problems in the tails for some distributions, with the difficulties being more persistent in the lower tail in a number of cases.

Dagum (1983) estimated a two-parameter lognormal distribution for 1978 U.S. family incomes. The distribution is outperformed by wide margins by the Dagum type III and type I as well as the Singh–Maddala distribution (four- and three-parameter models, respectively), and even the two-parameter gamma distribution does considerably better. In particular, the mean income is substantially overestimated.

For the French wages stratified by occupation for 1970–1978 the three-parameter model outperforms a three-parameter Weibull distribution as well as a four-parameter beta type I model, but the Dagum type II, the Singh–Maddala, and a Box–Cox-transformed logistic appear to be more appropriate for these data (Espinguet and Terraza, 1983).

Arguing that in most studies income distribution functions are put forth as approximate descriptive devices that are not meant to hold exactly, Ransom and Cramer (1983) suggested employing a measurement error model, viewing observed income as the sum of a systematic component and an independent $N(0, \sigma^2)$ error term. Utilizing models with systematic components following Pareto, lognormal, and gamma distributions, they found that the lognormal variant performs best in terms of chi-square statistics for U.S. family incomes for 1960. However, these goodness-of-fit tests still reject all three models.

McDonald (1984) estimated the lognormal distribution for 1970, 1975, and 1980 U.S. family incomes. However, the distribution is outperformed by 9 out of 10 alternative models (of gamma or beta type), for all three data sets. In McDonald and Xu (1995), the distribution is outperformed by all 10 alternative models, again mainly of beta and gamma type, for 1985 U.S. family incomes.

Kmietowicz (1984) used a bivariate lognormal model for the distribution of household size and income when analyzing a subsample consisting of 200 rural households from the Household Budget and Living Conditions Survey, Iraq, for the period 1971–1972. The model was also fitted to data from other household budget surveys, namely, Iraq 1971–1972 (urban), Iraq 1976 (rural and urban), and Kenya 1963–1964 (rural). In all cases, the distribution of household income per head follows the lognormal distribution more closely than the marginal distribution of household income.

Kmietowicz and Ding (1993) considered the distribution of household incomes per head in the Jiangsu province of China (the city of Shanghai is located at the southeastern tip of this province), for 1980, 1983, and 1986. The fit is quite poor for the 1980 data but somewhat better for 1983 incomes; however, chi-square goodness-of-fit tests reject the lognormal distribution as an appropriate model. For 1986 this is no longer the case and hence the lognormal distribution may be considered appropriate for these data. (Note the substantial economic changes in China beginning in 1982–1983.)

For Japanese incomes for 1963–1971 the distribution is outperformed by the Singh–Maddala, Fisk, beta, and gamma distributions, with only the one-parameter Pareto (II) yielding inferior results (Suruga, 1982). Compared to eight other distributions utilizing income data from the 1975 Japanese Income Redistribution Survey (in grouped form), the lognormal ranks 6–8th for various strata in terms of SSE and information criteria (Atoda, Suruga, and Tachibanaki, 1988). In a later study modeling individual incomes from the same source, Tachibanaki, Suruga, and Atoda (1997) employed six different distributions; here the lognormal is almost always the worst model.

Henniger and Schmitz (1989) considered the lognormal distribution when comparing various parametric models (including, among others, gamma and Singh–Maddala) for the UK Family Expenditure Survey for the period 1968–1983 to nonparametric fittings. However, for the whole population all parametric models are rejected; for subgroups models such as the Singh–Maddala or Fisk perform considerably better than the lognormal, in terms of goodness-of-fit tests.

Summarizing research on the distribution of income in Poland over 50 years, Kordos (1990) observed that the two-parameter lognormal distribution describes the Polish data until 1980 with a reasonable degree of accuracy. In particular, for the distribution of monthly wages in 1973 the lognormal model compares favorably with alternative beta type I, beta type II, and gamma fittings.

Bordley, McDonald, and Mantrala (1996) fit the lognormal model to U.S. family incomes for 1970, 1975, 1980, 1985, and 1990. For all five data sets the distribution is outperformed by 13 out of 15 considered distributions, mainly of beta and gamma type, by very wide margins; only the (one-parameter) exponential distribution does worse.

Creedy, Lye, and Martin (1997) estimated the two-parameter lognormal distribution for individual earnings from the 1987 U.S. Current Population Survey (March Supplement). The distribution is outperformed, by wide margins, by a generalized lognormal-type distribution; see (4.67) below, as well as the standard and a generalized gamma distribution.

Botargues and Petrecolla (1997, 1999a,b) fit the lognormal distribution to the labor incomes for the province of greater Buenos Aires for each year from 1992–1997. However, the model is outperformed by several other distributions, notably the Dagum models.

Wealth Data

Sargan (1957) considered British wealth data for 1911–1913, 1924–1930, 1935–1938, and 1946–1947. Graphical methods indicate a fairly good approximation to a lognormal distribution.

Chesher (1979) estimated a lognormal model for the distribution of wealth in Ireland (grouped into 26 classes) in 1966 over the population of individuals with a recorded estate size. It is clear that the lognormal distribution is superior to the Pareto distribution on these data, with χ^2 and likelihood improvements of about 93%. In view of the conventional wisdom that the Pareto distribution is an appropriate model for the upper tail, it is particularly noteworthy that the fit is "unexpectedly good in the upper tail" (p. 7).

Bhattacharjee and Krishnaji (1985) fit the lognormal distribution to Indian data of landholdings, for 14 states for 1961–1962. However, the distribution is outperformed by both the gamma and loggamma distributions.

Firm Sizes

In his pioneering research, Kalecki (1945) considered the size distribution of factories (size being defined as "number of workers") in the U.S. manufacturing industry in 1937, finding the agreement between actual and calculated series to be "fairly good."

Observing that the empirical Lorenz curve for British firm size data is roughly symmetric about the alternate diagonal of the unit square, Hart and Prais (1956) approximated the distribution of firm sizes by a lognormal distribution, the best-known distribution possessing this property. However, in the discussion of the Hart and Prais paper, their choice was criticized by Champernowne (1956) and Kendall (1956), both of whom provided general expressions for distributions with this property.

Quandt (1966a), in a study investigating the distribution of firm sizes (size being measured in terms of assets) in the United States, found the lognormal distribution to be more appropriate than the Pareto distributions of types I–III for his data. He considered the *Fortune* lists of the 500 largest firms in the United States in 1955 and 1960, and 30 samples representing industries according to four-digit S.I.C. classes. However, Pareto type I and II distributions appear to fit the two *Fortune* samples rather well.

More recently, Stanley et al. (1995) used the lognormal to model the size distribution of American firms (by sales), noting that the model overpredicts in the upper tails. Hart and Oulton (1997) studied the size distribution (by employment) of 50,441 independent UK firms in 1993 and arrived at the opposite conclusion—for UK firms there is excess mass in the upper tail compared to a lognormal benchmark model. Voit (2001) considered the size (defined in terms of annual sales) of 570 German firms over the period 1987–1997. He noted that for these data the lognormal lower tail decreases too fast toward the abscissa.

Insurance Losses

The lognormal distribution is favored by a number of studies for a diverse variety of types of insurance.

Benckert (1962) studied industrial and nonindustrial fire losses and business interruption and accident insurance as well as automobile third-party insurance in Sweden for 1948–1952.

Ferrara (1971) employed a three-parameter lognormal distribution for modeling industrial fire losses in Italy for the period 1963–1965.

Benckert and Jung (1974) studied fire insurance claims for four types of houses in Sweden for the period 1958–1969, concluding that for one class of buildings ("stone dwellings") the lognormal distribution provides a reasonable fit.

Considering automobile bodily injury loss data, Hewitt and Lefkowitz (1979) employed the two-parameter lognormal distribution as well as a lognormal-gamma mixture. The latter model performs considerably better on these data.

Hogg and Klugman (1983) fit the lognormal distribution to a small data set (35 observations) of hurricane losses and found that it fits about as well as the Weibull distribution. They also considered data for malpractice losses, for which (variants of) Pareto distributions are preferable to the lognormal distribution.

Cummins et al. (1990) fit the two-parameter lognormal distribution to aggregate fire losses. However, most of the distributions they considered (mainly of the gamma and beta type) seem to be more appropriate. The same authors also considered data on the severity of fire losses and fit the lognormal distribution to both grouped and individual observations. Again, most of the other distributions they considered do considerably better for these data.

Burnecki, Kukla, and Weron (2000) used the lognormal distribution when modeling property insurance losses and found that it outperforms the Pareto distribution for these data.

Overall, it would thus seem that the popular lognormal distribution is not the best choice for modeling income, firm sizes, and insurance losses.

4.10 GENERALIZED LOGNORMAL DISTRIBUTION

The material in this section has been collected from diverse sources and to the best of our knowledge appears here in unified form for the first time.

Vianelli (1982a,b, 1983) proposed a three-parameter generalized lognormal distribution. It is obtained as the distribution of $X = \exp Y$, where Y follows a generalized error distribution, with density

$$f(y) = \frac{1}{2r^{1/r}\sigma_r\Gamma(1+1/r)}\exp\left\{-\frac{1}{r\sigma_r^r}|y-\mu|^r\right\}, \quad -\infty < y < \infty, \tag{4.41}$$

where $-\infty < \mu < \infty$ is the location parameter, $\sigma_r = [E|Y-\mu|^r]^{1/r}$ is the scale parameter, and $r > 0$ is the shape parameter. Like many of the distributions discussed in this book, the generalized error distribution is known under a variety of names and it was (re)discovered several times in different contexts. For $r = 2$ we arrive at the normal distribution and $r = 1$ yields the Laplace distribution. The generalized error distribution is thus known as both a generalized normal distribution, in particular in the Italian literature (Vianelli, 1963), and a generalized Laplace distribution. The generalized form was apparently first proposed by Subbotin (1923) in a Russian publication. Box and Tiao (1973) called it the *exponential power distribution*, the name under which this distribution is presumably best known in statistical literature, and used the following parameterization of the p.d.f.:

$$f(y) = \frac{1}{2^{(3+\beta)/2}\sigma\Gamma[(3+\beta)/2]}\exp\left\{-\frac{1}{2\sigma}|y-\mu|^{2/(1+\beta)}\right\}, \quad -\infty < y < \infty, \tag{4.42}$$

where $-1 < \beta \leq 1$, $\sigma > 0$. Here $\beta = 0$ corresponds to the normal and $\beta = 1$ to the Laplace distribution.

The exponential power distribution has been frequently employed in robustness studies, and also as a prior distribution in various Bayesian models (see Box and Tiao, 1973, for several examples).

If we start from (4.41), the density of $X = \exp Y$ is

$$f(x) = \frac{1}{2xr^{1/r}\sigma_r\Gamma(1+1/r)}\exp\left\{-\frac{1}{r\sigma_r^r}|\log x-\mu|^r\right\}, \quad 0 < x < \infty. \tag{4.43}$$

Here e^μ is a scale parameter and σ_r, r are shape parameters. The effect of the new parameter r is illustrated in Figures 4.4 and 4.5.

Figure 4.5 suggests that the density becomes more and more concentrated on a bounded interval with increasing r. This is indeed the case: For $\mu = 0$, $r \to \infty$, the limiting form is the distribution of the exponential of a random variable following a uniform distribution on $[-1, 1]$, as may be seen from the following argument given

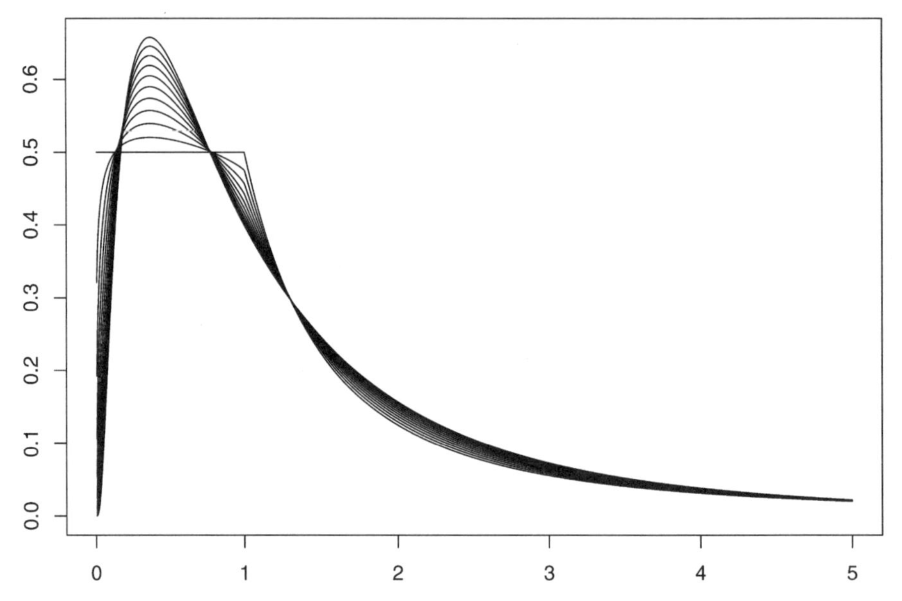

Figure 4 From the log-Laplace to the lognormal distribution: $\mu = 0$, $\sigma = 1$, and $r = 1(0.1)2$ (from the bottom).

by Lunetta (1963). The characteristic function of the generalized error distribution (4.41) is given by

$$\begin{aligned}\phi(t) &= \frac{1}{\sigma_r r^{1/r}\Gamma(1+1/r)}\int_0^\infty \exp\left\{-\frac{|x|^r}{r\sigma_r^r}\right\}\cos(tx)\,dx \\ &= \sum_{s=0}^{\infty}(-1)^s\frac{\Gamma[(2s+1)/r]t^{2s}}{r^{2s/r}\sigma_r^{2s}\Gamma(1/r)(2s)!}.\end{aligned} \tag{4.44}$$

Now,

$$\lim_{r\to\infty}\frac{\Gamma[(2s+1)/r]}{r^{2s/r}\sigma_r^{2s}\Gamma(1/r)} = \lim_{r\to\infty}\frac{\Gamma[1+(2s+1)/r]}{(2s+1)r^{2s/r}\sigma_r^{2s}\Gamma(1+1/r)} = \frac{1}{2s+1},$$

yielding

$$\lim_{r\to\infty}\phi(t) = \frac{1}{t}\sum_{s=0}^{\infty}(-1)^s\frac{t^{2s+1}}{(2s+1)!} = \frac{\sin t}{t},$$

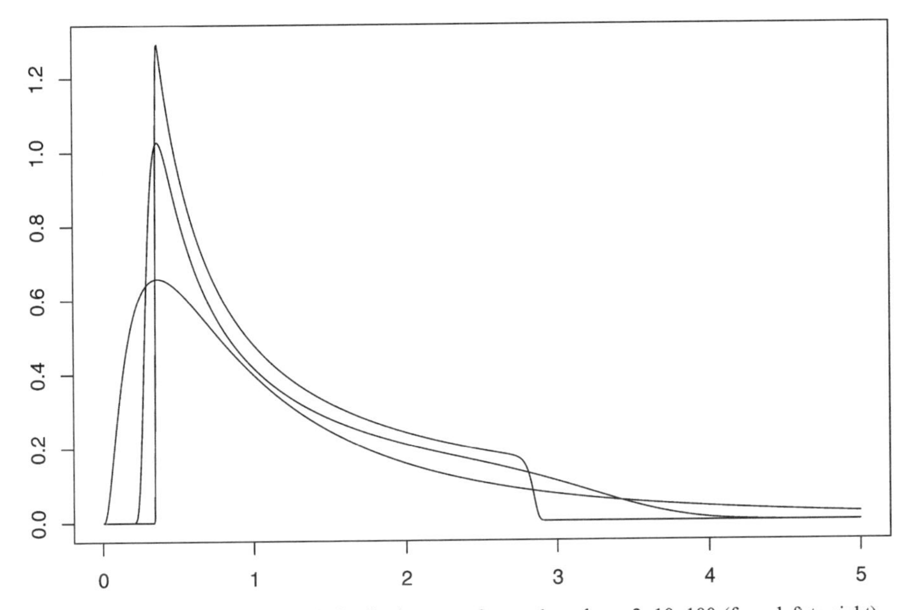

Figure 5 Beyond the lognormal distribution: $\mu = 0$, $\sigma = 1$, and $r = 2, 10, 100$ (from left to right).

which can be recognized as the characteristic function of a random variable distributed uniformly on the interval $[-1, 1]$. The resulting distribution therefore possesses the p.d.f.

$$f(x) = \frac{1}{2x}, \quad e^{-1} \leq x \leq e. \tag{4.45}$$

Interestingly, this is the p.d.f. of a doubly truncated Pareto-type variable, a distribution considered by Bomsdorf (1977) which was briefly mentioned in the preceding chapter, under the name of *prize-competition distribution*.

The case where $r = 1$ (the log-Laplace distribution) was proposed as an income distribution by Fréchet as early as 1939. Here closed forms for the c.d.f. and quantile function are available, namely,

$$F(x) = \begin{cases} \frac{1}{2}\exp\left(-\frac{\mu - \log x}{\sigma_1}\right), & \text{for } 0 < x < \exp\mu, \\ 1 - \frac{1}{2}\exp\left(-\frac{\log x - \mu}{\sigma_1}\right), & \text{for } x \geq \exp\mu, \end{cases}$$

and

$$F^{-1}(u) = \begin{cases} \exp\{\mu + \sigma_1 \log(2u)\}, & \text{for } 0 < u < 0.5, \\ \exp\{\mu - \sigma_1 \log[2(1-u)]\}, & \text{for } 0.5 \leq u < 1. \end{cases}$$

In the general case, the c.d.f. may be written in the form (Pollastri, 1987b)

$$F(x) = \begin{cases} \dfrac{\Gamma[1/r, B(x)]}{2\Gamma(1/r)}, & \text{for } x < \exp\mu, \\ \dfrac{1}{2}, & \text{for } x = \exp\mu, \\ \dfrac{1}{2} + \dfrac{\gamma[1/r, M(x)]}{2\Gamma(1/r)}, & \text{for } x > \exp\mu, \end{cases} \tag{4.46}$$

where $B(x) = [(\mu - \log x)/\sigma_r]^r/r$, $M(x) = [(\log x - \mu)/\sigma_r]^r/r$, and $\Gamma(\nu, x) = \int_x^\infty e^{-t}t^{\nu-1}\,dt$, $\gamma(\nu, x) = \int_0^x e^{-t}t^{\nu-1}\,dt$ are the incomplete gamma functions.

The moments of the generalized lognormal distribution are (see, e.g., Brunazzo and Pollastri, 1986)

$$E(X^k) = \frac{\exp(k\mu)}{\Gamma(1/r)} \sum_{i=0}^{\infty} \frac{(k\sigma_r)^{2i}}{(2i)!} r^{2i/r} \Gamma\left(\frac{2i+1}{r}\right). \tag{4.47}$$

For $r = 1$, that is, for the log-Laplace distribution, the infinite sum on the r.h.s. converges only for $|k\sigma_1| < 1$; in that case, it equals (Vianelli, 1982b)

$$E(X^k) = \frac{\exp(k\mu)}{1 - (k\sigma_1)^2}. \tag{4.48}$$

In particular, for $k = 1$ we get $E(X) = \exp\mu/(1 - \sigma_1^2)$. For generalized lognormal distributions with $r > 1$, all the moments exist. Brunazzo and Pollastri (1986) noted that the mean, variance, and standard deviation are increasing in σ_r and decreasing in r, whereas for Pearson's coefficient of skewness the opposite behavior is observed. Thus, the distribution becomes more symmetric as σ_r increases. Pollastri (1997) investigated the kurtosis of the distribution utilizing Zenga's (1996) kurtosis diagram and found that for fixed median and mean deviation from the median the kurtosis decreases as r and/or σ_r increase.

Since the exponential power distribution (4.41) is symmetric about μ, the median of the generalized lognormal distribution is given by

$$x_{\text{med}} = \exp(\mu) \tag{4.49}$$

and the mode equals

$$x_{\text{mode}} = \exp[\mu - \sigma_r^{r/(r-1)}] \quad \text{for } r > 1, \tag{4.50}$$

while

$$x_{\text{mode}} = \exp\mu \quad \text{for } r = 1,\ \sigma_1 < 1. \tag{4.51}$$

The distribution is thus unimodal, with a cusped mode in the log-Laplace case.

From (4.47), (4.49), and (4.50) it follows that the generalized lognormal distribution satisfies the mean-median-mode inequality, namely,

$$\frac{\exp \mu}{\Gamma(1/r)} \sum_{i=0}^{\infty} \frac{\sigma_r^{2i}}{(2i)!} r^{2i/r} \Gamma\left(\frac{2i+1}{r}\right) > \exp \mu > \exp[\mu - \sigma_r^{r/(r-1)}].$$

The entropy of the distribution is (Scheid, 2001)

$$E[-\log f(X)] = \frac{1}{r} - \log \frac{c}{\sigma_r} + \mu, \tag{4.52}$$

where $c = [2r^{1/r}\Gamma(1 + 1/r)]^{-1}$, which simplifies to

$$E[-\log f(X)] = 1 + \log(2\sigma_1) + \mu \tag{4.53}$$

for the log-Laplace distribution. It can be shown that the entropy is a decreasing function of r.

The Lorenz and Zenga curves may be obtained numerically in terms of the first-moment distribution; Pollastri (1987b) provided some illustrations. She observed that inequality, as measured by the Lorenz curve, is decreasing in r for fixed σ_r and increasing in σ_r for fixed r.

Expressions for inequality measures of the generalized lognormal are often somewhat involved. For the Gini coefficient there does not seem to be a simple expression for a general r. Pollastri (1987b) suggested evaluating numerically

$$G = 1 - \sum_{i=1}^{k} [F(x_i) - F(x_{i-1})][F_{(1)}(x_i) + F_{(1)}(x_{i-1})],$$

where $F(x_0) = F_{(1)}(x_0) = 0$. However, in the log-Laplace case there is a simple closed form for the Gini index (Moothathu and Christudas, 1992)

$$G = \frac{3\sigma_1^{-1}}{(4\sigma_1^{-2} - 1)}. \tag{4.54}$$

The formulas of Theil's inequality measures were derived by Scheid (2001). The (first) Theil coefficient is given by

$$T_1 = \frac{\exp(\mu)}{\Gamma(1/r)E(X)} \sum_{i=0}^{\infty} \frac{(\sigma_r)^{2i+2}}{(2i+1)!} r^{(2i+2)/r} \Gamma\left(\frac{2i+3}{r}\right)$$
$$- \log \{E(X)/\exp \mu\}, \tag{4.55}$$

and Theil's second measure equals

$$T_2 = \log[E(X)/\exp \mu]. \tag{4.56}$$

Again, in the log-Laplace case these expressions are simplified and reduced to

$$T_1 = \log(1 - \sigma_1^2) + \frac{2\sigma_1^2}{1 - \sigma_1^2} \tag{4.57}$$

and

$$T_2 = -\log(1 - \sigma_1^2), \tag{4.58}$$

respectively, provided $\sigma_1 < 1$.

Variants of the first measure with the median and mode replacing the mean, as suggested by Shimizu and Crow (1988) for the standard lognormal distribution, are (Scheid, 2001)

$$T_{\text{med}} = \frac{\sigma_r}{\Gamma(1/r)} r^{2/r} \sum_{i=0}^{\infty} \frac{(\sigma_r)^{2i+1}}{(2i+1)!} r^{2i/r} \Gamma\left(\frac{2i+3}{r}\right),$$

and, for $r > 1$,

$$T_{\text{mode}} = \frac{\sigma_r \exp[\sigma_r^{r/(r-1)}]}{\Gamma(1/r)} \sum_{i=0}^{\infty} \frac{(\sigma_r)^{2i}}{(2i)!} r^{2i/r} \left[\frac{\sigma_r}{2i+1} r^{2/r} \Gamma\left(\frac{2i+3}{r}\right) \right.$$
$$\left. + \sigma_r^{1/(r-1)} \Gamma\left(\frac{2i+1}{r}\right) \right].$$

Like the Gini coefficient, the Pietra coefficient can only be evaluated numerically. However, for a modified version where the mean deviation from the mean is replaced by the mean deviation from the median, Scheid (2001) provided the expression

$$P_{\text{med}} = \frac{E(|X - x_{\text{med}}|)}{2E(X)} = \frac{\sum_{i=0}^{\infty} [(\sigma_r)^{2i+1}/(2i+1)!\, r^{(2i+1)/r} \Gamma((2i+2)/r)]}{2\sum_{i=0}^{\infty} [(\sigma_r)^{2i}/(2i)!\, r^{2i/r} \Gamma((2i+1)/r)]}.$$

In the log-Laplace case this is simply

$$P_{\text{med}} = \frac{\sigma_1}{2},$$

implying that as in the lognormal case, inequality increases as the shape parameter σ_1 increases.

Finally, the variance of logarithms equals

$$\mathrm{VL}(X) = r^{2/r}\sigma_r^2 \frac{\Gamma(3/r)}{\Gamma(1/r)}. \tag{4.59}$$

It would seem that all the measures cited decrease with r and increase with σ_r, but a rigorous proof of this fact is still lacking.

Brunazzo and Pollastri (1986) suggested estimating the parameters via a method of moments estimation of the generalized normal parameters, that is, a method of moments estimation for the logarithms of the data. It is easy to see that μ and σ_r may be estimated by

$$\hat{\mu} = \frac{1}{n}\sum_{i=1}^{n} \log x_i \quad \text{and} \quad \hat{\sigma}_r = \left[\frac{1}{2}\sum_{i=1}^{n} |\log x_i - \overline{\log x}|^{\hat{r}}\right]^{1/\hat{r}},$$

once an estimate of r is available. To obtain such an estimate, one requires a certain ratio of absolute central moments of $\log X$. For the generalized normal distribution (Lunetta, 1963),

$$\mu_p' := E(|Y - \mu|^p) = \frac{1}{r^{p/r}\sigma_r^p} \frac{\Gamma[(p+1)/r]}{\Gamma(1/r)}.$$

Thus,

$$\beta_p := \frac{E(|Y - \mu|^{2p})}{E(|Y - \mu|^p)} = \frac{\Gamma(1/r)\Gamma[(2p+1)/r]}{\Gamma^2[(p+1)/r]}.$$

If we set $p = r$, this simplifies to

$$\beta_r = r + 1.$$

Thus, r may be estimated using the empirical counterpart of β_r. Brunazzo and Pollastri (1986) suggested solving the equation

$$\widehat{\mu_{2r}'} - (r+1)\widehat{\mu_r'}^2 = 0$$

by means of the *regula falsi*. [Using the parameterization of the exponential power distribution (4.42), for which the additional shape parameter β varies on a bounded set, Rahman and Gokhale (1996) suggested using the bisection method.]

Scheid (2001) considered the maximum likelihood estimation of the parameters. The gradient of the log-likelihood is given by

$$\frac{\partial \log L}{\partial \mu} = \frac{1}{\sigma_r}\sum_{i=1}^{n}\left|\frac{\log x_i - \mu}{\sigma_r}\right|^{r-1} \operatorname{sign}(\log x_i - \mu)$$

$$\frac{\partial \log L}{\partial \sigma_r} = -\frac{n}{\sigma_r} + \frac{1}{\sigma_r^{r+1}}\sum_{i=1}^{n} |\log x_i - \mu|^r$$

$$\frac{\partial \log L}{\partial r} = \frac{n \log r}{r^2} - \frac{n}{r^2} + \frac{n}{r^2}\psi\left(1 + \frac{1}{r}\right) + \frac{1}{r^2}\sum_{i=1}^{n}\left|\frac{\log x_i - \mu}{\sigma_r}\right|^r$$
$$- \frac{1}{r}\sum_{i=1}^{n}\left|\frac{\log x_i - \mu}{\sigma_r}\right|^{r-1}\log\left(\left|\frac{\log x_i - \mu}{\sigma_r}\right|^{r-1}\right),$$

where ψ denotes the digamma function. [Somewhat earlier, Bologna (1985) obtained the ML estimators in the log-Laplace case as well as their sampling distributions, and also the distributions of the sample median and sample geometric mean.]

The Fisher information matrix for the parameter $\theta := (\mu, \sigma_r, r)^\top$ can be shown to be (Scheid, 2001)

$$I(\theta) = \begin{bmatrix} \dfrac{(r-1)\Gamma(1-1/r)}{\sigma_r^2 r^{2/r}\Gamma(s)} & 0 & 0 \\ 0 & \dfrac{r}{\sigma_r^2} & -\dfrac{B}{r\sigma_r} \\ 0 & -\dfrac{B}{r\sigma_r} & \dfrac{s\psi'(s) + B^2 - 1}{r^3} \end{bmatrix}, \tag{4.60}$$

where $s := 1 + 1/r$ and $B := \log r + \psi(s)$.

This matrix coincides with the Fisher information of the generalized normal distribution, which had previously been derived by Agrò (1995); see also Rahman and Gokhale (1996). Both of these works use the term exponential power distribution.

An expression for the asymptotic covariances of the ML estimates is obtained by inversion of the Fisher information; thus, for a sample of size n we obtain

$$I^{-1}(\theta) = \frac{1}{n}\left\{\begin{matrix} \dfrac{\sigma_r^2 r^{2/r}\Gamma(s)}{(r-1)\Gamma(1-1/r)} & 0 & 0 \\ 0 & \dfrac{\sigma_r^2}{r}\left[1 + \dfrac{B^2}{s\psi'(s) - 1}\right] & \dfrac{rB\sigma_r}{s\psi'(s) - 1} \\ 0 & \dfrac{rB\sigma_r}{s\psi'(s) - 1} & \dfrac{r^3}{s\psi'(s) - 1} \end{matrix}\right\}.$$

For the generalized normal distribution, two simulation studies have been conducted in order to investigate the small sample behavior of the estimators.

Rahman and Gokhale (1996) found that the method of moments (MM) and ML estimators for μ and σ_r perform similarly for $r \leq 2$, whereas for $r > 2$ the ML estimator of r seems to perform better than its MM counterpart for small samples. For $r < 2$ the situation is reversed. Agrò (1995) noted that for samples of size $n \leq 100$ there is sometimes no well-defined optimum of the likelihood when $r \geq 3$ and that r is frequently overestimated in small samples for $r > 2$.

In both of these works, random samples were obtained using a rejection method following Tiao and Lund (1970). However, it is possible to generate simulated data using methods that exploit the structure of the generalized normal distribution. The following algorithm for the generation of samples from the generalized lognormal distribution makes use of a mixture representation utilizing the gamma (Ga) distribution and is adapted from Devroye (1986, p. 175):

- Generate $V \sim U[-1, 1]$ and $W \sim \text{Ga}(1 + 1/r, 1)$.
- Compute $Y := r^{1/r}\sigma_r V W^{1/r} + \mu$.
- Obtain $X = \exp(Y)$.

Jakuszenkow (1979) and Sharma (1984) studied the estimation of the variance of a generalized normal distribution. (In the terminology of inequality measurement, this is the variance of logarithms of the generalized lognormal distribution.) Since the variance of the distribution is a multiple of σ_r^2, one may equivalently study the estimation of σ_r^2. Sharma showed that the estimator

$$\frac{\Gamma[(n+2)/r](\sum_{i=1}^{n} |x_i|^k)^{2/k}}{\Gamma[(n+4)/k]}$$

is Lehmann-unbiased for the loss function $L(\theta) = (\hat{\theta} - \theta)\theta^{-2}$ and also admissible, for fixed r. Thus, the result may be best perceived as pertaining to the familiar special cases of the generalized lognormal family, the lognormal and log-Laplace distributions.

Further results on estimation are available in the log-Laplace case (where $r = 1$). Moothathu and Christudas (1992) considered the UMVU estimation of log-Laplace characteristics when $\mu = 0$. They noted, that the statistic $T = \sum_{i=1}^{n} |\log x_i|$ follows a gamma distribution and is complete as well as sufficient for σ_1. An unbiased estimator of σ_1^p, $p = 1, 2, \ldots,$ is given by

$$\hat{\sigma}_1^p = \frac{T^p}{(n)_p},$$

where $(n)_p$ denotes Pochhammer's symbol for the forward factorial function.

Furthermore, they showed that the UMVUE of the Gini coefficient is given by

$$\hat{G} = \frac{3T}{n} {}_1F_2\left[1; \frac{n+1}{2}, \frac{n+2}{2}; \frac{T^2}{16}\right],$$

where ${}_1F_2$ is a generalized hypergeometric function.

We conclude our brief survey of the generalized lognormal distribution by reporting on some empirical applications. The distribution was fitted to Italian and German income data with mixed success. Brunazzo and Pollastri (1986) used the distribution for approximating Italian data of 1948 and obtained a shape parameter $r = 1.4476$, considerably below the lognormal benchmark value of $r = 2$. Scheid (2001) fit the distribution to 1993 German household incomes (sample size: 40,000). Estimating by the method of moments and maximum likelihood, she found that the distribution improves upon the two-parameter lognormal, but the estimate of r is only slightly below 2. Although both likelihood ratio and score tests confirm the significance of the difference, the standard as well as the generalized lognormal distributions are empirically rejected using nonparametric goodness-of-fit tests.

Inoue (1978) postulated a stochastic process giving rise to the log-Laplace distribution and fit the distribution to British data for the period 1959–1960 by the method of maximum likelihood. He found that the fit is more satisfactory than for the lognormal for this period.

Vianelli (1982a, 1983) briefly considered a family of generalized lognormal distributions with bounded support. The p.d.f. is of the form

$$f(x) = \frac{r}{2\sigma_r(rq)^{1/r}B(1/r, q+1)x}\left(1 - \frac{1}{\sigma_r^r}|\log x - \mu|\right),$$

where $\mu e^{-\sigma_r(rq)^{1/r}} \leq x \leq \mu e^{\sigma_r(rq)^{1/r}}$. The logarithm of a random variable following this distribution with $r = 2$ can be viewed a Pearson type II distribution. For $b = \sigma_r(rq)^{1/r} \to \infty$, we get the generalized lognormal distribution (4.43).

4.11 AN ASYMMETRIC LOG-LAPLACE DISTRIBUTION

The preceding section presented a one-parameter family of generalizations for the lognormal distribution. Interestingly, an asymmetric variant of one of its members, the log-Laplace distribution (a generalized lognormal distribution with $r = 1$), appears in a recent dynamic model of economic size phenomena proposed by Reed (2001a,b, 2003).

He started from a continuous-time model with a varying but size-independent growth rate. The probably most widely known dynamic model possessing this property is a stochastic version of simple exponential growth, geometric Brownian motion, that is defined in terms of the stochastic differential equation

$$dX_t = \mu X_t\, dt + \sigma X_t\, dB_t, \tag{4.61}$$

where B_t denotes a standard Brownian motion. Hence, the proportional increase in X during dt has a systematic drift component $\mu\, dt$ and a stochastic diffusion component $\sigma\, dB_t$. This is essentially a continuous-time version of Gibrat's law of proportionate effect.

The novelty of Reed's approach lies in the assumption that the time of observation T is not fixed—this would lead to the well-known lognormal case—but random. An economic interpretation is that even if the evolution of each individual income follows a geometric Brownian motion (Gibrat's law of proportionate effect) when observing the income distribution at a fixed point in time, we may not know for how long a person has lived. If different age groups are mixed and the distribution of time in the workforce of any individual follows an exponential distribution, the observed distribution should be that of the state of the geometric Brownian motion stopped after an exponentially distributed time. What does this distribution look like?

For a geometric Brownian motion with a fixed initial state x_0, the conditional state at time T is lognormally distributed

$$Y|T := \log(X|T) \sim N\left[x_0 + \left(\mu - \frac{\sigma^2}{2}\right)T,\ \sigma^2 T\right].$$

Hence, $Y|T$ possesses the m.g.f.

$$m_{Y|T}(t) = \exp\left\{x_0 t + \left(\mu - \frac{\sigma^2}{2}\right)t + \frac{\sigma^2 t^2}{2}\right\}.$$

Assuming that T itself follows an exponential distribution, $T \sim \text{Exp}(\lambda)$, with m.g.f.

$$m_T(t) = \frac{\lambda}{\lambda - t},$$

we obtain for the m.g.f. of the unconditional state variable Y

$$m_Y(t) = E(e^{Yt}) = E_T[E(e^{Y|T\cdot t}|T)] = \frac{\lambda e^{x_0 t}}{\lambda + (\mu - \sigma^2/2)t - \sigma^2 t^2}, \tag{4.62}$$

which may be rewritten in the form

$$m_Y(t) = \frac{\lambda e^{x_0 t}\alpha\beta}{(\alpha - t)(\beta + t)}, \tag{4.63}$$

where α, $-\beta$ are the roots of the quadratic equation defined by the denominator of (4.62). These parameters are therefore functions of the drift and diffusion constants of the underlying geometric Brownian motion and of the scale parameter λ of the exponentially distributed random time T. Expression (4.63) is recognized as the m.g.f. of an asymmetric Laplace distribution (see, e.g., Kotz, Kozubowski, and

Podgórski, 2001), and so the unconditional distribution of the size variable $X = \exp(Y)$ randomly stopped at T is given by

$$f(x) = \begin{cases} \dfrac{x_0 \alpha\beta}{\alpha + \beta}\left(\dfrac{x}{x_0}\right)^{\beta - 1}, & x < x_0, \\ \dfrac{\alpha\beta}{x_0(\alpha + \beta)}\left(\dfrac{x}{x_0}\right)^{-\alpha - 1}, & x \geq x_0, \end{cases} \tag{4.64}$$

a density that exhibits power-law behavior in both tails. This is noteworthy, since the underlying geometric Brownian motion is essentially a multiplicative generative model and hence in view of the law of proportionate effect a lognormal distribution would be expected (see Section 4.2). Thus, a seemingly minor modification—introducing a random observational time—yields a power-law behavior. This type of effect was noticed some 20 years earlier by Montroll and Schlesinger (1982, 1983) who showed that a mixture of lognormal distributions with a geometric weighting distribution would have essentially a lognormal main part but a Pareto-type distribution in the upper tail.

Because of the power-law behavior in both tails, Reed referred to (4.64) as a *double-Pareto distribution*; in view of its genesis, it could also be called a "log-asymmetric Laplace distribution."

A generalization of the above model assumes that the initial state X_0 is also random, following a lognormal distribution. This yields an unconditional distribution to which Reed (2001b) referred as the *double-Pareto-lognormal distribution*. He estimated this four-parameter model for U.S. household incomes of 1997, Canadian personal earnings in 1996, 6-month household incomes in Sri Lanka for 1981, and Bohemian personal incomes in 1933 (considered earlier by Champernowne, 1952), for all of which the fit is excellent.

It is worth mentioning that the distribution (4.64) appears in a model of underreported income discussed by Hartley and Revankar (1974); see also Hinkley and Revankar (1977). In an underreporting model the goal is to make an inference about the distribution of the true income X_* when only a random sample from observable income X is available. It is therefore necessary to relate the p.d.f. of X to the parameters of the p.d.f. of X_*. Suppose the true but unobservable incomes X_* follow a Pareto type I distribution (3.2)

$$f(x_*) = \frac{\alpha}{x_0}\left(\frac{x_*}{x_0}\right)^{-\alpha - 1}, \quad x_0 \leq x_*,$$

and assume that observable income X is given by

$$X = X_* - U,$$

where U is the underreporting factor. It is natural to assume that $0 < U < X_*$. Hartley and Revankar (1974) postulated that the proportion of X_* that is underreported, denoted by

$$W_* = \frac{U}{X_*},$$

is distributed independently of X_* with the p.d.f.

$$f(w) = \beta(1-w)^{\beta-1}, \quad 0 \leq w \leq 1, \beta > 0,$$

a special case of the beta distribution. It is not difficult to show that the observable income X has the p.d.f. (4.64).

4.12 RELATED DISTRIBUTIONS

Since the t distribution can be viewed as a generalization of the normal distribution, it is not surprising that the log-t distribution has also been suggested as a model for the size distribution of incomes (Kloek and van Dijk, 1977) or of insurance losses (Hogg and Klugman, 1983). Its p.d.f. is

$$f(x) = \frac{\nu^{\nu/2}}{B(1/2, \nu/2)x} \cdot \left[\nu + \frac{(\log x - \log \mu)}{\sigma^2}\right]^{-(\nu+1)/2}, \quad x > 0, \tag{4.65}$$

where $\mu \in \mathbb{R}$, $\sigma, \nu > 0$. As in the case of the t distribution, no closed-form expression of the c.d.f. is available. From the properties of the t distribution, the variance of logarithms is given by

$$\text{VL}(X) = \frac{\nu\sigma^2}{\nu - 2}, \tag{4.66}$$

provided $\nu > 2$. Apparently it has not been appreciated in the econometrics literature how heavy the tails of this distribution are. Kleiber (2000b) pointed out that the log-t distribution does not have a single finite moment, that is, $E(X^k) = \infty$ for all $k \in \mathbb{R}\backslash\{0\}$. However, the Lorenz curve and most of the standard inequality measures only exist when the mean is finite. Specifically, for (4.65) the variance of logarithms is the only inequality measure among the common ones that exists, provided ν is sufficiently large. (In the case where $\nu = 1$, i.e., the log-Cauchy case, even the variance of logarithms is infinite.)

Hogg and Klugman (1983) presented the following interesting mixture representation for the log t distribution: Suppose X has a lognormal distribution, parameterized in the form

$$f(x \mid \theta) = \frac{1}{\theta}\sqrt{\frac{\theta}{2\pi}}\exp\left[\frac{-\theta(\log x - \mu)^2}{2}\right], \quad x > 0,$$

that is, $\log X$ has a normal distribution with mean μ and variance $1/\theta$, and $\theta \sim \mathrm{Ga}(\nu, \lambda)$. Then

$$
\begin{aligned}
f(x) &= \int_0^\infty \left\{ \frac{1}{\theta}\sqrt{\frac{\theta}{2\pi}} \exp\left[-\theta \frac{(\log x - \mu)^2}{2} \right] \right\} \left[\frac{\lambda^\nu \theta^{\nu-1} e^{-\lambda\theta}}{\Gamma(\nu)} \right] d\theta \\
&= \frac{\lambda^\nu \Gamma(\nu + 1/2)}{\sqrt{2\pi}\Gamma(\nu) x [\lambda + (\log x - \mu)^2]^{\nu+1/2}}.
\end{aligned}
$$

Thus, the log-t distribution may be considered a shape mixture of lognormal variates with inverse gamma weights.

Kloek and van Dijk (1977) fit a three-parameter log-t distribution to Australian family disposable incomes for the period 1966–1968, disaggregated by age, occupation, education, and family size. Although for about one half of the samples they considered, "one may doubt whether it is worthwhile to introduce the extra parameter [namely, ν]" (p. 447), for other cases the fit is considerably better. Using Cox tests (Cox, 1961), they found that the lognormal distribution is rejected when compared with the log-t, but not vice versa. Overall, they concluded that the log-t distribution appears to be a useful improvement over the lognormal.

Cummins et al. (1990) applied the log-t distribution to aggregate fire losses, a data set that seems to be better modeled by simpler distributions such as the inverse exponential or inverse gamma distribution.

Recently, Azzalini and Kotz (2002) fit a log-skewed-t distribution to U.S. family income data for 1970(5)1990 with rather encouraging but preliminary results.

Other generalized lognormal distributions—not to be confused with the distribution discussed in Section 4.10 above—were considered by Bakker and Creedy (1997, 1998) and Creedy, Lye, and Martin (1997). Their distributions arise as the stationary distribution of a certain stochastic model and possesses the p.d.f.

$$
f(x) = \exp\{\theta_1(\log x)^3 + \theta_2(\log x)^2 + \theta_3 \log x + \theta_4 x - \eta\}, \quad 0 < x < \infty, \tag{4.67}
$$

where $\exp(\eta)$ is the normalizing constant. Clearly, the two-parameter lognormal distribution is obtained for $\theta_1 = \theta_4 = 0$. (Observe that the gamma distribution is also a special case, arising for $\theta_1 = \theta_2 = 0$.)

Creedy, Lye, and Martin (1997) estimated this generalized lognormal distribution for individual earnings from the 1987 U.S. Current Population Survey (March Supplement), for which the model does about as well as a generalized gamma distribution and much better than the standard gamma and lognormal distributions. When applied to New Zealand wages and salaries for 1991, classified by age groups and sex, the distribution performs again consistently better than the two-parameter lognormal and about as well as the gamma distribution, in terms of chi-square criteria. However, in nine out of ten cases for males and in six out of ten cases for females it is outperformed by a generalized gamma distribution with the same number of parameters (Bakker and Creedy, 1997, 1998).

Saving (1965) used a S_B-type distribution (Johnson, 1949; see also Section 2.4) as a model for firm sizes that is a four-parameter lognormal-type distribution on a bounded domain.

Two additional distributions closely related to the lognormal, the Benini and Benktander type I distributions, will be discussed in Chapter 7.

CHAPTER FIVE

Gamma-type Size Distributions

For our purposes, gamma-type distributions comprise all distributions that are members of the generalized gamma family introduced by Luigi Amoroso in the 1920s, including the classical gamma and Weibull distributions and simple transformations of them, such as inverted or exponentiated forms.

The literature dealing with these models is rather substantial, notably in engineering and more recently in medical applications, and detailed accounts of the gamma and Weibull distributions are, for instance, available in Chapters 17 and 21 of Johnson, Kotz, and Balakrishnan (1994). There is a book-length treatment of estimation for the gamma distribution, written by Bowman and Shenton (1988).

Our exposition therefore focuses on the "size aspects" and log-gamma and log-Gompertz distributions, two distributions whose applications (as of today) appear to be mainly in connection with size phenomena.

5.1 GENERALIZED GAMMA DISTRIBUTION

In the American literature the generalized gamma distribution is most often referred to as the Stacy (1962) distribution although by now it is acknowledged that Amoroso's (1924–1925) paper in *Annali di Matematica* was probably the first work in which the generalized gamma distribution appeared. To this, we add the less known fact that D'Addario in the 1930s dealt with this generalization of the gamma distribution. In a report by A. C. Cohen (1969) entitled *A Generalization of the Weibull Distribution*, this distribution was rediscovered again! It is not out of the question that a more thorough search would locate a source for this distribution during the years 1940–1960.

Statistical Size Distributions in Economics and Actuarial Sciences, By Christian Kleiber and Samuel Kotz.
ISBN 0-471-15064-9

5.1.1 Definition and Interrelations

The classical gamma distribution is defined by

$$f(z) = \frac{1}{b^p \Gamma(p)} z^{p-1} e^{-z/b}, \quad z > 0, \tag{5.1}$$

where p, $b > 0$. If a power transformation X^a, $a > 0$, of some random variable X follows this distribution, the p.d.f. of X is

$$f(x) = \frac{a}{\beta^{ap} \Gamma(p)} x^{ap-1} e^{-(x/\beta)^a}, \quad x > 0. \tag{5.2}$$

Here $\beta = b^{1/a}$ is a scale and a, p are shape parameters. This was introduced by Amoroso (1924–1925) as the family of generalized gamma distributions. Amoroso originally considered a four-parameter variant defined by $X - \mu$, $\mu \in \mathbb{R}$, but we shall confine ourselves to the three-parameter version (5.2). We use the notation $X \sim \mathrm{GG}(a, \beta, p)$. It is sometimes convenient to allow for $a < 0$ in (5.2); one then simply replaces a by $|a|$ in the numerator. For clarity, we shall always refer to a generalized gamma distribution as the distribution with the p.d.f. (5.2) and $a > 0$; the variant with $a < 0$ will be called an *inverse generalized gamma distribution* and denoted as $\mathrm{InvGG}(a, \beta, p)$.

There are a number of alternative parameterizations: Amoroso used $(s, \gamma, p) \equiv (1/a, 1/\beta, p)$, Stacy (1962) employed $(a, \beta, d) \equiv (a, \beta, ap)$, whereas Taguchi (1980) suggested $(\alpha, \beta, h) \equiv (ap, \beta, 1/p)$. We shall use (5.2), which was also employed by McDonald (1984) and Johnson, Kotz, and Balakrishnan (1994).

The generalized gamma distribution is a fairly flexible family of distributions; it includes many distributions supported on the positive halfline as special or limiting cases:

- The gamma distribution is obtained for $a = 1$; hence, if $X \sim \mathrm{GG}(a, \beta, p)$, then $X^{1/a} \sim \mathrm{Ga}(b, p)$. In particular, the chi-squared distribution with ν degrees of freedom is obtained for $a = 1$ and $p = \nu/2$.
- The inverse gamma (or Vinci) distribution is obtained for $a = -1$.
- $p = 1$, $a > 0$ yields the Weibull distribution.
- $p = 1$, $a < 0$ yields the inverse Weibull distribution (to be called the log-Gompertz distribution in this chapter).
- $a = p = 1$ ($a = -1, p = 1$) yields the (inverse) exponential distribution.
- $a = 2, p = 1/2$ yields the half-normal distribution. This caused Cammillieri (1972) to refer to the generalized gamma distribution as the generalized seminormal distribution. More generally, all positive even powers and all

positive powers of the modulus of a normal random variable (with mean zero) follow a generalized gamma distribution.

- As $a \to 0$, $p \to \infty$, $\beta \to \infty$ but $a^2 \to 1/\sigma^2$ and $\beta p^{1/a} \to \mu$, the distribution tends to a lognormal LN(μ, σ^2).
- For $a \to 0$, $p \to \infty$, with $ap \to r$, $r > 0$, the distribution tends to a power function distribution PF(r, β). Since the power function distribution is the inverse Pareto distribution [see (3.38)], one directly determines that for $a \to 0, p \to -\infty$, with $ap \to -r$, the Pareto distribution Par(β, r) is also a limiting case.

Figure 5.1 summarizes the interrelations between the distributions that are included in the present chapter.

We should note that the preceding list comprises several of the most popular lifetime distributions. The generalized gamma distribution is also useful for discriminating among these models.

5.1.2 The Generalized Gamma Distribution as an Income Distribution

Esteban (1986) proposed characterizing income distributions in terms of their *income share elasticity*. If $\xi(x, x+h)$ denotes the share of total income earned by individuals with incomes in the interval $[x, x+h]$, we can write $\xi(x, x+h) =$

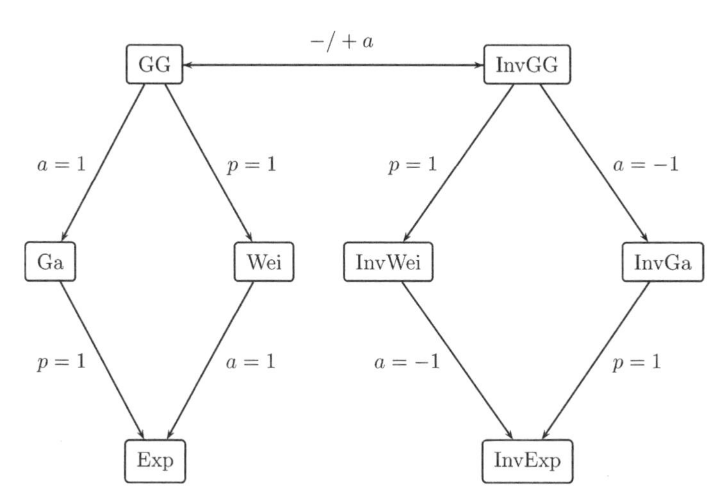

Figure 1 Gamma-type distributions and their interrelations: generalized gamma distribution (GG), inverse generalized gamma distribution (InvGG), gamma distribution (Ga), Weibull distribution (Wei), inverse Weibull (=log-Gompertz) distribution (InvWei), inverse gamma distribution (InvGa), exponential distribution (Exp), and inverse exponential distribution (InvExp).

$F_{(1)}(x+h) - F_{(1)}(x)$ and define the elasticity $\eta(x, f)$ of this quantity

$$\eta(x, f) = \lim_{h \to 0} \frac{d \log \xi(x, x+h)}{d \log x} \tag{5.3}$$

$$= \lim_{h \to 0} \frac{d \log \int_x^{x+h} t f(t)\, dt}{E(X) d \log x} \tag{5.4}$$

$$= \lim_{h \to 0} \frac{x(x+h) f(x+h) - x^2 f(x)}{E(X) \xi(x, x+h)} \tag{5.5}$$

$$= 1 + \frac{x f'(x)}{f(x)}. \tag{5.6}$$

It indicates the rate of change for the first-moment distribution (the distribution of income shares) at each income level. Since f is a density, the income share elasticity characterizes an income distribution. It is therefore possible to characterize the generalized gamma distribution in terms of $\eta(x, f)$: Suppose that, for a distribution supported on $(0, \infty)$,

- $\lim_{x \to \infty} \eta(x, f) = -\alpha$, for some $\alpha > 0$ (a type of weak Pareto law).
- There is at least one interior mode, that is, $f'(m) = 0$ for some $m \in (0, \infty)$.
- $\eta(x, f)$ exhibits a constant rate of decline, that is, either $\eta'(x, f) = 0$ or

$$\frac{d \log \eta'(x, f)}{d \log x} = -(1 + \varepsilon)$$

  for some $\varepsilon > -1$.

Upon integration, the third assumption can be rephrased as requiring either

$$\eta(x, f) = -\alpha + \frac{\delta}{\log x}, \quad \text{if } \varepsilon = 0,$$

or

$$\eta(x, f) = -\alpha + \frac{\delta}{|\varepsilon| x^{\varepsilon}}, \quad \text{if } \varepsilon \neq 0,$$

where α and δ are constants of integration.

If we combine the first and third assumption, it follows that both α and ε must be positive. If there is to be a unique maximum m, we must further have $\eta(m, f) = 1$ and therefore

$$m = \left[\frac{\delta}{(1 + \alpha)\varepsilon} \right]^{1/\varepsilon}.$$

Hence, $\delta > 0$. A reparameterization yields the desired elasticity

$$\eta(x, f) = -\alpha + (1+\alpha)\left(\frac{x}{m}\right)^{-\varepsilon}, \quad \alpha > 1, \varepsilon > 0, m > 0. \tag{5.7}$$

On the other hand, for the generalized gamma distribution (5.2) we have

$$\eta(x, f) = ap + a\left(\frac{x}{\beta}\right)^{a}. \tag{5.8}$$

A direct comparison shows that (5.7) defines a generalized gamma distribution with $-\varepsilon = a$ and $\alpha = -ap$, an inverse generalized gamma distribution in our terminology.

This shows that the (inverse) generalized gamma distribution can be derived from three salient features of an income distribution, all of which can be verified from empirical data.

5.1.3 Moments and Other Basic Properties

Like the c.d.f. of the standard gamma distribution (5.1) (see below), the c.d.f. of the generalized gamma distribution can be expressed in terms of Kummer's confluent hypergeometric function

$${}_1F_1(c_1; c_2; x) = \sum_{n=0}^{\infty} \frac{(c_1)_n}{(c_2)_n} \frac{x^n}{n!}, \tag{5.9}$$

where $(c)_n = c(c+1)(c+2)\cdots(c+n-1)$ is Pochhammer's symbol, in the form

$$F(x) = \frac{e^{-(x/\beta)^a}(x/\beta)^{ap}}{\Gamma(p+1)} {}_1F_1\left[1; p+1; \left(\frac{x}{\beta}\right)^a\right], \quad x \geq 0. \tag{5.10}$$

Equivalently, it can be expressed in terms of an incomplete gamma function ratio

$$F(x) = \frac{1}{\Gamma(p)} \int_0^z t^{p-1} e^{-t} dt, \quad x \geq 0, \tag{5.11}$$

where $z = (x/\beta)^a$. Here $\gamma(\nu, z) = \int_0^z t^{\nu-1} e^{-t} dt$ is often called an *incomplete gamma function*, although in the statistical literature this name is sometimes also used in connection with (5.11).

The moments of the distribution (5.2) are given by

$$E(X^k) = \frac{\beta^k \Gamma(p + k/a)}{\Gamma(p)}. \tag{5.12}$$

Hence, the first moment is

$$E(X) = \frac{\beta\Gamma(p + 1/a)}{\Gamma(p)} \tag{5.13}$$

and the variance equals

$$\text{var}(X) = \beta^2 \left\{ \frac{\Gamma(p+2/a)\Gamma(p) - [\Gamma(p+1/a)]^2}{[\Gamma(p)]^2} \right\}. \tag{5.14}$$

The expressions for the skewness and kurtosis coefficients are rather lengthy and therefore not given here. It is however interesting that there is a value $a = a(p)$ for which the shape factor $\sqrt{\beta_1} = 0$. For $a < a(p)$, $\sqrt{\beta_1} < 0$; for $a > a(p)$, $\sqrt{\beta_1} > 0$. This property of the generalized gamma distribution is inherited by the Weibull distribution discussed below. Figures 5.2 and 5.3 depict some generalized gamma densities, including left-skewed examples.

As noted in the preceding chapter, the best known distribution that is not determined by the sequence of its moments (despite all the moments being finite) is the lognormal distribution. Pakes and Khattree (1992) showed that the generalized gamma distribution provides a further example of a distribution possessing this somewhat pathological and unexpected property. Specifically, the distribution is determined by the moments only if $a > \frac{1}{2}$, whereas for $a \leq \frac{1}{2}$, any distribution with p.d.f.

$$f(x)\{1 - \epsilon \sin(2\pi ap + x^a \tan 2\pi a)\} \tag{5.15}$$

has the same moments for $-1 \leq \epsilon < 1$.

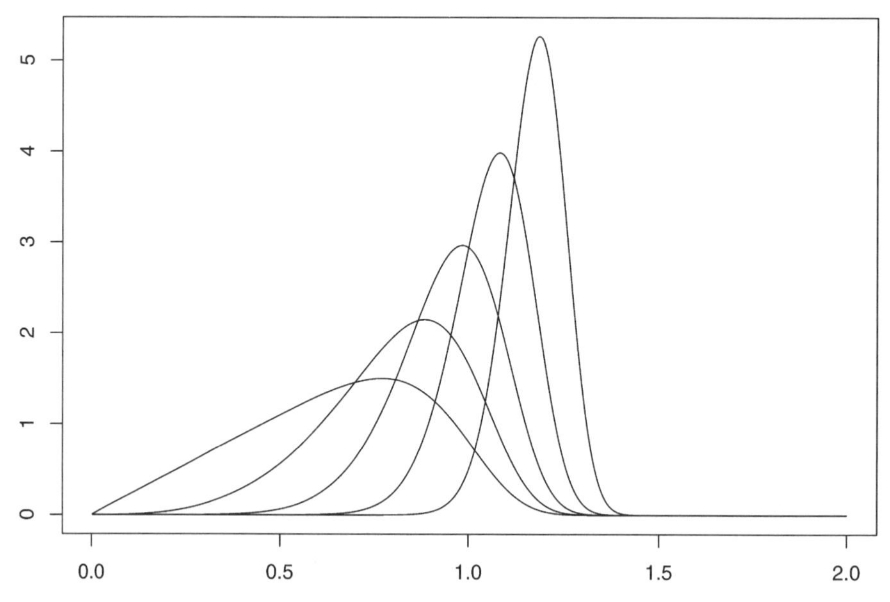

Figure 2 Generalized gamma densities: $a = 8$, $\beta = 1$, and $p = 0.25, 0.5, 1, 2, 4$ (from left to right).

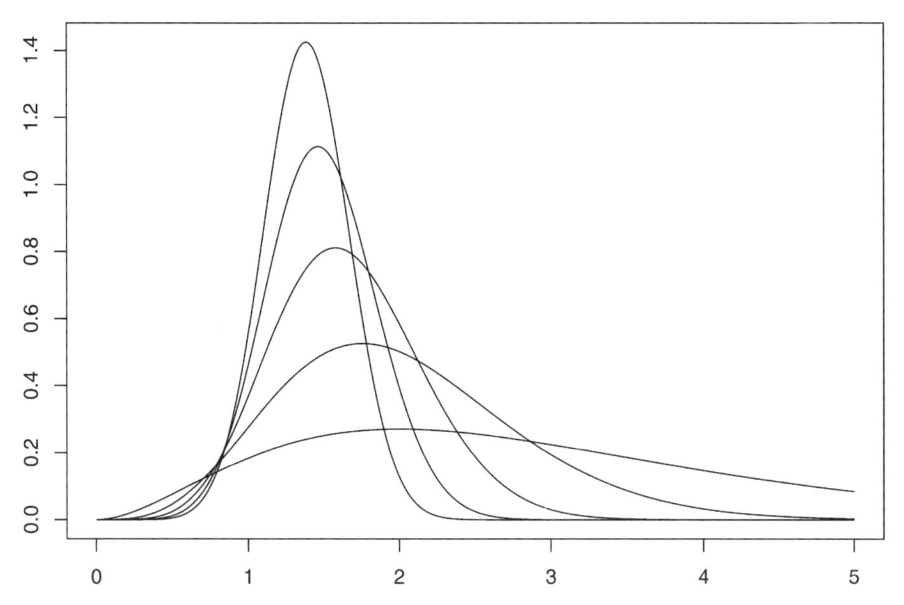

Figure 3 Generalized gamma densities: $p = 4$, $\beta = 1$, and $a = 1, 1.5, 2, 2.5, 3$ (from right to left).

The mode of the generalized gamma distribution occurs at

$$x_{\text{mode}} = \beta\left(p - \frac{1}{a}\right)^{1/a}, \quad \text{for } ap > 1. \tag{5.16}$$

Otherwise, the distribution is zeromodal with a pole at the origin if $ap < 1$.

The generalized gamma distribution allows for a wide array of shapes of the hazard rate. The situation is best analyzed utilizing general results due to Glaser (1980). He considered the reciprocal hazard rate

$$g(x) = \frac{1}{r(x)} = \frac{1 - F(x)}{f(x)}$$

whose derivative is

$$g'(x) = g(x)\vartheta(x) - 1,$$

where

$$\vartheta(x) = -\frac{f'(x)}{f(x)}.$$

The shape of $r(x)$ now depends on the behavior of ϑ'. It is not difficult to see that $\vartheta'(x) > 0$ ($\vartheta'(x) < 0$), for all $x > 0$, implies an increasing (decreasing) hazard rate.

If ϑ' changes signs, with $\vartheta'(x_0) = 0$ for some $x_0 > 0$, and $\vartheta'(x) < 0$ for $x < x_0$ and $\vartheta'(x) > 0$ for $x > x_0$, we have an increasing hazard rate if $\lim_{x\to 0} f(x) = 0$ and a $\bigcup$-shaped hazard rate if $\lim_{x\to 0} f(x) = \infty$. Similarly, we obtain decreasing and $\bigcap$-shaped hazard rates if the inequalities in the preceding conditions are reversed.

For the generalized gamma distribution the function ϑ is

$$\vartheta(x) = \frac{1 - ap}{x} + \frac{ax^{a-1}}{\beta^a}.$$

Hence,

$$\vartheta'(x) = \frac{x^a[a(a-1)] + (ap-1)\beta^a}{x^2\beta^a}.$$

This shows that the overall shape of the hazard rate ultimately depends on the signs of $a(a-1)$ and $ap - 1$. The salient features now follow easily from the preceding conditions; Table 5.1 summarizes the results.

A closer look reveals that $\bigcup$- and $\bigcap$-shaped hazard rates are only possible if neither a nor p equals 1; hence, these cases cannot occur with the gamma or Weibull distributions.

5.1.4 Lorenz Curve and Inequality Measures

The quantile function of the generalized gamma distribution is not available in closed form; hence, we must use the representation of the Lorenz curve in terms of the first moment distribution. From Butler and McDonald (1989) we know that the kth moment distribution is

$$F_{(k)}(x; a, \beta, p) = F\left(x; a, \beta, p + \frac{k}{a}\right), \quad x \geq 0, \tag{5.17}$$

and therefore of the same form as the underlying distribution. Utilizing (2.6), we obtain

$$\{[u, L(u)]\} = \left\{\left[F(x; a, \beta, p), F\left(x; a, \beta, p + \frac{1}{a}\right)\right] \middle| x \in (0, \infty)\right\}. \tag{5.18}$$

Table 5.1 Hazard Rates of Generalized Gamma Distributions

Sign of $a(a-1)$	Sign of $ap - 1$	Shape of $r(x)$
$-$	$-$	decreasing
$-$	$+$	$\bigcap$-shaped
$+$	$-$	$\bigcup$-shaped
$+$	$+$	increasing

Source: Glaser (1980), McDonald and Richards (1987).

Regarding the Lorenz ordering, Taillie (1981) asserted (without detailed derivation) the result

$$X_1 \geq_L X_2 \Longleftrightarrow a_1 \leq a_2 \quad \text{and} \quad a_1 p_1 \leq a_2 p_2. \tag{5.19}$$

A detailed proof using a density crossing argument was later provided by Wilfling (1996a).

For the Gini coefficient McDonald (1984) derived the expression

$$G = \frac{1}{2^{2p+1/a} B(p, p+1/a)} \left\{ \left(\frac{1}{p}\right) {}_2F_1\left(1, 2p + \frac{1}{a}; p+1; \frac{1}{2}\right) - \left(\frac{1}{p+1/a}\right) {}_2F_1\left(1, 2p + \frac{1}{a}; p+1+\frac{1}{a}; \frac{1}{2}\right) \right\}. \tag{5.20}$$

Special cases of this result were already known to Amoroso (1924–1925).

5.1.5 A Compound Generalized Gamma Distribution

Starting from (5.2), compound gamma distributions can be constructed by assigning (joint) distributions to the parameters a, β, p. If the parameter β itself follows a three-parameter inverse generalized gamma distribution (with the *same* parameter a as the structural distribution),

$$f(\beta) = \frac{|a|}{\Gamma(q)} \beta^{aq-1} e^{-\beta^a} \quad (a < 0),$$

the resulting compound distribution has the p.d.f. (Malik, 1967; Ahuja, 1969)

$$f(x) = \frac{a x^{ap-1}}{B(p, q)(1 + x^a)^{p+q}}, \quad x > 0, \tag{5.21}$$

which is the density of a power transformation of a random variable following a Pearson type VI distribution (or beta distribution of the second kind). This family will be studied in greater detail in the following chapter; for the moment it should be noted that equation (5.21) provides the link between Chapters 5 and 6.

5.1.6 Estimation

In early work with the generalized gamma distribution, there were significant problems in developing inference procedures. The essential difficulty is the estimation of the additional (compared to the classical gamma distribution) shape parameter a. In fact, if a is known, one can apply the transformation X^a and use the methods appropriate to gamma distributions (as described in, e.g., Bowman and Shenton, 1988).

Much of the difficulty with the model arises because distributions with rather different sets of parameters can look very much alike. For example, the work of Johnson and Kotz (1972) showed that for certain values $a < 0$ (an inverse generalized gamma distribution in our terminology) two generalized gamma distributions exist for certain constellations of the shape factors $\sqrt{\beta_1}$ and β_2. Consequently, it will not be possible to estimate such distributions by moments alone.

Maximum likelihood estimation is also not straightforward. Unfortunately, the likelihood function is in general not unimodal, nor does it necessarily exhibit a maximum. There are two main approaches for obtaining MLEs. The first consists of the direct maximization of the likelihood; see Lawless (1980) who used a parameterization of the distribution of $Y = \log X$ proposed by Prentice (1974). Here the likelihood is maximized over a subset of the parameters, with the remaining parameters temporarily held fixed at some initial values. This is followed by a heuristic interpolation scheme that attempts to refine further the values of the fixed parameters. This approach is not very efficient computationally and guarantees at most a local maximum of the likelihood function. It also relies to some extent on the judgment of the statistician when determining appropriate values for the parameters that are held fixed.

The second approach, proposed by Hager and Bain (1970), suggests solving a scalar nonlinear equation derived from the likelihood equations

$$-n\hat{p} + \sum_{i=1}^{n} \left(\frac{X_i}{\hat{\beta}} \right)^{\hat{a}} = 0, \tag{5.22}$$

$$\frac{n}{\hat{a}} + \hat{p} \sum_{i=1}^{n} \log\left(\frac{X_i}{\hat{\beta}} \right) - \sum_{i=1}^{n} \left(\frac{X_i}{\hat{\beta}} \right)^{\hat{a}} \log\left(\frac{X_i}{\hat{\beta}} \right) = 0, \tag{5.23}$$

$$-n\psi(\hat{p}) + \hat{a} \sum_{i-1}^{n} \left(\frac{X_i}{\hat{\beta}} \right) = 0. \tag{5.24}$$

The first equation (5.22) yields an expression for the scale parameter in terms of the shape parameters, $\hat{\beta} \equiv \hat{\beta}(\hat{a}, \hat{p}) = [\sum_{i=1}^{n} X_i^{\hat{a}}/(n\hat{p})]^{1/\hat{a}}$. Upon substituting this expression into the second equation (5.23), we get

$$\hat{p} \equiv \hat{p}(\hat{a}) = \left\{ \hat{a} \left[\left(\sum_{i=1}^{n} \frac{\log X_i}{n} \right) - \frac{(\sum_{i=1}^{n} X_i^{\hat{a}} \log X_i)}{(\sum_{i=1}^{n} X_i^{\hat{a}})} \right] \right\}^{-1}. \tag{5.25}$$

Substituting this into the remaining equation (5.24) gives us an equation in $\hat{a}$

$$\phi(\hat{a}) := -\psi(\hat{p}) + \frac{\hat{a}}{n} \sum_{i=1}^{n} \log X_i - \log\left(\sum_{i=1}^{n} X_i^{\hat{a}} \right) + \log(n\hat{p}) = 0, \tag{5.26}$$

where $\hat{p}$ is determined by (5.25). When trying to solve the preceding equations, for example, Hager and Bain (1970) reported persistent divergence using an unstabilized Newton algorithm.

Wingo (1987b) argued that these problems stem from an inappropriate use of zero-finding algorithms striving for fast local convergence that often diverge upon application to highly nonlinear problems such as the present one. Also, (5.26) is defined only for $a > 0$; hence, any iterative numerical procedure must assure that only positive iterates are obtained. Wingo (1987a) showed that $\hat{a} = 0$ is a double root of (5.26), which explains to some extent the numerical problems reported in the earlier literature.

Wingo recommended (1987b) a derivative-free numerical root isolation method developed by Jones, Waller, and Feldman (1978) that assures globally optimal maximum likelihood estimators for the parameters. The procedure is able to locate all of the zeros of the likelihood equations or indicate that none exist. The method determines all of the real zeros of $\phi(\hat{a})/\hat{a}^2 = 0$ on a sufficiently large interval [$\phi(\hat{a})$ being defined in (5.26)], if any, and obtains estimates of p and β from (5.25) and (5.22). For the resulting sets of parameter estimates, the likelihoods are compared and the parameter set with the largest likelihood is selected. Wingo applied this method to three data sets containing zero, one, and two solutions of $\phi(\hat{a})/\hat{a}^2 = 0$, respectively. [In Wingo (1987a) it is also conjectured, based on extensive computational tests, that $\phi(\hat{a}) = 0$ never has more than two positive zeros.]

The Fisher information on $\theta = (a, b, p)^\top$ is given by

$$I(\theta) = \begin{pmatrix} \frac{1}{a^2}\{1 + \psi(p)[2 + \psi(p)] + p\psi'(p)\} & -\frac{1 + p\psi(p)}{b} & -\frac{\psi(p)}{a} \\ -\frac{1 + p\psi(p)}{b} & \frac{a^2 p}{\beta^2} & \frac{1}{\beta} \\ -\frac{\psi(p)}{a} & \frac{1}{\beta} & \psi'(p) \end{pmatrix}, \tag{5.27}$$

from which the asymptotic covariance matrix of $\sqrt{n}(\hat{a}, \hat{\beta}, \hat{p})^\top$ can be obtained by inversion.

In our context, Kloek and van Dijk (1978) reported that for their data the asymptotic correlation matrix of generalized gamma parameter estimates is nearly singular, with highly correlated estimates $\hat{a}$ and $\hat{p}$. This once again underlines the problems associated with parameter estimation in connection with this distribution.

5.1.7 Empirical Results

Incomes and Wealth

Amoroso (1924–1925) fit a four-parameter generalized gamma distribution to Prussian incomes of 1912. Some 50 years later Bartels (1977) applied the three-parameter version to 1969 fiscal incomes for three regions in the Netherlands and found that it does better than the gamma and Weibull special cases, but not as good as the log-logistic and Champernowne distributions.

Analyzing 1973 Dutch earnings, Kloek and van Dijk (1978) determined the generalized gamma distribution to be superior to the gamma and lognormal models but also reported on numerical problems. They preferred other three-parameter models such as the Champernowne and log-t distributions.

McDonald (1984) estimated the generalized gamma distribution for 1970, 1975, and 1980 U.S. family incomes. It outperforms eight other distributions and is inferior to only the GB2 and Singh–Maddala distributions (see the following chapter).

Atoda, Suruga, and Tachibanaki (1988) considered grouped data from the Japanese Income Redistribution Survey for 1975, stratified by occupation. Among the distributions they employed, the Singh–Maddala appears to be the most appropriate for the majority of strata. It is noteworthy that when a model is selected via information criteria such as the AIC, the generalized gamma is always inferior to one of its special cases, the Weibull and gamma distributions. In a later study employing individual data from the same source, the generalized gamma was sometimes the best distribution in terms of likelihood but only marginally better than its Weibull and gamma special cases (Tachibanaki, Suruga, and Atoda, 1997).

In a comprehensive study employing 15 income distribution models of the beta and gamma type, Bordley, McDonald, and Mantrala (1996) fit the generalized gamma distribution to U.S. family incomes for 1970, 1975, 1980, 1985, and 1990. It is outperformed by the (G)B2, (G)B1, Dagum, and Singh–Maddala distributions—see the following chapter—but does significantly better than all two-parameter models considered.

In an application using 1984–1993 German household incomes, the generalized gamma distribution is revealed as inappropriate model for these data (Brachmann, Stich, and Trede, 1996). Specifically, it does not provide an improvement over the two-parameter gamma distribution. The data seem to require a more flexible model such as the GB2 and Singh–Maddala distribution.

Actuarial Losses

In the actuarial literature, Cummins et al. (1990) considered 16 loss distributions when modeling the Cummins and Freifelder (1978) fire loss data. They found that an inverse generalized gamma provides an excellent fit but contains too many parameters. It emerges that one- or two-parameter special cases such as the inverse gamma, inverse Weibull, and inverse exponential distributions are already sufficiently flexible. For the Cummins and Freifelder (1978) severity data the Singh–Maddala distribution is preferable.

5.1.8 Related Distributions

A "quadratic elasticity" distribution with p.d.f.

$$f(x) = \frac{1}{\delta(p, \beta, \gamma)} x^{p-1} e^{-\beta x - \gamma^2 x^2}, \quad x > 0, \tag{5.28}$$

was considered by Bordley, McDonald, and Mantrala (1996). Here the normalizing constant is given by

$$\delta(p, \beta, \gamma) = (2\beta^2)^{-p/2}\Gamma(p)e^{\beta^2/(8\gamma^2)}D_{-p}\left(\frac{\beta}{\sqrt{2}\gamma}\right), \tag{5.29}$$

where D_{-p} is a parabolic cylinder function. (An alternative but more lengthy expression occurs in terms of the confluent hypergeometric function ${}_1F_1$.)

The quadratic elasticity distribution was originally motivated by Bordley and McDonald (1993) in connection with the estimation of income elasticity in an aggregate demand model for specific car lines. The gamma distribution is associated with a linear income share elasticity [see (5.8)], whereas the data considered by Bordley and McDonald showed signs of being slightly quadratic. The income share elasticity of the new distribution is

$$\eta(x, f) = (p - 1) - \beta x - 2\gamma^2 x^2. \tag{5.30}$$

Clearly, the gamma distribution is the special case where $\gamma = 0$ and a generalized gamma with $a = 2$ is obtained for $\beta = 0$. The moments of (5.28) are given by

$$E(X^k) = \frac{\delta(p + k, \beta, \gamma)}{\delta(p, \beta, \gamma)}, \tag{5.31}$$

where δ is defined in (5.29).

This model was fitted to U.S. family incomes for 1970, 1975, 1980, 1985, and 1990 by Bordley, McDonald, and Mantrala (1996); the performance was found to be intermediate between gamma and generalized gamma but inferior to beta-type distributions with a comparable number of parameters.

A further generalization along these lines is a member of the so-called *generalized exponential family* used by Bakker and Creedy (1997, 2000) and Creedy, Lye, and Martin (1997). Its p.d.f. is given by

$$f(x) = \exp\{\theta_1 \log x + \theta_2 x + \theta_3 x^2 + \theta_4 x^3 - \eta\}, \quad x \geq 0. \tag{5.32}$$

Here $\exp(\eta)$, is the normalizing constant. A generalized gamma distribution with $a = 2$ [using the notation of (5.2)] is obtained for $\theta_2 = 0 = \theta_4$ and the quadratic elasticity distribution (5.28) is the special case where $\theta_4 = 0$. An important feature of this model is that it can accommodate multimodality.

Creedy, Lye, and Martin (1997) estimated this generalized gamma-type distribution for individual earnings from the 1987 U.S. Current Population Survey (March Supplement), for which it does about as well as a generalized lognormal distribution and considerably better than the standard gamma and lognormal distributions. When fit to New Zealand wages and salaries for 1991, classified by age groups and sex, the distribution is superior to generalized lognormal, lognormal, and gamma fittings in nine of ten cases for males and in six of ten cases for females, in terms of chi-square criteria (Bakker and Creedy, 1997, 1998). In a further application of this model to male individual incomes (before tax, but including transfer payments) from the New Zealand Household Expenditure Surveys for each year over the period 1985–1994, Bakker and Creedy (2000) found the coefficient on x^3 to be statistically insignificant and therefore confined themselves to the special case where $\theta_4 = 0$, which is the quadratic elasticity model (5.28). Their data exhibit bimodality that they attribute to the inclusion of transfer payments in their measure of income. Following earlier work by, for example, Metcalf (1969) (mentioned in the preceding chapter) they investigated how macroeconomic variables influence the parameters of distribution over time. It emerges that the rate of unemployment is the primary influence on the shape of the distribution.

5.2 GAMMA DISTRIBUTION

The gamma (more precisely, the Pearson type III) distribution is certainly among the five most popular distributions in applied statistical work when unimodal and positive data are available. In economic and engineering applications it has two rivals—lognormal and Weibull. It is hard to state categorically which one is the frontrunner. Lancaster (1966) asserted that both Laplace (in the 1836 third edition of his *Théorie analytique des probabilités*) and Bienaymé in 1838 (in his *Mémoires de l'Academie de Sciences de l'Institute de France*) obtained the gamma distribution.

However, these references pertain to normal sampling theory and therefore essentially to the history of χ^2 distributions. For a gamma distribution with a general (i.e., not limited to half-integers) shape parameter, an early reference predating Pearson's (1895) seminal work on asymmetric curves may be attributed to De Forest (1882–1883), as pointed out by Stigler (1978).

Here we shall briefly sketch the basic properties of the gamma distribution and concentrate on aspects more closely related to size and income distributions.

5.2.1 Definition, History, and Basic Properties

The pioneering work marking the initial use of the gamma distribution as an income distribution is due to the French statistician Lucien March, who in 1898 fit the gamma distribution to various French, German, and U.S. earnings distributions. March was inspired by Pearson's work on asymmetric curves. As mentioned above, some 25 years later Amoroso in 1924 introduced a generalized gamma distribution, and another 50 years later the standard gamma distribution resurfaced as a size distribution almost simultaneously but independently in the cybernetics (Peterson and von Foerster, 1971) and econometrics (Salem and Mount, 1974) literatures.

The p.d.f. of the gamma distribution is

$$f(x) = \frac{1}{\beta^p \Gamma(p)} x^{p-1} e^{-x/\beta}, \quad x > 0, \tag{5.33}$$

where $p, \beta > 0$, with p being a shape and β a scale parameter.

As mentioned in the preceding section, the gamma distribution includes the exponential (for $p = 1$) and chi-square distributions (for $p = \nu/2$, ν an integer) as special cases.

From (5.12), the moments of the gamma distribution are given by

$$E(X^k) = \beta^k \frac{\Gamma(p+k)}{\Gamma(p)}. \tag{5.34}$$

Hence, the mean is

$$E(X) = \beta p, \tag{5.35}$$

and the variance equals

$$\mathrm{var}(X) = \beta^2 p. \tag{5.36}$$

The coefficient of variation is therefore of the simple form

$$\mathrm{CV} = \frac{1}{\sqrt{p}}. \tag{5.37}$$

The shape factors are

$$\sqrt{\beta_1} = \frac{2}{\sqrt{p}} \tag{5.38}$$

and

$$\beta_2 = 3 + \frac{6}{p}. \tag{5.39}$$

Hence, in the gamma case we have the relation

$$\sqrt{\beta_1} = 2 \cdot \mathrm{CV}. \tag{5.40}$$

The mode is at $\beta(p-1)$, for $p > 1$, and at zero otherwise.

A basic property of gamma variables is their closure under addition: Suppose that $X_i \sim \mathrm{Ga}(p_i\ \beta)$, $i = 1, 2$, are independent. Then

$$X_1 + X_2 \sim \mathrm{Ga}(p_1 + p_2, \beta). \tag{5.41}$$

The mean excess function is not available in simple closed form; we have the expansion

$$e(x) = 1 + \frac{p-1}{x} + \frac{(p-1)(p-2)}{x^2} + O(x^{-3}), \tag{5.42}$$

which shows that $e(x)$ eventually decreases for large x for $p > 1$ and increases for $p < 1$.

Similarly, the hazard rate is

$$r(x) = \left\{1 + \frac{p-1}{x} + \frac{(p-1)(p-2)}{x^2} + O(x^{-3})\right\}^{-1}, \tag{5.43}$$

which is an increasing function for large x for $p > 1$ and a decreasing function for $p < 1$.

5.2.2 The Angle Process

In a series of papers in the sociological literature, Angle (1986a,b, 1990, 1992, 1993a, 2000) discussed a stochastic model whose long-term wealth distribution appears to be well approximated by a gamma distribution.

His generating mechanism is motivated by the *surplus theory of social stratification* from sociology and anthropology (e.g., Lenski, 1966) that encompasses two concepts of wealth: (1) *subsistence wealth*, which is the wealth necessary to keep producers of wealth alive and to cover the long-term costs of production, and (2) *surplus wealth*, which is the difference between subsistence wealth and total wealth (net product). The central concept of this theory, inequality resulting from contagious competition, is perhaps as old as the adage "the rich get richer, the poor get poorer."

We confine ourselves here to the perhaps most elementary version of the surplus theory that may be described by the following propositions:

1. When people are able to produce a surplus, some of it will become fugitive and leave the possession of its producers.
2. Wealth confers, on those who possess it, the ability to extract wealth from others. Each person's ability to do this in a general competition for surplus wealth depends on his or her own surplus wealth; specifically, the rich tend to take the surplus away from the poor.

Angle's *inequality process* formalizes these propositions in what is known in mathematical physics as an *interacting particle system* with binary interactions. (Binary interactions are employed because no others are specified by the surplus theory.) The process describes a competition between random pairs of individuals for each other's wealth in which the richer individual has a fixed probability p of winning ($.5 < p < 1$). Also, the loser in a random encounter loses a fixed proportion ω of wealth ($0 < \omega < 1$). ω and p are the parameters of the process.

The transition equations for the wealth of a random pair of individuals i, j are

$$X_{i,t} = X_{i,t-1} + D_t \omega X_{j,t-1} - (1 - D_t)\omega X_{i,t-1}, \tag{5.44}$$

$$X_{j,t} = X_{j,t-1} + (1 - D_t)\omega X_{i,t-1} - D_t \omega X_{j,t-1}, \tag{5.45}$$

where $X_{i,t}$ is i's surplus wealth after an encounter with j, $X_{i,t-1}$ is i's surplus wealth before the encounter, and D_t is a sequence of identically distributed Bernoulli variables with $P(D_t = 1) = p$.

Proposition 1 states only that surplus wealth will be fugitive in encounters between members of a population and is implemented in (5.44) and (5.45). Proposition 2 states that people with greater surplus wealth will tend to win encounters. This can be implemented by specifying D_t to be

$$D_t = \begin{cases} 1 \text{ with probability } p, & \text{if } X_{i,t-1} \geq X_{j,t-1}, \\ 0 \text{ with probability } 1-p, & \text{if } X_{i,t-1} < X_{j,t-1}, \end{cases} \tag{5.46}$$

with $.5 < p < 1$.

There are several generalizations of this basic mechanism: Angle (1990, 1992) introduced coalitions among the wealth holders, whereas in 1999 he allowed for *random* shares of lost wealth ω. Angle found, by means of extensive simulation studies, that all these variants of the inequality process generate income or wealth distributions that are well approximated by gamma distributions, although a rigorous proof of this fact appears to be unavailable at present. He further conjectured (1999) that the shape parameter p of this approximating gamma distribution is related to the parameters of the process as $p = (1 - \omega)/\omega$. No doubt the Angle process deserves further scrutiny.

5.2.3 Characterizations

A characterization of the gamma distributions in terms of maximum entropy among all distributions supported on $[0, \infty)$ is as follows: If both the arithmetic and geometric means are prescribed, the maximum entropy p.d.f. is the gamma density (Peterson and von Foerster, 1971; see also Kapur, 1989, pp. 56–57).

Another useful characterization occurs by the following property: If X_1 and X_2 are independent positive random variables and

$$X_1 + X_2 \quad \text{and} \quad \frac{X_1}{X_1 + X_2}$$

are also independent, then each X_i is gamma with a common β but possibly different p. This characterization is due to Lukacs (1965) and has been extended by Marsaglia (1989) relaxing the positivity condition.

In Chapter 2 we saw that the coefficient of variation is, up to a monotonic transformation, a member of the family of generalized entropy measures of inequality. A characterization of the gamma distribution in terms of the (sample) coefficient of variation should therefore be of special interest in our context. Hwang and Hu (1999) showed that if $X_1, \ldots, X_n$ $(n \geq 3)$ are i.i.d. random variables possessing a density, the independence of the sample mean $\bar{X}_n$ and the sample coefficient of variation $\mathrm{CV}_n = S_n/\bar{X}_n$ characterizes the gamma distribution. This result can be refined by replacing the sample standard deviation S_n in the numerator of CV_n with other measures of dispersion. Specifically, we may use Gini's mean difference $\Delta_n = \sum_{i=1}^n \sum_{j=1}^n |X_i - X_j|/[n(n-1)]$ and obtain that independence of $\bar{X}_n$ and $\Delta_n/\bar{X}_n$ characterizes the gamma distribution under the previously stated assumptions (Hwang and Hu, 2000). In our context the latter result is probably best remembered as stating that independence of the sample mean and the sample Gini index G_n characterizes the gamma distribution, since one of the many representations of this inequality measure is $G_n = \Delta_n/(2\bar{X}_n)$.

5.2.4 Lorenz Curve and Inequality Measures

Specializing from (5.17), we see that the kth moment distribution is given by

$$F_{(k)}(x; \beta, p) = F(x; \beta, p+k), \quad x > 0. \tag{5.47}$$

Hence the gamma distribution provides a further example of a distribution that is closed with respect to the formation of moment distributions. This yields the parametric expression for the Lorenz curve

$$\{[u, L(u)]\} = \{[F(x; \beta, p), F(x; \beta, p+1)] | x \in (0, \infty)\}. \tag{5.48}$$

From (5.19) moreover we get

$$X_1 \geq_L X_2 \Longleftrightarrow p_1 \leq p_2. \tag{5.49}$$

Hence, the Lorenz order is linear within the family of two-parameter gamma distributions.

Expressions for inequality measures are not as cumbersome as those of the generalized gamma distribution. Specifically, the Gini coefficient is given by (McDonald and Jensen, 1979)

$$G = \frac{\Gamma(p+1/2)}{\Gamma(p+1)\sqrt{\pi}}. \tag{5.50}$$

The Pietra coefficient can be written in the two forms (McDonald and Jensen, 1979; Pham and Turkkan, 1994)

$$P = \left(\frac{p}{e}\right)^p \frac{1}{\Gamma(p+2)} {}_1F_1(2; p+2; p) = \left(\frac{p}{e}\right)^p \frac{1}{\Gamma(p+1)}, \tag{5.51}$$

where ${}_1F_1$ is the confluent hypergeometric function.

Theil's entropy measure T_1 is given by (Salem and Mount, 1974)

$$T_1 = \frac{1}{p} + \psi(p) - \log p. \tag{5.52}$$

As was to be expected from (5.49), all three coefficients are decreasing in p.

5.2.5 Estimation

Regarding estimation, we shall again be brief since detailed accounts for the estimation of two- and three-parameter gamma distributions are available in Bowman and Shenton (1988) and Cohen and Whitten (1988), among other sources.

The likelihood equations for a simple random sample of size n are

$$\sum_{i=1}^{n} \log X_i - n \log \hat{\beta} - n\psi(\hat{p}) = 0, \tag{5.53}$$

$$\sum_{i=1}^{n} X_i - n\hat{p}\hat{\beta} = 0. \tag{5.54}$$

These can be solved iteratively, and indeed procedures for estimation in the gamma distribution are nowadays available in many statistical software packages.

The Fisher information on $\theta = (\beta, p)^\top$ is given by

$$I(\theta) = \begin{bmatrix} \dfrac{p}{\beta^2} & \dfrac{1}{\beta} \\ \dfrac{1}{\beta} & \psi'(p) \end{bmatrix}, \tag{5.55}$$

from which the asymptotic covariance matrix of $\sqrt{n}(\hat{\beta}, \hat{p})^\top$ can be obtained by inversion or direct computation.

Equation (5.53) shows that the score function of the gamma distribution is unbounded, implying that the MLE is very sensitive to outliers and other aberrant observations. Victoria-Feser and Ronchetti (1994) and Cowell and Victoria-Feser (1996) demonstrated by simulation for complete as well as truncated data that parameter estimates and implied inequality measures for a gamma population can indeed be severely biased when a nonrobust estimator such as the MLE is used. They suggested employing robust methods such as an optimal bias-robust estimator (OBRE) (see Section 3.6 for an outline of the basic ideas behind this estimator). If only grouped data are available, Victoria-Feser and Ronchetti (1997) proposed a type of minimum Hellinger distance estimator that enjoys better robustness properties than the classical MLEs for grouped data.

McDonald and Jensen (1979) studied the sampling behavior of method of moment estimators and MLEs of the Theil, Gini, and Pietra coefficients. They noted that the knowledge of the sample arithmetic mean, geometric mean, and sample variance are sufficient to calculate both the MLEs and method of moment estimators of the Gini, Pietra, and Theil coefficients of inequality and provided a table that facilitates these calculations.

The optimal grouping of data from a gamma population was considered by Schader and Schmid (1986). The situation is somewhat more difficult than in the Pareto or lognormal cases (see Sections 3.6 and 4.6, respectively), since the optimal class boundaries now depend on the shape parameter p and must therefore be derived separately for each value of this parameter. Schader and Schmid reported that there is always a unique set of optimal class boundaries in the gamma case.

Table 5.2 provides these boundaries $z_1^*, \ldots, z_{k^*-1}^*$ based on the least number of classes k^* for which the loss of information is less than or equal to a given value of γ, for $\gamma = 0.1$, 0.05, 0.025, and 0.01 and for $p = 0.5$, 1, 2 and $\beta = 1$. (See Section 3.6 for further details about this problem.)

From the table, optimal class boundaries a_j^* for a gamma distribution with parameters p and β can be obtained upon setting $a_j^* = \beta z_j^*$.

5.2.6 Empirical Results

Incomes and Wealth

March (1898) in his pioneering contribution fit the gamma model to wage distributions for France, Germany, and the United States, stratified by occupation.

Salem and Mount (1974) applied the gamma distribution to U.S. pre-tax personal incomes for 1960–1969, concluding that the gamma provides a better fit than the lognormal distribution. Specifically, the Gini coefficients implied by the estimated gamma distributions mostly fall within the feasible bounds defined by Gastwirth (1972), whereas those implied by lognormal fittings do not.

Kloek and van Dijk (1977) employed the gamma distribution when analyzing data originating from the Australian survey of consumer expenditures and finances

Table 5.2 Optimal Class Boundaries for Gamma Data

p	γ	k^*	$z_1^*, \ldots, z_{k^*-1}^*$
0.5	0.1	5	0.0005 0.0127 0.1002 0.5119
	0.05	7	0.0001 0.0021 0.0161 0.0742 0.2603 0.8293
	0.025	10	0.0000 0.0003 0.0023 0.0102 0.0342 0.0946 0.2313 0.5313 1.2486
	0.01	16	0.0000 0.0000 0.0002 0.0007 0.0024 0.0065 0.0153 0.0325 0.0638 0.1179 0.2083 0.3574 0.6057 1.0401 1.9170
1	0.1	5	0.0351 0.1792 0.5416 1.4073
	0.05	7	0.0143 0.0712 0.2036 0.4607 0.9398 1.9215
	0.025	10	0.0054 0.0265 0.0742 0.1611 0.3041 0.5289 0.8804 1.4543 2.5316
	0.01	17	0.0012 0.0059 0.0164 0.0350 0.0643 0.1070 0.1665 0.2468 0.3531 0.4923 0.6744 0.9148 1.2370 1.6872 2.3615 3.5437
2	0.1	5	0.3479 0.8515 1.6372 3.0202
	0.05	7	0.2174 0.5113 0.9195 1.4884 2.3246 3.7469
	0.025	10	0.1314 0.3017 0.5241 0.8996 1.1640 1.6233 2.2347 3.1071 4.5599
	0.01	17	0.0615 0.1386 0.2350 0.3505 0.4859 0.6428 0.8238 1.0326 1.2746 1.5574 1.8920 2.2956 2.7965 3.4460 4.3535 5.8377

Source: Schader and Schmid (1986).

for the period 1966–1968 (see Podder, 1972). Here the fit is not impressive, the distribution being outperformed by the Champernowne, log-t, and lognormal distributions, two of which have one additional parameter. For 1973 Dutch earnings (Kloek and van Dijk, 1978), the distribution does also not do well, notably in the middle income range, and considerable improvements are possible utilizing three- and four-parameter models such as the Champernowne and log-t distributions.

Suruga (1982) considered Japanese incomes for 1963–1971. Here the gamma distribution is outperformed by the Singh–Maddala, Fisk, and beta distributions but does somewhat better than the lognormal and one-parameter Pareto (II) distribution. In an extension of Suruga's work, Atoda, Suruga, and Tachibanaki (1988) considered grouped data from the Japanese Income Redistribution Survey for 1975, stratified by occupation. Although among the distributions they employed the Singh–Maddala appears to be the most appropriate for the majority of strata, it is remarkable that when a model is selected via information criteria (such as the AIC), the gamma distribution is often preferred over the more flexible generalized gamma distribution. In a later study employing individual data from the same source the gamma is again often indistinguishable from the generalized gamma distribution (Tachibanaki, Suruga, and Atoda, 1997).

Dagum (1983) fit a gamma distribution to 1978 U.S. family incomes. The (four- and three-parameter) Dagum type III and type I as well as the Singh–Maddala distribution perform considerably better; however, the gamma distribution outperforms the lognormal by wide margins.

Ransom and Cramer (1983) considered a measurement error model, viewing observed income as the sum of a systematic component and an independent $N(0, \sigma^2)$ error term. For competing models with systematic components following Pareto or lognormal distributions, they found that the gamma variant is rejected by chi-square goodness-of-fit tests for U.S. family incomes for 1960 and 1969.

McDonald (1984) estimated the gamma model for 1970, 1975, and 1980 U.S. family incomes. The distribution is outperformed by three- and four-parameter families such as the (generalized) beta, generalized gamma, and Singh–Maddala distributions, but is superior to all other two-parameter models, notably the lognormal and Weibull distributions (an exception being the 1980 data for which the Weibull does somewhat better). Also, the improvements achieved by employing the generalized gamma are quite small for the 1970 and 1975 data.

Bhattacharjee and Krishnaji (1985) found that the landholdings in 17 Indian states for 1961–1962 appear to be more adequately approximated by a gamma distribution with a decreasing density ($p < 1$) than by a lognormal distribution. A log-gamma distribution provides a comparable fit.

Bordley and McDonald (1993) employed the gamma distribution for the estimation of the income elasticity in an aggregate demand model for automotive data. It turns out that the income-elasticity estimates provided by the gamma are fairly similar to those associated with distributions providing good approximations to U.S. income distribution, such as the generalized beta distribution of the second kind (GB2; see Chapter 6).

Angle (1993b, 1996) found that personal income data from the 1980–1987 U.S. Current Population Survey as well as from the Luxembourg Income Study (for eight countries in the 1980s) stratified by levels of education are well approximated by gamma distributions, with shape parameters that moreover increase with educational attainment. Somewhat surprisingly, it turns out that the aggregate distribution can also be fitted by a gamma distribution, although finite mixtures of gamma distributions are known in general not to follow this distribution.

Victoria-Feser and Ronchetti (1994) fit a gamma distribution to incomes of households on income support using the 1979 UK Family Expenditure Survey. They employed the MLE as well as an optimal B-robust estimator (OBRE) (see above) and concluded that the latter provides a better fit because it gives more importance to the majority of data.

In the study of Brachmann, Stich, and Trede (1996) utilizing 1984–1993 German household incomes, the gamma distribution emerges as the best two-parameter model. However, only the more flexible GB2 and Singh–Maddala distributions seem to be appropriate for these data.

Bordley, McDonald, and Mantrala (1996) fit the gamma distribution to U.S. family incomes for 1970, 1975, 1980, 1985, and 1990. Although it is outperformed by all three- and four-parameter models, notably the GB2, Dagum, and Singh–Maddala distributions, it turns out to be the best two-parameter model except for 1980 and 1985, where the Weibull distribution does equally well.

Creedy, Lye, and Martin (1997) estimated the gamma distribution for individual earnings from the 1987 U.S. Current Population Survey (March Supplement). The performance is comparable to, although slightly worse than, the one provided by a generalized gamma and a generalized lognormal distribution. The standard lognormal distribution does much worse for these data.

From these studies it would seem that the gamma distribution is perhaps the best two-parameter model for approximating the size distribution of personal income.

Actuarial Losses

In the actuarial literature Ramlau-Hansen (1988) fit a gamma distribution to windstorm losses for a portfolio of single-family houses and dwellings in Denmark for the period 1977–1981. However, there are problems with his choice because the empirical skewness and coefficient of variation are very close, whereas for a gamma distribution the former necessarily equals twice the latter [see (5.40)].

Cummins et al. (1990), who employed 16 loss distributions when modeling the Cummins and Freifelder (1978) fire loss data, found the gamma distribution to be among the worst distributions they considered. Specifically, the data seem to require a model with much heavier tails, such as an inverse gamma distribution.

5.3 LOG-GAMMA DISTRIBUTION

If there is a lognormal distribution that enjoys wide applicability, why should we deprive ourselves of a log-gamma distribution? A logarithmic transform is often a sensible operation to smooth the data, especially if we confine ourselves to the interval $(1, \infty)$.

5.3.1 Definition and Basic Properties

If Y follows a two-parameter gamma distribution, the random variable $X = \exp Y$ is said to possess a log-gamma distribution. The density of X is therefore

$$f(x) = \frac{\beta^p}{\Gamma(p)} x^{-\beta-1} \{\log(x)\}^{p-1}, \quad 1 \le x, \tag{5.56}$$

where $p > 0$, $\beta > 0$. Here both parameters β, p are shape parameters. Many distributional properties of this model are given in Taguchi, Sakurai, and Nakajima (1993), who also discuss a bivariate form. [The parameterization (5.56) differs slightly from the one used in connection with the gamma distribution in that, compared to (5.33), we use $1/\beta$ instead of β. (5.56) appears to be the standard parameterization of this distribution in the literature.]

Despite its genesis, the log-gamma distribution is perhaps best considered a generalized Pareto distribution since the classical Pareto type I with a unit scale is the special case where $p = 1$. (The parameter β now plays the role of Pareto's α.) Although the Pareto type I has a decreasing density, the log-gamma is more flexible in that it allows for unimodal densities. Specifically, the density is decreasing for $p \le 1$, whereas for $p > 1$ an interior mode exists. The parameter β determines the shape in the upper income range, whereas p governs the lower tail.

In order to enhance the flexibility of this distribution, it is sometimes desirable to introduce a scale parameter, yielding

$$f(x) = \frac{\beta^p b^\beta}{\Gamma(p)} x^{-\beta-1} \left\{\log\left(\frac{x}{b}\right)\right\}^{p-1}, \quad 0 < b \le x, \tag{5.57}$$

where $p > 0$, $\beta > 0$, and $b > 0$ is the scale. A further generalization along the lines of the generalized gamma distribution is the generalized log-gamma—or rather log-"generalized gamma"—distribution given by the p.d.f. (Taguchi, Sakurai, and Nakajima, 1993)

$$f(x) = \frac{a\beta^{ap}}{x\Gamma(p)} \left\{\log \frac{x}{b}\right\}^{ap-1} \exp\left[-\left(\beta \log \frac{x}{b}\right)^a\right], \quad 0 < b \le x, \tag{5.58}$$

where $a, b, p, \beta > 0$. The logarithm of a random variable with this density follows a four-parameter generalized gamma distribution. In our presentation of the basic properties of the log-gamma distribution, we shall confine ourselves to the two-parameter case (5.56).

The moments exist for $k < \beta$, in which case they are given by

$$E(X^k) = \left(\frac{\beta}{\beta - k}\right)^p. \tag{5.59}$$

Note that the latter expression does not contain any terms involving the gamma function. This reflects the fact that the log-gamma distribution is closely related to the classical Pareto distribution, whose moments are obtained for $p = 1$.

From (5.59) we get

$$E(X) = \left(\frac{\beta}{\beta - 1}\right)^p \tag{5.60}$$

and

$$\text{var}(X) = \left(\frac{\beta}{\beta - 2}\right)^p - \left(\frac{\beta}{\beta - 1}\right)^{2p}, \quad \beta > 2. \tag{5.61}$$

The mode and geometric mean are given by

$$x_{\text{mode}} = \exp\left\{\frac{p - 1}{\beta + 1}\right\} \tag{5.62}$$

and

$$x_{\text{geo}} = \exp\left(\frac{p}{\beta}\right), \tag{5.63}$$

respectively. We see that $x_{\text{mode}} < x_{\text{geo}} < E(X)$.

The basic reproductive property of the distribution is a direct consequence of the well-known reproductive property of the gamma distribution (5.41). Since the latter is closed under addition, if the scale parameters are equal, we determine that the log-gamma distribution is closed with respect to the formation of products: if $X_i \sim \log \text{Ga}(p_i, \beta)$, $i = 1, 2$, are independent, we have

$$X_1 \cdot X_2 \sim \log \text{Ga}(p_1 + p_2, \beta). \tag{5.64}$$

Regarding estimation, we can be brief, because in view of its genesis, we may translate this problem to that of parameter estimation for a two-parameter gamma distribution and make use of the extensive literature devoted to that topic. Specifically, the Fisher information on $\theta = (\beta, p)^\top$ is identical to the Fisher information $I(\tilde{\theta})$ of the classical gamma distribution with parameters $\tilde{\theta} = (1/\beta, p)^\top$; see (5.55). (Recall that for the log-gamma distribution the p.d.f. was reparameterized via $\beta \to 1/\beta$.) The Fisher information $I(\theta)$ of the reparameterization is thus obtained using the relation $I(\theta) = JI(\tilde{\theta})J^\top$, where J is the Jacobian of the inverse transformation (see, e.g., Lehmann and Casella, 1998, p. 125), and therefore given by

$$I(\theta) = \begin{bmatrix} \dfrac{p}{\beta^2} & -\dfrac{1}{\beta} \\ \dfrac{-1}{\beta} & \psi'(p) \end{bmatrix}. \tag{5.65}$$

5.3.2 Lorenz Curve and Inequality Measures

The Lorenz curve of the log-gamma distribution is not available in a simple closed form. It can be expressed in terms of a moment distribution in the form

$$\left\{(u, v) \mid u = \frac{\gamma(p, \beta y)}{\Gamma(p)}, \ v = \frac{\gamma[p, (\beta - 1)y]}{\Gamma(p)}, \text{ where } y = \log x\right\}. \tag{5.66}$$

Here $\gamma(\cdot, \cdot)$ is the incomplete gamma function.

The Gini coefficient of the two-parameter log-gamma distribution is (Bhattacharjee and Krishnaji, 1985)

$$G = 1 - 2B_a(p, p), \quad a = \frac{\beta - 1}{2\beta - 1}. \tag{5.67}$$

This shows that, for a fixed p, the Gini coefficient decreases with increasing β, with a (positive) lower bound that depends on the second shape parameter p.

5.3.3 Empirical Results

Bhattacharjee and Krishnaji (1985) showed that the landholdings in 17 Indian states for 1961–1962 are better approximated by a log-gamma than by a lognormal distribution. A gamma distribution with a decreasing density provides a comparable fit.

In the actuarial literature, Hewitt and Lefkowitz (1979) estimated a two-component gamma-loggamma mixture for automobile bodily injury loss data. This mixture model does considerably better than a two-parameter lognormal distribution. Also, Ramlau-Hansen (1988) reported that the log-gamma distribution provides a satisfactory fit when modeling fire losses for a portfolio of single-family houses and dwellings in Denmark for the period 1977–1981. His estimated tail index β (essentially Pareto's alpha) falls in the vicinity of 1.4.

5.4 INVERSE GAMMA (VINCI) DISTRIBUTION

5.4.1 Definition and Basic Properties

Some authors, especially those dealing with reliability applications such as Barlow and Proschan (1981), call this distribution the *inverted* gamma. Indeed, if Z is a gamma variable, then $X = 1/Z$ is a variable with the density

$$f(x) = \frac{\beta^p}{\Gamma(p)} x^{-p-1} e^{-(\beta/x)}, \quad x > 0, \tag{5.68}$$

where $p, \beta > 0$. (As in the preceding section, we have reparameterized the density using $\beta \to 1/\beta$.)

The distribution is encountered in Bayesian reliability applications. It is also hidden among the Pearson curves, specifically Pearson V, and Vinci (1921) should be credited for his income distribution applications.

Distribution (5.68) is a special case of the inverse generalized gamma distribution ($a = -1$); for $p = 1$ we obtain a special case of the log-Gompertz distribution to be discussed below. The case where $p = 1/2$, more widely known as the *Lévy distribution*, is of special interest in probability theory: It is one of the few stable distributions for which an expression of the density in terms of elementary functions is available and arises as the distribution of first-passage times in Brownian motion (see, e.g., Feller, 1971).

Since the density (5.68) is regularly varying (at infinity) with index $-p-1$, the moments exist only for $k < p$. They are given by [compare with (5.12)]

$$E(X^k) = \beta^k \frac{\Gamma(p)}{\Gamma(p-k)}. \tag{5.69}$$

The mode occurs at $(p + 1)\beta$.

As mentioned above, the classical gamma distribution arises as the maximum entropy distribution under the constraints of a fixed first moment and a fixed geometric mean. Similarly, the inverse gamma distribution is the maximum entropy distribution if the first moment and harmonic mean are prescribed (Ord, Patil, and Taillie, 1981).

Regular variation of the p.d.f. also gives us the basic asymptotic property of the hazard rate and mean excess function

$$r(x) \in \mathrm{RV}_\infty(-1) \tag{5.70}$$

and

$$e(x) \in \mathrm{RV}_\infty(1). \tag{5.71}$$

5.4.2 Inequality Measurement

From (5.49) we know that the gamma distributions are Lorenz-ordered with respect to the shape parameter p, with diminishing inequality being associated with increasing p. Unfortunately, this does not translate directly into a corresponding result for the inverse gamma distribution. In order to obtain such a result, a stronger ordering concept than the Lorenz ordering is required. As mentioned in Chapter 2, the star-shaped ordering implies Lorenz ordering, and van Zwet (1964) showed that the gamma distribution is ordered according to the convex transform ordering and therefore also in the sense of the weaker star-shaped ordering. Since the star-shaped ordering is closed under inversion, in the sense that (e.g., Taillie, 1981)

$$X_1 \geq_* X_2 \Longleftrightarrow \frac{1}{X_1} \geq_* \frac{1}{X_2}, \tag{5.72}$$

these results translate into

$$X_1 \geq_L X_2 \Longleftrightarrow p_1 \leq p_2, \tag{5.73}$$

for the inverse gamma distributions, provided $p > 1$ [in order to assure the existence of $E(X)$ and therefore that of the Lorenz curve].

5.4.3 Estimation

In view of the genesis of the distribution, estimation can proceed by considering the "inverted" data $1/x_i$, $i = 1, \ldots, n$, and using methods appropriate for the gamma distribution. Alternatively, the likelihood equations

$$\frac{np}{\beta} - \sum_{i=1}^{n} \frac{1}{x_i} = 0, \tag{5.74}$$

$$n \log \beta - n\psi(p) - \sum_{i=1}^{n} \log x_i = 0 \tag{5.75}$$

can be solved directly.

From (5.74) and (5.75) we get the Fisher information on $\theta = (\beta, p)^\top$

$$I(\theta) = \begin{bmatrix} \dfrac{p}{\beta^2} & -\dfrac{1}{\beta} \\ -\dfrac{1}{\beta} & \psi'(p) \end{bmatrix}. \tag{5.76}$$

5.4.4 Empirical Results

Although the Vinci distribution was proposed as an income distribution some 80 years ago, we have not been able to track down fittings to income data in the literature available to us. In the actuarial literature, Cummins et al. (1990) used the inverse gamma distribution for approximating the fire loss experiences of a major university. The distribution turns out to be one of the best two-parameter models; in fact, the data are appropriately modeled by the one-parameter special case where $a = 1$, an inverse exponential distribution.

5.5 WEIBULL DISTRIBUTION

All that Waloddi Weibull, a Swedish physicist, did in his pioneering reports No.'s 151 and 153 for the Engineering Academy in 1939 was to add a "small a" to the c.d.f. of the exponential distribution, and what a difference it did cause! Nowadays we refer to this operation as *Weibullization*. The Weibull distribution has no doubt received maximum attention in the statistical and engineering literature of the last ten years and is still going strong. In economics it is probably less prominent, but D'Addario (1974) noticed its potentials for income data and Hogg and Klugman (1983) for insurance losses.

Even the strong evidence that the Weibull distribution is indeed due to Weibull is shrouded in minor controversy. Rosin and Rammler in 1933 used this distribution in their paper "The laws governing the fineness of powdered coal." Some Russian sources insist that it should be called Weibull–Gnedenko (or preferably Gnedenko–Weibull!) since it turns out to be one of the three types of extreme value limit distributions established rigorously by Gnedenko in his famous paper in the *Annals of Mathematics*, published during World War II. And the French would argue that this is nothing else but Fréchet's distribution, who initially identified it in 1927 to be an extreme-value distribution in his "Sur la loi de probabilité de l'écart maximum." Here we shall follow the same pattern as in the gamma distribution section and concentrate on the income and size applications of the Weibull distribution, some of them quite recent.

5.5.1 Definition and Basic Properties

Being a generalized gamma distribution with $p = 1$, the Weibull density is given by

$$f(x) = \frac{a}{\beta}\left(\frac{x}{\beta}\right)^{a-1} e^{-(x/\beta)^a}, \quad x > 0, \tag{5.77}$$

where $a, \beta > 0$. Unlike the c.d.f. of the classical and generalized gamma distributions, the c.d.f. of the Weibull distribution is available in terms of elementary functions; it is simply

$$F(x) = 1 - e^{-(x/\beta)^a}, \quad x > 0. \tag{5.78}$$

We note that irrespective of the value of a,

$$F(\beta) = 1 - e^{-1}.$$

From (5.78) we see that even the quantile function is available in closed form

$$F^{-1}(u) = \beta\{-\log(1-u)\}^{1/a}, \quad 0 < u < 1, \tag{5.79}$$

a property that facilitates the derivation of Lorenz-ordering results (see below). From (5.79) we determine that the median of the distribution is

$$x_{\text{med}} = \beta(\log 2)^{1/a}. \tag{5.80}$$

A simple argument leading to a Weibull distribution as the distribution of fire loss amount was given by Ramachandran (1974). Under the two assumptions that (1) the hazard rate of the fire duration T is given by $\lambda(t) = \exp(at)$, $a > 0$, and (2) that the resulting damage X is an exponential function of the duration, $X = x_0 \exp(kT)$, for some x_0, $k > 0$, the c.d.f. is given by (5.78).

The inverse Weibull distribution, that is, the distribution of $1/X$ for $X \sim \text{Wei}(a, \beta)$, is discussed in the following section, under the name of the log-Gompertz distribution.

From (5.12), the moments of the Weibull distribution are

$$E(X^k) = \beta^k \Gamma\left(1 + \frac{k}{a}\right), \tag{5.81}$$

specifically

$$E(X) = \beta \Gamma\left(1 + \frac{1}{a}\right), \tag{5.82}$$

and

$$\text{var}(X) = \beta^2 \left\{ \Gamma\left(1 + \frac{2}{a}\right) - \left[\Gamma\left(1 + \frac{1}{a}\right)\right]^2 \right\}. \tag{5.83}$$

The mode is at $\beta(1 - 1/a)^{1/a}$ if $a > 1$; otherwise, the distribution is zero-modal, with a pole at the origin if $a < 1$.

The somewhat peculiar skewness properties of the Weibull distribution have been studied by several authors; see for example, Cohen (1973), Rousu (1973), or Groeneveld (1986). As is the case with the generalized gamma distribution, there is a value a for which the shape factor $\sqrt{\beta_1} = 0$. Unlike in the generalized gamma case, this value does not depend on other parameters of the distribution; it equals (to four decimals) $a_0 = 3.6023$. For $a < a_0$ we have $\sqrt{\beta_1} > 0$, while for $a > a_0$

we have $\sqrt{\beta_1} < 0$. Since empirical size distributions are heavily skewed to the right, it would seem that $a < a_0$ is the relevant range in our context.

The mean excess function is

$$e(x) = e^{(x/\beta)^a} \int_x^\infty e^{-(t/\beta)^a} dt, \quad x > 0, \tag{5.84}$$

which is asymptotic to x^{1-a} (Beirlant and Teugels, 1992). [Incidentally, $e(x) \propto x^{1-a}$, $0 < a \leq 1$, defines the Benktander type II distribution, a loss distribution that will be discussed in Section 7.4 below.]

The Weibull hazard rate

$$r(x) = \frac{a}{\beta}\left(\frac{x}{\beta}\right)^{a-1}, \quad x > 0, \tag{5.85}$$

is a decreasing function when the shape parameter a is less than 1, a constant when a equals 1 (the exponential distribution), and an increasing function when a is greater than 1. The simple and flexible form of the hazard rate may explain why the Weibull distribution is quite popular in lifetime studies.

A useful property of the Weibull order statistics is distributional closure of the minima. Specifically,

$$\begin{aligned} f_{1:n}(x) &= n\{(1 - F(x)\}^{n-1} f(x) \\ &= \frac{na}{\beta}\left(\frac{x}{\beta}\right)^{a-1} \exp\left\{-n\left(\frac{x}{\beta}\right)^a\right\}, \quad x > 0. \end{aligned} \tag{5.86}$$

Hence, $X_{1:n} \sim \text{Wei}(a, \beta n^{-1/a})$.

5.5.2 Lorenz Curve and Inequality Measurement

Lorenz-ordering relations are easily obtained using the star-shaped ordering (see Section 2.1.1). Specifically, we have for $X_i \sim \text{Wei}(a_i, 1)$, $i = 1, 2$, using (5.79)

$$\frac{F_1^{-1}(u)}{F_2^{-1}(u)} = \frac{\{-\log(1-u)\}^{1/a_1}}{\{-\log(1-u)\}^{1/a_2}}, \quad 0 < u < 1,$$

which is seen to be increasing in u if and only if $a_1 \leq a_2$. Since the star-shaped ordering implies the Lorenz ordering, we get (Chandra and Singpurwalla, 1981)

$$X_1 \geq_L X_2 \Longleftrightarrow a_1 \leq a_2. \tag{5.87}$$

Hence, the Lorenz order is linear within the family of two-parameter Weibull distributions. The result could, of course, also have been obtained directly from (5.19).

In view of (5.86), the Gini coefficient is most easily derived using the representation in terms of order statistics (2.22), yielding

$$G = 1 - \frac{E(X_{1:2})}{E(X)} = 1 - 2^{-1/a}, \tag{5.88}$$

which is decreasing in a.

5.5.3 Estimation

Parameter estimation for the Weibull distribution is discussed in many sources, notably in texts on engineering statistics. See also Cohen and Whitten (1988) and Johnson, Kotz, and Balakrishnan (1994).

Briefly, the ML estimators satisfy the equations

$$\hat{a} = \left[\left\{ \sum_{i=1}^{n} x_i^{\hat{a}} \log x_i \right\} \left\{ \sum_{i=1}^{n} x_i^{\hat{a}} \right\}^{-1} - \frac{1}{n} \sum_{i=1}^{n} \log x_i \right]^{-1}, \tag{5.89}$$

$$\hat{\beta} = \left\{ \frac{1}{n} \sum_{i=1}^{n} x_i^{\hat{a}} \right\}^{1/\hat{a}}. \tag{5.90}$$

Here (5.89) is solved for $\hat{a}$; the result is then substituted in (5.90).

We note that only if $a > 2$, are we in a situation where the regularity conditions for ML estimation are satisfied. In this case, the Fisher information on $\theta = (a, \beta)^\top$ is given by

$$I(\theta) = \begin{bmatrix} \dfrac{6(C-1)^2 + \pi^2}{6a^2} & \dfrac{C-1}{\beta} \\ \dfrac{C-1}{\beta} & \dfrac{a^2}{\beta^2} \end{bmatrix}, \tag{5.91}$$

where $C = 0.577216\ldots$ is Euler's constant. If $a < 1$, the density is unbounded.

5.5.4 Empirical Results

As noted in the introduction to this chapter, the Weibull distribution was apparently used only sporadically as an income or size distribution and most applications are of comparatively recent date.

Incomes

Bartels (1977) fit the distribution to 1969 fiscal incomes in three regions of the Netherlands. Here the model seems to provide a rather unsatisfactory fit; it is outperformed by (generalized) gamma and Champernowne distributions and variants of it. For the French wages stratified by occupation for 1970–1978 the three-parameter distribution is also not satisfactory, being outperformed by the Dagum type II, a Box–Cox-transformed logistic, the Singh–Maddala, and the three-parameter lognormal distributions (Espinguet and Terraza, 1983). However, it does better than a four-parameter beta type I distribution for these data.

McDonald (1984) applied the Weibull distribution for 1970, 1975, and 1980 U.S. family incomes. The distribution is outperformed by three- and four-parameter models such as the (generalized) beta, generalized gamma, and Singh–Maddala distributions, but is superior to all other two-parameter models—notably the lognormal and gamma distributions—for the 1980 data, where it ranks fourth out of 11 models considered.

Atoda, Suruga, and Tachibanaki (1988) considered grouped income data from the Japanese Income Redistribution Survey for 1975, stratified by occupation. Although among the distributions they employed, the Singh–Maddala often appears to be the most appropriate, the Weibull does only slightly worse than the more flexible generalized gamma distribution, in one case even better when the selection criterion is the AIC. In a later study employing individual data from the same source, the Weibull distribution is again comparable to the generalized gamma distribution for one stratum, and only slightly worse for the remaining ones (Tachibanaki, Suruga, and Atoda, 1997).

Bordley, McDonald, and Mantrala (1996) fit the Weibull distribution to U.S. family incomes for 1970, 1975, 1980, 1985, and 1990. It is outperformed by all three- and four-parameter models—notably the GB2, Dagum, and Singh–Maddala distributions—and by the two-parameter gamma distribution for three data sets. For the remaining years it is comparable to the gamma distribution.

Brachmann, Stich, and Trede (1996), in their study of German household incomes over the period 1984–1993, found the Weibull distribution to perform better than the lognormal, but not nearly as well as the gamma distribution. However, only the GB2 and Singh–Maddala distribution seem to provide a satisfactory fit for these data.

Actuarial Losses

In the actuarial literature, Hogg and Klugman (1983) fit the Weibull distribution to a small data set (35 observations) of hurricane losses and found that it performs about as well as the lognormal distribution. They also considered data for malpractice losses, for which beta type II and Lomax (Pareto type II) distributions are preferable.

In the Cummins et al. (1990) study employing 16 loss distributions, the Weibull distribution does not provide an adequate fit to the Cummins and Freifelder (1978) fire loss data. Specifically, the data seem to require a model with heavier tails such as an inverse Weibull distribution.

Nonetheless, from these works it is clear that the Weibull distribution often does considerably better than the more popular lognormal distribution. Among the two-parameter models it appears to be comparable to the gamma distribution.

5.6 LOG-GOMPERTZ DISTRIBUTION

The Gompertz distribution was introduced some 120 years before the Weibull one by Benjamin Gompertz in 1825 in the *Philosophical Transactions of the Royal Society*, to fit mortality tables. It is usually defined for positive values. Nowadays it is used in actuarial statistics and competing risks, in the early 1970s it attracted attention in *Applied Statistics* (Garg, Rao, and Redmond, 1970; Prentice and El Shaarawi, 1973). When defined over the real line, the Gompertz c.d.f. is

$$F(y) = \exp(-ae^{-y/\beta}), \quad -\infty < y < \infty, \tag{5.92}$$

where a, $\beta > 0$, which is a type I extreme value distribution.

5.6.1 Definition and Basic Properties

The log-Gompertz distribution appears to be used mainly in income and size distributions and was noticed by Dagum (1980c) in this connection. It is a member of Dagum's (1980c, 1990a, 1996) generating system; see Section 2.4.

From (5.92), the c.d.f. of $X = \exp Y$ is

$$F(x) = \exp\left\{-\left(\frac{x}{\beta}\right)^{-a}\right\}, \quad x > 0, \tag{5.93}$$

where $\beta > 0$ and $a > 0$. This yields the log-Gompertz density

$$f(x) = a\beta^a x^{-a-1} e^{-(x/\beta)^{-a}}, \quad x > 0, \tag{5.94}$$

which is easily recognized as the p.d.f. of an inverse Weibull distribution. The case where $a = 1$, the inverse exponential distribution, is also a special case of the inverse gamma (Vinci) distribution discussed in Section 5.4.

As in the Weibull case, the quantile function is available in closed form, being

$$F^{-1}(u) = \beta\{(-\log u)^{-1/a}\}, \quad 0 < u < 1. \tag{5.95}$$

The median therefore occurs at

$$x_{\text{med}} = \beta(\log 2)^{-1/a}. \tag{5.96}$$

The moments exist only for $k < a$; in that case, they are given by

$$E(X^k) = \beta^k \Gamma\left(1 - \frac{k}{a}\right). \tag{5.97}$$

Specifically,

$$E(X) = \beta\Gamma\left(1 - \frac{1}{a}\right) \tag{5.98}$$

and

$$\operatorname{var}(X) = \beta^2\left\{\Gamma\left(1 - \frac{2}{a}\right) - \Gamma^2\left(1 - \frac{1}{a}\right)\right\}. \tag{5.99}$$

The mode is at

$$x_{\text{mode}} = \beta\left(\frac{a}{a+1}\right)^{1/a}. \tag{5.100}$$

We see that in contrast to the Weibull distribution, there is always an interior mode.

The hazard rate and mean excess function were studied by Erto (1989) in an Italian publication dealing with lifetime applications. The hazard rate is

$$r(x) = \frac{a\beta^a x^{-a-1} e^{-(x/\beta)^{-a}}}{1 - \exp\{-(x/\beta)^{-a}\}}, \quad x \geq 0. \tag{5.101}$$

It is noteworthy that, irrespectively of the value of the shape parameter a, $r(0) = 0$ and $\lim_{x\to\infty} r(x) = 0$; in fact, $r(x)$ is a unimodal function (similar to the lognormal hazard rate). There is no simple expression for the abscissa of the mode, but it can be bounded: From the derivative of $\log r(x)$ we obtain the first-order condition

$$\exp\left\{\left(-\frac{x}{\beta}\right)^{-a}\right\}\frac{1}{x} = \frac{1}{x} - \frac{a}{\beta(a+1)}\left(\frac{x}{\beta}\right)^{-a-1},$$

which is of the form

$$u(x) = v(x).$$

It is not difficult to see that both functions $u(x)$ and $v(x)$ are increasing up to a point

$$x_n = \beta a^{1/a},$$

which is to the right of the mode (5.100), and decreasing thereafter. Also, $\lim_{x\to 0} u(x) = 0$ and $v(x_{\text{mode}}) = 0$. This yields

$$u(x_{\text{mode}}) > v(x_{\text{mode}}) = 0, \quad u(x_n) < v(x_n), \quad x_{\text{mode}} < x_n,$$

from which it follows that the point of intersection of $u(x)$ and $v(x)$—defining the mode of $r(x)$—is contained in the interval (x_{mode}, x_n).

Since the log-Gompertz density is regularly varying at infinity, we determine that the mean excess function is asymptotically linearly increasing [see (2.67)]

$$e(x) \in \mathrm{RV}_\infty(1). \tag{5.102}$$

5.6.2 Estimation

Being the inverse Weibull distribution, parameter estimation for the log-Gompertz distribution proceeds most easily by considering the reciprocal observations $1/x_i$, $i = 1, \ldots, n$, and using methods appropriate for Weibull data.

The relationship with the Weibull distribution also yields the Fisher information on $\theta = (a, \beta)^\top$

$$I(\theta) = \begin{bmatrix} \dfrac{6(C-1)^2 + \pi^2}{6a^2} & \dfrac{1-C}{\beta} \\ \dfrac{1-C}{\beta} & \dfrac{a^2}{\beta^2} \end{bmatrix}, \tag{5.103}$$

which coincides up to the sign of the off-diagonal elements with (5.91).

Erto has suggested a simple estimator utilizing a linearization of the survival function. He proposed estimating parameters in the equation

$$\log\left(\frac{1}{x}\right) = \frac{1}{a}\log\log\left[\frac{1}{1-F(x)}\right] + \log b \tag{5.104}$$

by least squares.

5.6.3 Inequality Measurement

As in the Weibull case, Lorenz ordering relations are easily obtained using the star-shaped ordering (see Section 2.1.1). Specifically, we have for $X_i \sim \mathrm{logGomp}(a_i, 1)$, $i = 1, 2$, using (5.95),

$$\frac{F_1^{-1}(u)}{F_2^{-1}(u)} = \frac{(-\log u)^{-1/a_1}}{(-\log u)^{-1/a_2}}, \quad 0 < u < 1,$$

which is seen to be increasing in u if and only if $a_1 \leq a_2$. Since the star-shaped ordering implies the Lorenz ordering—provided $a_i > 1$, so that $E(X_i) < \infty$—we have

$$X_1 \geq_L X_2 \Longleftrightarrow a_1 \leq a_2. \tag{5.105}$$

Hence, the log-Gompertz family is another family of distributions within which the Lorenz ordering is linear.

5.6.4 Empirical Results

Cummins et al. (1990), in their comprehensive study employing 16 loss distributions, fit the log-Gompertz (under the name of inverse Weibull) distribution to the annual fire loss experiences of a major university. The distribution turns out to be the best two-parameter model; however, the data are appropriately modeled by the one-parameter special case where $a = 1$, an inverse exponential distribution.

CHAPTER SIX

Beta-type Size Distributions

Beta distributions (there are two kinds of this distribution) are members of the celebrated Pearson system and have been widely utilized in all branches of sciences—both soft and hard. They are intrinsically related to the incomplete beta function ratio

$$I_x(p, q) = \frac{1}{B(p, q)} \int_0^x u^{p-1}(1-u)^{q-1}\, du, \quad 0 \leq x \leq 1, \tag{6.1}$$

and the incomplete beta function

$$B_x(p, q) = \int_0^x u^{p-1}(1-u)^{q-1}\, du, \quad 0 \leq x \leq 1. \tag{6.2}$$

A historical account of these functions was provided by Dutka (1981), who traced them to a letter from Isaac Newton to Henry Oldenberg in 1676. Needless to say, as Thurow (1970) put it, "using a beta distribution is not meant to imply that God is a beta generating function."

6.1 (GENERALIZED) BETA DISTRIBUTION OF THE SECOND KIND

For our purposes, the pivotal distribution in this family is the so-called generalized beta distribution of the second kind (hereafter referred to as GB2). We should note the contributions of McDonald (1984) and his associates in the development of GB2 distributions as an income distribution and in unifying the various research activities in closely related fields. We could mention as an example the multiauthor paper by Cummins et al. (1990), which is a combination of two manuscripts by McDonald

Statistical Size Distributions in Economics and Actuarial Sciences, By Christian Kleiber and Samuel Kotz.
ISBN 0-471-15064-9

and Pritchett (the Brigham Young University School) and Cummins and Dionne (the University of Pennsylvania School) dealing with applications of the GB2 family of distributions to insurance losses submitted independently to *Insurance: Mathematics and Economics* in early 1988 and resulted in a unified treatment.

The interrelations between particular cases of the GB2 distributions and other distributions known in the literature are somewhat confusing, but a natural consequence of the independent uncoordinated research that has been so prevalent the last 20–30 years. One indication is the discovery that the distribution proposed by Majumder and Chakravarty (1990) is simply a reparameterization of the GB2 distribution. This observation escaped researchers for at least five years in spite of the fact that the papers appeared in not unrelated journals: one oriented toward applications and the other of a more theoretical bent.

6.1.1 Definition and Interrelations

The c.d.f. of the GB2 distribution may be introduced using an alternative expression for the incomplete beta function ratio that is obtained upon setting $u := t/(1+t)$ in (6.1)

$$I_z(p, q) = \frac{1}{B(p, q)} \int_0^z \frac{t^{p-1}}{(1+t)^{p+q}}\, dt, \quad z > 0. \tag{6.3}$$

Introducing additional scale and shape parameters b and a and setting $z := (x/b)^a$, we get a distribution with c.d.f.

$$F(x) = I_z(p, q), \text{ where } z = \left(\frac{x}{b}\right)^a, \quad x > 0, \tag{6.4}$$

with corresponding density

$$f(x) = \frac{a x^{ap-1}}{b^{ap} B(p, q)[1 + (x/b)^a]^{p+q}}, \quad x > 0. \tag{6.5}$$

Here all four parameters a, b, p, q are positive, b is a scale, and a, p, q are shape parameters. It is not difficult to see that the GB2 density is regularly varying at infinity with index $-aq - 1$ and regularly varying at the origin with index $-ap - 1$; thus, all three shape parameters control the tail behavior of the model. Nonetheless, these three parameters are not on an equal footing: If the distribution of $Y = \log X$, with density

$$f(y) = \frac{a e^{ap(y - \log b)}}{B(p, q)[1 + e^{a(y - \log b)}]^{p+q}}, \quad -\infty < y < \infty, \tag{6.6}$$

is considered, a turns out to be a scale parameter, whereas p and q are still shape parameters.

We see that the larger the value of a, the thinner the tails of the density (6.5) are, whereas the relative values of p and q are important in determining the skewness of the distribution of $\log X$. Figures 6.1–6.3 illustrate the effect of the three shape parameters; each graph keeps two parameters constant and varies the remaining one. (Note that there is considerable variation in the shape of the density for small a and p in Figures 6.1 and 6.2.)

As an income distribution, (6.5) was proposed by McDonald (1984) and independently as a model for the size-of-loss distribution in actuarial science by Venter (1983), who called it a *transformed beta distribution*. A decade earlier, it was briefly discussed by Mielke and Johnson (1974) in a meteorological application as a generalization of two distributions that are included in the following sections under the names of the Singh–Maddala and Dagum distributions, respectively. It may also be considered a generalized F distribution, and it appears under this name in, for example, Kalbfleisch and Prentice (1980). The distribution is further referred to as a Feller–Pareto distribution by Arnold (1983), who introduced an additional location parameter, and it was rediscovered, in a different parameterization, in the econometrics literature by Majumder and Chakravarty (1990). McDonald and Mantrala (1993, 1995) observed that the Majumder–Chakravarty model is equivalent to the GB2 distribution. We shall use McDonald's (1984) notation below.

The case where $a = 1$, that is, the beta distribution of the second kind (B2) with p.d.f. is a member of the Pearson system of distributions (see Chapter 2), namely,

$$f(x) = \frac{x^{p-1}}{b^p B(p, q)[1 + x/b]^{p+q}}, \quad x > 0, \tag{6.7}$$

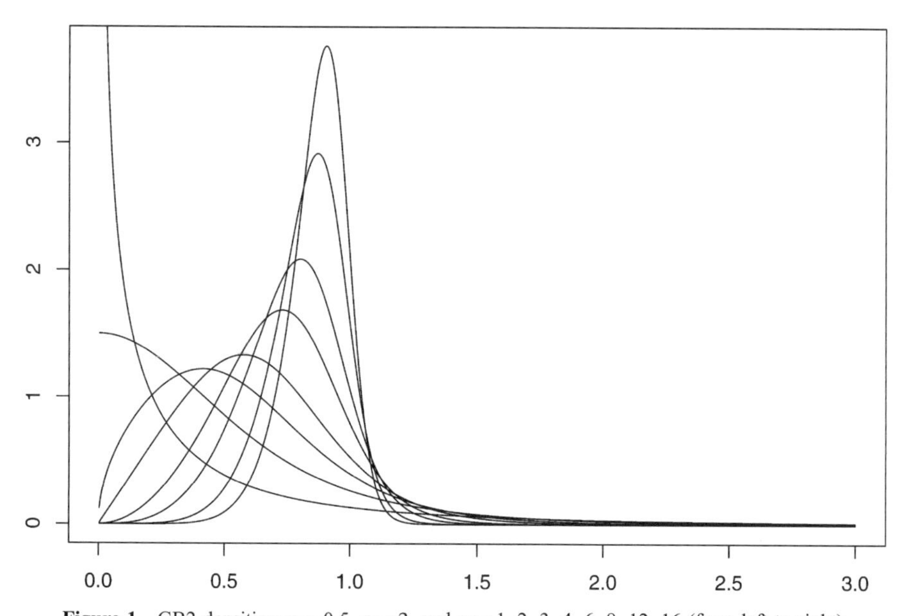

Figure 1 GB2 densities: $p = 0.5$, $q = 2$, and $a = 1, 2, 3, 4, 6, 8, 12, 16$ (from left to right).

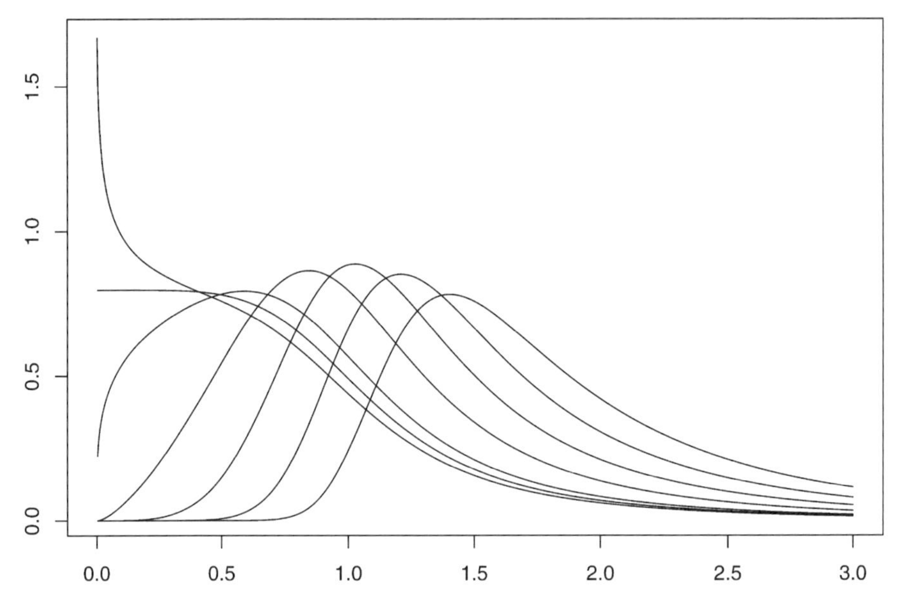

Figure 2 GB2 densities: $a = 5$, $q = 0.5$, and $p = 0.17, 0.25, 0.5, 1, 2, 4$ (from left to right).

the Pearson VI distribution. As an income distribution, it was proposed somewhat earlier than the GB2, in an application to Finnish data (Vartia and Vartia, 1980).

The GB2 can also be considered a generalized log-logistic distribution: Setting $a = b = p = q = 1$ in (6.6), we get

$$f(y) = \frac{e^y}{(1 + e^y)^2}, \quad -\infty < y < \infty,$$

which is the density of a standard logistic distribution (see, e.g., Johnson, Kotz, and Balakrishnan, 1995, Chapter 23). Specifically, the distribution of (6.6) is a skewed generalized logistic distribution, symmetry being attained only for $p = q$. McDonald and Xu (1995) referred to (6.6) as the density of an *exponential GB2* distribution.

Recently, Parker (1999a,b) derived GB2 and B2 earnings distributions from microeconomic principles (a neoclassical model of optimizing firm behavior), thereby providing some rationale as to why such distributions may be observed. In his model the shape parameters p and q are functions of the output-labor elasticity and the elasticity of income returns with respect to human capital, thus permitting some insight into the potential causes of observed inequality trends.

The GB2 model is most useful for unifying a substantial part of the size distributions literature. It contains a large number of income and loss distributions as special or limiting cases: The Singh–Maddala distribution is obtained for $p = 1$, the Dagum distribution for $q = 1$, the beta distribution of the second kind (B2) for

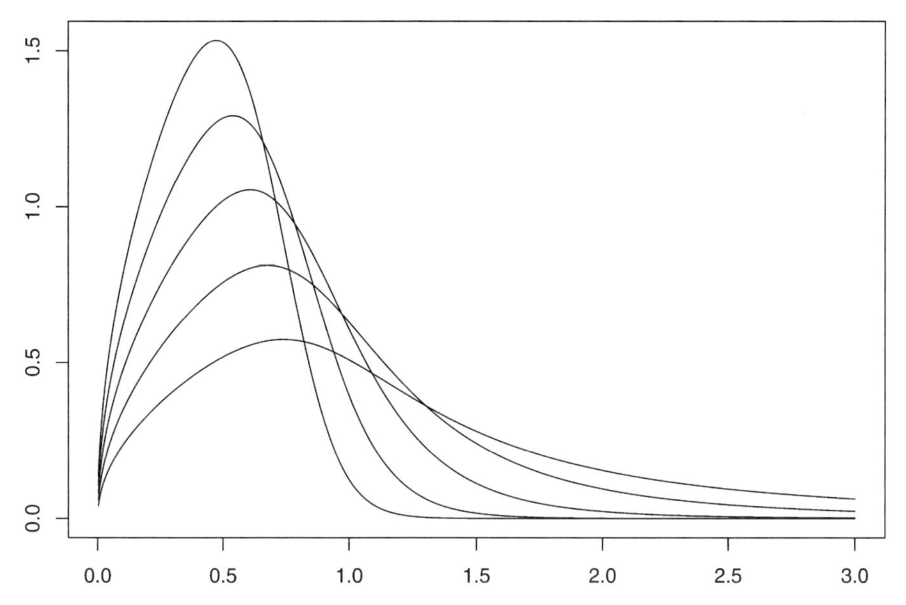

Figure 3 GB2 densities: $a = 5$, $p = 0.3$, and $q = 0.25, 0.5, 1, 2, 4$ (from right to left).

$a = 1$, the Fisk (or log-logistic) distribution for $p = q = 1$, and the Lomax (or Pareto type II) distribution for $a = p = 1$. Figure 6.4 illustrates these interrelations. [For completeness (and symmetry!) we include an inverse form of the Lomax distribution, although there appears to be hardly any work dealing explicitly with this distribution.] Apart from the B2 distribution, which is included in the present section, these models will be discussed in greater detail in the following sections. Furthermore, the generalized gamma distribution (see Chapter 5) emerges as a limiting case upon setting $b = q^{1/a}\beta$ and letting $q \to \infty$. Consequently, the gamma and Weibull distributions are also limiting cases of the GB2, since both are special cases of the generalized gamma distribution.

6.1.2 Moments and Other Basic Properties

In (6.4) the GB2 distribution was introduced via the incomplete beta function ratio. Utilizing the relation with Gauss's hypergeometric function ${}_2F_1$ (e.g., Temme, 1996), we obtain

$$I_z(p, q) = \frac{z^p}{pB(p, q)}\, {}_2F_1(p, 1 - q; p + 1; z), \quad 0 \le z \le 1, \tag{6.8}$$

where

$${}_2F_1(a_1, a_2; b; x) = \sum_{n=0}^{\infty} \frac{(a_1)_n (a_2)_n}{(b)_n} \frac{x^n}{n!}, \quad |x| < 1. \tag{6.9}$$

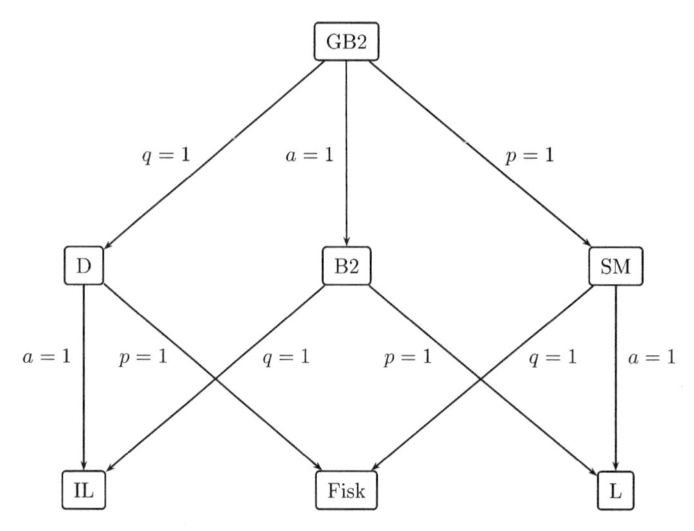

Figure 4 Beta-type size distributions and their interrelations: generalized beta distribution of the second kind (GB2), Dagum distribution (D), beta distribution of the second kind (B2), Singh–Maddala distribution (SM), inverse Lomax distribution (IL), Fisk (log-logistic) distribution (Fisk), Lomax distribution (L).

The c.d.f. can also be expressed in the form (McDonald, 1984)

$$F(x) = \frac{[(x/b)^a/(1+(x/b)^a)]^p}{pB(p,q)}\ {}_2F_1\left(p, 1-q; p+1; \left[\frac{(x/b)^a}{1+(x/b)^a}\right]\right), \quad x > 0. \tag{6.10}$$

If an expression for the survival function is required, we may use the complementary relation

$$I_{1-z}(p,q) = 1 - I_z(q,p)$$

in conjunction with (6.4).

The mode of the GB2 distribution occurs at

$$x_{\text{mode}} = b\left(\frac{ap-1}{aq+1}\right)^{1/a}, \quad \text{if } ap > 1, \tag{6.11}$$

and at zero otherwise.

As noted above, the density of the GB2 distribution is regularly varying at infinity with index $-aq-1$ and also regularly varying at the origin with index $-ap-1$. This implies that the moments exist only for $-ap < k < aq$. They are

$$E(X^k) = \frac{b^k B(p+k/a, q-k/a)}{B(p,q)} = \frac{b^k \Gamma(p+k/a)\Gamma(q-k/a)}{\Gamma(p)\Gamma(q)}. \tag{6.12}$$

Note that this last expression is, when considered a function of k, equal to the moment-generating function $m_Y(\cdot)$ of $Y = \log X$, in view of the relation $m_Y(k) = E(e^{kY}) = E(X^k)$. This point of view is useful for the computation of the moments of $\log X$, which are required for deriving, for example, the Fisher information matrix of the GB2 distribution and its subfamilies.

It is easy to see that the GB2 is closed under power transformations:

$$X \sim \text{GB2}(a, b, p, q) \Longrightarrow X^r \sim \text{GB2}\left(\frac{a}{r}, b^r, p, q\right), \quad r > 0. \tag{6.13}$$

Also, Venter (1983) observed that the GB2 distribution is closed under inversion, in the sense that

$$X \sim \text{GB2}(a, b, p, q) \Longleftrightarrow \frac{1}{X} \sim \text{GB2}\left(a, \frac{1}{b}, q, p\right). \tag{6.14}$$

Thus, it is sometimes convenient to allow for $a < 0$ in (6.5), one then simply replaces a by $|a|$ in the numerator.

The hazard rate of the GB2 distributions can exhibit a wide variety of shapes. Considering the special case where $p = q = 1$, it is clear that it can at least be monotonically decreasing as well as unimodal. Since a detailed analysis is rather involved, we refer the interested reader to McDonald and Richards (1987).

In view of the density being regularly varying and general results for the mean excess function of such distributions (see Chapter 2), it follows that the mean excess function is asymptotically linearly increasing

$$e(x) \in \text{RV}_\infty(1). \tag{6.15}$$

6.1.3 Characterizations and Representations

It is possible to characterize the B2 distribution in terms of maximum entropy among all distributions supported on $[0, \infty)$: If both $E[\log X]$ and $E[\log(1 + X)]$ are prescribed, then the maximum entropy p.d.f. is the B2 density. Hence, this distribution is characterized by the geometric means of X and $1 + X$ (Kapur, 1989, p. 66).

An extension of the classical maximum entropy approach was considered by Leipnik (1990), who used a relative maximum entropy principle that leads to several income distributions included in the present chapter, such as the GB2, GB1 (see Section 6.5), and Singh–Maddala distributions. The idea here is that income recipients are affected by ordinal as well as cardinal considerations, and therefore subjectively reduce their incomes by multiplication with a subjective, but not individualized, reduction factor $\phi[x, 1 - F(x)]$, which depends on the actual income x as well as the income status, here measured by the proportion $1 - F(x)$ of income receivers earning more than a preassigned income x. This leads to an adjusted income $\xi(x) = x\phi[x, 1 - F(x)]$ that is perhaps best interpreted as utility. A relative income entropy density f is now defined in terms of the p.d.f. f and the marginal subjective income $\partial\xi[x, F(x)]/\partial x$ and determined by the maximization of $\int f(x)\,dx$ under constraints on $E\xi(X)$ and $E\xi^2(X)$, the first two moments of ξ with respect to F. The resulting nonlinear differential equation (according to Leipnik of a type that is not much studied outside of hydrology and astrophysics) is sometimes solvable and leads, under appropriate specifications of the adjustment function ϕ, to income distributions of the generalized beta type.

As mentioned in the preceding chapter, Malik (1967) and Ahuja (1969) both showed that if $X_1 \sim \mathrm{GG}(a, 1, p)$, $X_2 \sim \mathrm{GG}(a, 1, q)$, and X_1 and X_2 are independent (note the identical shape parameter a!),

$$\frac{X_1}{X_2} \sim \mathrm{GB2}(a, 1, p, q). \tag{6.16}$$

The relation (6.16) is perhaps more familiar in the form

$$\left(\frac{X_1}{X_2}\right)^{1/a} \sim \mathrm{GB2}(a, 1, p, q). \tag{6.17}$$

where now $X_1 \sim \mathrm{Ga}(1, p)$, $X_2 \sim \mathrm{Ga}(1, q)$, and X_1 and X_2 are independent, which is a generalization of the well-known relation between the standard gamma and B2 distributions

$$\frac{X_1}{X_2} \sim \mathrm{B2}(1, p, q). \tag{6.18}$$

These relations can be exploited to obtain random samples from the GB2 distribution: There is a large number of gamma random number generators (see Devroye, 1986), from which generalized gamma samples can be obtained by a power transformation. Simulating independent data from two generalized gamma distributions with the required shape parameters we arrive at GB2 samples via the above relation.

All this can be further rephrased, utilizing the familiar relation between the gamma distribution and the classical beta distribution (Pearson type I). If $X_1 \sim \mathrm{Ga}(1, p)$, $X_2 \sim \mathrm{Ga}(1, q)$, and X_1 and X_2 are independent, the random variable

$$W := \frac{X_1}{X_1 + X_2} \tag{6.19}$$

follows a beta type I distribution with parameters p and q. In view of (6.18), it is therefore possible to define the B2 distribution via

$$X \sim W^{-1} - 1,$$

as was done by, for example, Feller (1971, p. 50), who notes that the resulting distribution is sometimes named after Pareto. This caused Arnold (1983) to refer to the GB2 distribution, which can evidently be expressed as

$$X \sim b[W^{-1} - 1]^{1/a} \tag{6.20}$$

as the *Feller–Pareto distribution.*

The representation (6.16) can also be considered a mixture representation, namely, as a scale mixture of generalized gamma distributions with inverse generalized gamma weights (Venter, 1983; McDonald and Butler, 1987). Recall that the generalized gamma distribution has the density

$$f(x) = \frac{a}{\theta^{ap}\Gamma(p)} x^{ap-1} e^{-(x/\theta)^a}.$$

Now, if the parameter $\theta \sim \text{InvGG}(a, b, q)$, then, in an obvious notation,

$$\text{GG}(a, \theta, p) \bigwedge_{\theta} \text{InvGG}(a, b, q) = \text{GB2}(a, b, p, q). \tag{6.21}$$

Evidently, this is just a restatement of (6.16). For the beta distribution of the second kind, that is, the case where $a = 1$, the above representation simplifies to

$$\text{Ga}(\theta, p) \bigwedge_{\theta} \text{InvGa}(b, q) = \text{B2}(b, p, q), \tag{6.22}$$

where InvGa denotes an inverse gamma (or Vinci) distribution.

Thus, the GB2 distribution and its subfamilies have a theoretical justification as a representation of incomes arising from a heterogeneous population of income receivers, or, in actuarial terminology, as a representation of claims arising from a heterogeneous population of exposures. This argument was used by Parker (1997) when utilizing a B2 distribution (6.22) as the distribution of self-employment income; he argued that individuals are heterogeneous with regard to entrepreneurial characteristics.

Table 6.1 presents the mixture representations for all the distributions discussed in the present chapter (McDonald and Butler, 1987).

6.1.4 Lorenz Curves and Inequality Measures

The Lorenz curve of the GB2 distribution, which exists whenever $aq > 1$, cannot be obtained directly from the Gastwirth representation since the quantile function is not available in closed form. However, the Lorenz curve is available in terms of the first-moment distribution in the following manner: Butler and McDonald (1989) observed that the normalized incomplete moments, that is, the c.d.f.'s of the higher-order moment distributions, can be expressed as the c.d.f.'s of

Table 6.1 Mixture Representations for Beta-type Size Distributions

Distribution	Structural Distribution	Mixing Distribution
GB2(x; a, b, p, q)	GG(x; a, θ, p)	InvGG(θ; a, b, q)
B2(x; b, p, q)	Ga(x; θ, p)	InvGa(θ; b, q)
SM(x; a, b, q)	Wei(x; a, θ)	InvGG(θ; a, b, q)
Dagum(x; a, b, p)	GG(x; a, θ, p)	InvWei(θ; a, b)
Fisk(x; a, b)	Wei(x; a, θ)	InvWei(θ; a, b)
Lomax(x; b, q)	Exp(x; θ)	InvGa(θ; b, p)

Source: McDonald and Butler (1987).

GB2 distributions with different sets of parameters. Namely, for $X \sim \text{GB2}(a, b, p, q)$ and

$$F_{(k)}(x) = \frac{\int_0^x t^k f(t)\,dt}{E(X^k)}, \quad 0 < x < \infty,$$

we have

$$F_{(k)}(x) = F\left(x; a, b, p + \frac{k}{a}, q - \frac{k}{a}\right), \quad 0 < x < \infty. \tag{6.23}$$

This closure property is of practical importance in that computer programs used to evaluate the distribution function of the GB2 distribution can also be used to evaluate the higher-order moment distributions by merely changing the values of the parameters. In particular, theoretical Lorenz curves of GB2 distributions can be obtained by plotting $F_{(1)}(x)$ against $F(x)$, $0 < x < \infty$. This closure property is special to the GB2 and B2 distributions; it does not extend to any of the subfamilies discussed in the following sections.

Using the ratio representation (6.16), Kleiber (1999a) showed that for $X_i \sim \text{GB2}(a_i, b_i, p_i, q_i)$, $i = 1, 2$,

$$a_1 \le a_2, \quad a_1 p_1 \le a_2 p_2 \quad \text{and} \quad a_1 q_1 \le a_2 q_2 \tag{6.24}$$

together imply $X_1 \ge_L X_2$. This generalizes earlier results obtained by Wilfling (1996b). However, the condition is not necessary. The necessary conditions for Lorenz dominance are

$$a_1 p_1 \le a_2 p_2 \quad \text{and} \quad a_1 q_1 \le a_2 q_2. \tag{6.25}$$

This result was first obtained by Wilfling (1996b); for an alternative approach see Kleiber (2000a). It is worth noting that, although a full characterization of Lorenz order within the GB2 family is currently unavailable, the conditions (6.24) and (6.25) are strong enough to yield complete characterizations for all the subfamilies considered in the following sections.

Sarabia, Castillo, and Slottje (2002) obtained some Lorenz ordering results for nonnested models. In particular, if $X \sim \text{GG}(\tilde{a}, \tilde{b}, \tilde{p})$, $Y \sim \text{GB2}(a, b, p, q)$, and $aq > 1$, $\tilde{a} \geq a$, $\tilde{a}\tilde{p} \geq a$, then $Y \geq_L X$.

As noted in the previous section, the GB2 distribution has heavy tails; hence, only a few of the moments exist. This implies that inequality measures such as the generalized entropy measures only exist for sensitivity parameters within a certain range (Kleiber, 1997) that is often rather narrow in practice.

The above relation (6.23) also provides a simple way to obtain the Pietra index of inequality, namely,

$$P = F(\mu) - F_{(1)}(\mu),$$

where μ is the mean of X (Butler and McDonald, 1989).

The Gini coefficient of the GB2 is available in McDonald (1984) as a lengthy expression involving the generalized hypergeometric function ${}_3F_2$

$$\begin{aligned} G = {} & \frac{2B(2p + 1/a, 2q - 1/a)}{pB(p, q)B(p + 1/a, q - 1/a)} \left\{ \frac{1}{p} {}_3F_2\left[1, p + q, 2p + \frac{1}{a}; p + 1, 2(p + q); 1\right] \right. \\ & \left. - \frac{1}{p + 1/a} {}_3F_2\left[1, p + q, 2p + \frac{1}{a}; p + \frac{1}{a} + 1, 2(p + q); 1\right] \right\}. \end{aligned}$$

For the B2 subfamily the expression is less cumbersome and equals (McDonald, 1984)

$$G = \frac{2B(2p, 2q - 1)}{pB^2(p, q)}.$$

This is decreasing in both p and q.

6.1.5 Estimation

There are comparatively few results concerning parameter estimation for this somewhat complicated four-parameter distribution. Venter (1983) provided a brief discussion of ML estimation. The log-likelihood for a complete random sample of size n equals

$$\begin{aligned} \log L = {} & n \log \Gamma(p + q) + n \log a + (ap - 1) \sum_{i=1}^{n} \log x_i - nap \log b \\ & - n \log \Gamma(p) - n \log \Gamma(q) - (p + q) \sum_{i=1}^{n} \log\left[1 + \left(\frac{x_i}{b}\right)^a\right]. \end{aligned}$$

Denoting as above the derivative of $\log\Gamma(\cdot)$, the digamma function, by $\psi(\cdot)$ we obtain the following four equations from the partial derivatives with respect to a, b, p, q (in this order):

$$\frac{n}{a}+p\sum_{i=1}^{n}\log\left(\frac{x_i}{b}\right)=(p+q)\sum_{i=1}^{n}\log\left(\frac{x_i}{b}\right)\left[\left(\frac{b}{x_i}\right)^{a}+1\right]^{-1}, \tag{6.26}$$

$$np=(p+q)\sum_{i=1}^{n}\left[1+\left(\frac{b}{x_i}\right)^{a}\right]^{-1}, \tag{6.27}$$

$$n\psi(p+q)+a\sum_{i=1}^{n}\log\left(\frac{x_i}{b}\right)=n\psi(p)+\sum_{i=1}^{n}\log\left[1+\left(\frac{x_i}{b}\right)^{a}\right], \tag{6.28}$$

$$n\psi(p+q)=n\psi(q)+\sum_{i=1}^{n}\log\left[1+\left(\frac{x_i}{b}\right)^{a}\right]. \tag{6.29}$$

Venter noted that the first and second equations are linear in p and q; hence, they can easily be solved to yield p and q as functions of a and b. Substituting the resulting expressions into the third and fourth equations, one obtains two nonlinear equations in two unknowns that can be solved by, for example, Newton–Raphson iteration.

Recently, Brazauskas (2002) obtained the Fisher information matrix of the GB2 distributions. This family is a regular family (in terms of ML estimation); hence, we may use the expression

$$I(\theta)=\left[-E\left(\frac{\partial^2\log L}{\partial\theta_i\partial\theta_j}\right)_{i,j}\right]=:\begin{pmatrix} I_{11} & I_{12} & I_{13} & I_{14}\\ I_{21} & I_{22} & I_{23} & I_{24}\\ I_{31} & I_{32} & I_{33} & I_{34}\\ I_{41} & I_{42} & I_{43} & I_{44}\end{pmatrix}. \tag{6.30}$$

For $\theta=(a, b, p, q)^{\top}$, it follows from (6.28) and (6.29) that the second derivatives of the log-likelihood with respect to p and q are constants; hence, I_{33}, $I_{34}=I_{43}$, and I_{44} are easily obtained. The remaining derivations are rather tedious and will not be given here. The elements of $I(\theta)$ are

$$I_{11}=\frac{1}{a^2(1+p+q)}\left\{1+p+q+pq\left[\psi'(q)+\psi'(p)\right.\right.$$
$$\left.\left.+\left(\psi(q)-\psi(p)+\frac{p-q}{pq}\right)^2-\frac{p^2+q^2}{p^2q^2}\right]\right\}, \tag{6.31}$$

$$I_{21}=I_{12}=\frac{p-q-pq[\psi(p)-\psi(q)]}{b(1+p+q)}, \tag{6.32}$$

$$I_{22} = \frac{a^2pq}{b^2(1+p+q)}, \tag{6.33}$$

$$I_{23} = I_{32} = \frac{aq}{b(p+q)}, \tag{6.34}$$

$$I_{24} = I_{42} = -\frac{ap}{b(p+q)}, \tag{6.35}$$

$$I_{31} = I_{13} = -\frac{q[\psi(p) - \psi(q)] - 1}{a(p+q)}, \tag{6.36}$$

$$I_{14} = I_{41} = -\frac{p[\psi(q) - \psi(p)] - 1}{a(p+q)}, \tag{6.37}$$

$$I_{33} = \psi'(p) - \psi'(p+q), \tag{6.38}$$

$$I_{34} = I_{43} = -\psi'(p+q), \tag{6.39}$$

$$I_{44} = \psi'(q) - \psi'(p+q). \tag{6.40}$$

It is worth noting that these results were essentially obtained by Prentice (1975) some 25 years earlier. (He worked with the distribution of $\log X$ that leads to the same Fisher information, although using a different parameterization from which the equivalence is not easily recognized.) Prentice also discussed a reparameterization that is useful for discriminating between special and limiting cases of the GB2.

6.1.6 Empirical Results

Incomes and Wealth

Vartia and Vartia (1980) fit a four-parameter shifted B2 distribution (under the name of *scaled and shifted F distribution*) to the 1967 distribution of taxed income in Finland. Using ML and method-of-moments estimators, they found that their model fits systematically better than the two- and three-parameter lognormal distributions.

McDonald (1984) estimated the GB2 and B2 distributions for 1970, 1975, and 1980 U.S. family incomes. The GB2 outperforms 10 other distributions (mainly of the beta and gamma type), whereas the B2 fit is comparable to the fit obtained using gamma or generalized gamma distributions but is inferior to the Singh–Maddala distribution for each one of the data sets considered.

Butler and McDonald (1989) gave GB2 parameter estimates for U.S. Caucasian family incomes from 1948–1980, using maximum likelihood estimators applied to grouped data. They found shape parameters $a \in (2.8, 8.4)$, roughly decreasing over time, $p \in (0.17, 0.6)$, and $q \in (0.4, 1.7)$, with larger values of p and q pertaining to the post-1966 data. This shows that the tail indexes ap, aq are roughly constant over time, while there are considerable changes in the center of the distribution.

Majumder and Chakravarty (1990) considered the GB2 distribution when modeling U.S. income data for 1960, 1969, 1980, and 1983. The distribution ranks second when compared to their own distribution. However, there are some contradictory results in their paper, and in a reassessment of their findings emphasizing that the Majumder–Chakravarty distribution is merely a reparametrization of the GB2 distribution, McDonald and Mantrala (1993, 1995) fit this model to 1970, 1980, and 1990 U.S. family incomes using two different methods of estimation and alternative groupings of the data. The distribution provides the best fit among all the models considered. Majumder and Chakravarty's paper provides us with an important lesson in fitting income data and signals the dangers and pitfalls when data are not properly scrutinized.

In a study of the aggregate demand for specific car lines, Bordley and McDonald (1993) employed the GB2 distribution for the direct estimation of income elasticity from population income distribution. Their results are consistent with those of traditional econometric studies of automotive demand.

McDonald and Xu (1995) studied 1985 U.S. family incomes. Out of 11 distributions considered—mainly of the gamma and beta type—the GB2 ranks first (being observationally equivalent to the more general GB distribution, which is briefly mentioned in Section 6.5 below) in terms of likelihood.

In a very comprehensive already mentioned study employing 15 income distribution models (again of the beta and gamma type), Bordley, McDonald, and Mantrala (1996) fit the GB2 distribution to U.S. family incomes for 1970, 1975, 1980, 1985, and 1990. For all but the 1970 data (where the five-parameter GB distribution—see Section 6.5 below—provides a slight improvement) the GB2 outperforms every other distribution, with the improvements relative to the nested models being statistically significant. Bordley, McDonald, and Mantrala concluded that the GB2 distribution is the best-fitting four-parameter distribution for these data.

Brachmann, Stich, and Trede (1996) fit the GB2 (and many of its subfamilies and limiting cases) to individual household incomes from the German Socio-Economic Panel (SOEP) for the years 1984–1993. A comparison with nonparametric density estimates shows that only the GB2 and, to a lesser extent, the Singh–Maddala distribution to be discussed in the following section are satisfactory. However, the means and Gini coefficients appear to be slightly underestimated.

Parker (1997) considered a three-parameter B2 distribution when studying UK self-employment incomes from 1976–1991. Starting from a dynamic economic model of self-employment income, he found that incomes above some threshold $\gamma < 0$ follow a two-parameter B2 distribution. He estimated parameters by modified method-of-moments estimators (MMME), as advocated by Cohen and Whitten (1988) for models containing a threshold parameter, finding that incomes became progressively more unequally distributed, a fact he attributed to an increase of heterogeneity among the self-employed.

Actuarial Losses

In actuarial science, the already mentioned paper by Cummins et al. (1990) used the GB2 family to model the annual fire losses experienced by a major university (Cummins and Freifelder, 1978). It turns out that, in terms of likelihood, the full flexibility of the GB2 is not required and a one-parameter limiting case, the inverse exponential distribution, agrees very well with the four-parameter GB2. The same authors also considered data on the severity of fire losses and fit the GB2 to both grouped and individual observations. Here three-parameter special and limiting cases of the GB2 such as the Singh–Maddala and inverse generalized gamma distributions are selected.

Cummins et al. (1990) provided an important lesson about unnecessary overparameterization. In summary, the GB2 family is indeed an attractive, flexible, elegant, and ingenious family but it involves four parameters. The experience collected by statistical applications during the last 100 years teaches us that four-parameter distributions can sometimes be almost omnipotent and do not allow us to penetrate to the crux of the matter: in our case the mechanism and factors that determine the size distributions. One (almost) becomes nostalgic about Pareto's simple model proposed at the end of the nineteenth century about 90 years before McDonald's pioneering effort. Perhaps the leap forward is just too great? (Evidently, without the disastrous consequences of the Chinese leap forward in the late 1950s.)

6.1.7 Extensions

Zandonatti (2001) presented a generalization of the GB2 following Feller's definition of a GB2 distribution; cf. (6.20). The distribution is defined by (we omit an additional location parameter)

$$X = b[W^{-1/\theta} - 1]^{1/a},$$

where W denotes a standard beta(p, q) variable, and possesses the density

$$f(x) = \frac{a\theta x^{a-1}}{b^a B(p,q)}\left[1 + \left(\frac{x}{b}\right)^a\right]^{-\theta p}\left\{1 - \left[1 + \left(\frac{x}{b}\right)^a\right]^{-\theta}\right\}^{q-1}. \tag{6.41}$$

We may call the distribution given by (6.41) the Zandonatti distribution. Clearly, for $\theta = 1$ we obtain (6.5).

6.2 SINGH–MADDALA DISTRIBUTION

The Singh–Maddala distribution introduced by Singh and Maddala in 1975 and in a more polished form in 1976 has received special attention in the literature of income distributions. Although it was discovered under the above name before the GB2 distribution, it would be convenient to treat it as a special case of the GB2. In the recent econometric literature, it is often compared with the Dagum distribution (see the following section), perhaps due to the similarity of their c.d.f.'s.

6.2.1 Definition and Motivation

The Singh–Maddala distribution is a special case of the GB2 distribution, with $p = 1$. Its density is

$$f(x) = \frac{aqx^{a-1}}{b^a[1 + (x/b)^a]^{1+q}}, \quad x > 0, \tag{6.42}$$

where all three parameters a, b, q are positive. Here b is a scale parameter and a, q are shape parameters; q only affects the right tail, whereas a affects both tails. We shall use the notation $\text{GB2}(a, b, 1, q) \equiv \text{SM}(a, b, q)$.

This distribution was independently rediscovered many times in several loosely related areas. Consequently, it is known under a variety of names: It seems that it was first considered by Burr (1942), where it appears as the twelfth example of solutions of a differential equation defining the Burr system of distributions [see Kleiber, (2003a) for a recent survey of this family]. It is therefore usually called the Burr XII distribution, or—being the most popular Burr distribution—simply *the Burr distribution*. Kakwani (1980b, p. 24) reported that it was proposed as an income distribution as early as 1958 in an unpublished paper presented at a meeting of the Econometric Society by Sargan. It is also known as the Pareto (IV) distribution (Arnold, 1983), the beta-*P* distribution (Mielke and Johnson, 1974), or as a generalized log-logistic distribution (e.g., El-Saidi, Singh, and Bartolucci, 1990). This distribution is among the most commonly used models for the distribution of personal incomes. In actuarial science it is usually called the Burr distribution (e.g., Hogg and Klugman, 1983, 1984).

The special case where $a = q$ is sometimes called the paralogistic distribution (Klugman, Panjer, and Willmot, 1998). Further special cases are known as the Fisk (1961) or log-logistic, for $q = 1$, and the Lomax (1954) distribution, for $a = 1$. They will be discussed in greater detail in the following sections.

Unlike the c.d.f. of the GB2 distribution, the c.d.f. of the Singh–Maddala distribution is available in closed form; it is given by the pleasantly simple expression

$$F(x) = 1 - \left[1 + \left(\frac{x}{b}\right)^a\right]^{-q}, \quad x > 0. \tag{6.43}$$

We directly determine that the quantile function is equally straightforward

$$F^{-1}(u) = b[(1 - u)^{-1/q} - 1]^{1/a}, \quad \text{for } 0 < u < 1. \tag{6.44}$$

Hence, the Singh–Maddala distribution is one of the few distributions for which the density, c.d.f., and quantile function all have simple closed forms. This may partly explain why it was rediscovered so many times in various contexts.

Singh and Maddala (1976) derived their distribution by considering the hazard rate $r(x)$ of income. They observed that, although decreasing failure rate (DFR) distributions are unlikely to be observed when the underlying variable is time, when the variable is income, "... a priori plausibility on theoretical reasoning for DFR after a point is obvious" (p. 964), in that "income may help in earning more. The ability to make more money might increase with one's income." [A similar idea may be found some 50 years earlier in Hagstrœm (1925) in connection with the classical Pareto distribution.]

Singh and Maddala then introduced the hazard rate of $z := \log x$, that is,

$$r^*(z) = \frac{\partial F(z)}{\partial z} \frac{1}{1 - F(z)},$$

a quantity they called the *proportional failure rate* (PFR). It measures, at any income, the odds against advancing further to higher incomes, in a proportional sense. For the classical Pareto distribution (3.2) we have $r^*(z) = \alpha$. We know from Chapter 3 that this is also the slope of the survival function of income in the Pareto diagram. This suggests that $r^*(z)$ must be asymptotically constant. Singh and Maddala further assumed that $r^*(z)$ grows with z first with an increasing, then a decreasing, rate. Defining $y := -\log(1 - F)$, $y' > 0$, $y'' > 0$, they started from the differential equation

$$y'' = a \cdot y'(\alpha - y'),$$

where a is a constant. This may be rearranged, yielding

$$\frac{y''}{y'} + \frac{y''}{\alpha - y} = a\alpha,$$

which integrates to

$$\log y' - \log(\alpha - y') = a\alpha z + c_1,$$

where c_1 is the constant of integration. From this we get

$$\frac{y'}{\alpha - y'} = e^{a\alpha z + c_1},$$

or

$$y' = \frac{\alpha e^{a\alpha z + c_1}}{1 + e^{a\alpha z + c_1}}.$$

Further integration yields

$$\log y = \frac{1}{a}\log(1 + e^{a\alpha z + c_1}) + c_2,$$

where c_2 is another constant of integration. Substituting $-\log(1 - F)$ for y, $\log x$ for z, and rearranging, we obtain

$$F(x) = 1 - \frac{c}{(b + x^{a\alpha})^{1/a}},$$

where $c = (-c_2 - c_1)/\alpha$ and $b = 1/e^{c_1}$. The boundary condition $F(0) = 0$ results in $c = b^{1/a}$. Thus,

$$F(x) = 1 - \frac{b^{1/a}}{(b + x^{a\alpha})^{1/a}},$$

or

$$F(x) = 1 - \frac{1}{(1 + a_1 x^{a_2})^{a_3}}, \tag{6.45}$$

where $a_1 = 1/b$, $a_2 = a\alpha$, and $a_3 = 1/a$, which is the Singh–Maddala c.d.f. given above with $a_2 = a$, $a_3 = q$, and $a_1 = b^{a_2}$.

This derivation shows that the Singh–Maddala distribution is characterized by a proportional failure rate

$$r^*(z) = \frac{aq \cdot e^{az - \log b}}{1 + e^{az - \log b}},$$

a three-parameter logistic function with respect to income power $z - \log x$. Singh and Maddala's approach was criticized by Cramer (1978) who found the analogy with failure rates unconvincing, writing that "it is not clear what the DFR property of income distributions means, but since it applies to distributions that hold at a given time all references to the passage of time and to the process whereby individuals move from one income level to the next are inappropriate." Nonetheless, these reliability properties seem to capture some salient features of empirical income distributions.

The parameterization utilized in (6.45) is quite common in applications; a variant with $\tilde{a}_1 = 1/b^a$ was employed, for example, by Schmittlein (1983). However, in the size distributions area the parameterization based on (6.42) is often more convenient, because there b is a scale parameter and hence it is eliminated when scale-free quantities such as the Lorenz curve and common inequality measures are considered. We shall use this variant in the subsequent discussion.

6.2.2 Moments and Other Basic Properties

The kth moment exists for $-a < k < aq$; it equals [compare (6.12)]

$$E(X^k) = \frac{b^k B(1 + k/a, q - k/a)}{B(1, q)} = \frac{b^k \Gamma(1 + k/a)\Gamma(q - k/a)}{\Gamma(q)}. \tag{6.46}$$

In particular,

$$E(X) = \frac{bB(1 + 1/a, q - 1/a)}{B(1, q)} = \frac{b\Gamma(1 + 1/a)\Gamma(q - 1/a)}{\Gamma(q)} \tag{6.47}$$

and

$$\operatorname{var}(X) = \frac{b^2\{\Gamma(q)\Gamma(1 + 2/a)\Gamma(q - 2/a) - \Gamma^2(1 + 1/a)\Gamma^2(q - 1/a)\}}{\Gamma^2(q)}. \tag{6.48}$$

Hence, the coefficient of variation is (McDonald, 1981)

$$\mathrm{CV} = \sqrt{\frac{\Gamma(q)\Gamma(1 + 2/a)\Gamma(q - 2/a)}{\Gamma^2(1 + 1/a)\Gamma^2(q - 1/a)} - 1} \tag{6.49}$$

and the shape factors are

$$\sqrt{\beta_1} = \frac{\Gamma^2(q)\lambda_3 - 3\Gamma(q)\lambda_2\lambda_1 + 2\lambda_1^3}{[\Gamma(q)\lambda_2 - \lambda_1^2]^{3/2}} \tag{6.50}$$

and

$$\beta_2 = \frac{\Gamma^3(q)\lambda_4 - 4\Gamma^2(q)\lambda_3\lambda_1 + 6\Gamma(q)\lambda_2\lambda_1^2 - 3\lambda_1^4}{[\Gamma(q)\lambda_2 - \lambda_1^2]^2}, \tag{6.51}$$

where we have set $\lambda_i = \Gamma(1 + i/a)\Gamma(q - i/a)$, $i = 1, 2, 3, 4$. These expressions are rather unwieldy.

From (6.11), the mode of the Singh–Maddala distribution is at

$$x_{\text{mode}} = b\left(\frac{a - 1}{aq + 1}\right)^{1/a}, \quad \text{if } a > 1, \tag{6.52}$$

and at zero otherwise. Thus, the mode is seen to be decreasing with q, reflecting the fact that the right tail becomes lighter as q increases.

The closed forms of the c.d.f and the quantile function of the Singh–Maddala distribution also permit convenient manipulations with the characteristics of order statistics. In particular,

$$E(X_{1:n}) = \frac{b\Gamma(1+1/a)\Gamma(nq-1/a)}{\Gamma(nq)}. \tag{6.53}$$

This follows from the closure property (Arnold and Laguna, 1977; Arnold, 1983)

$$X \sim \mathrm{SM}(a, b, q) \Longrightarrow X_{1:n} \sim \mathrm{SM}(a, b, nq) \tag{6.54}$$

and the expression for the moments (6.46).

The moments of $X_{k:n}$, $2 \leq k \leq n$ can now be generated using recurrence relations, yielding (Tadikamalla, 1977)

$$E(X_{i:n}^k) = i\binom{n}{i} b^k q \sum_{j=0}^{i-1} (-1)^j \binom{i-1}{j} B\left(qn - qi + q - \frac{k}{a} + qj, 1 + \frac{k}{a}\right).$$

However, expectation of minima is all we need for the derivation of the generalized Gini coefficients below.

Arnold and Laguna (1977) provided tables of $E(X_{k:n})$, $1 \leq k \leq n$, for $n = 1(1)10$ and parameter values of $b = 1$, $q = 0.5(0.5)5.0$, and $a^{-1} = 0.1(0.1)1.0$.

Since the c.d.f. is available in a simple closed form, this is also the case for the hazard rate, which is given by

$$r(x) = \frac{aqx^{a-1}}{b^a\{1 + (x/b)^a\}}, \quad x > 0. \tag{6.55}$$

The general shape of this function depends on the value of the shape parameter a: For all $a > 0$ the hazard rate is eventually decreasing. For $a > 1$ we have a unimodal function, whereas for $a \leq 1$ it is decreasing for all $x > 0$. [The special case where $a = 2$ (which is associated with a unimodal hazard rate) is discussed by Greenwich (1992) in some detail.] Note that the parameter q is only a scale factor in (6.55) and does not determine the shape of the function.

From general results for the mean excess function of regularly varying distributions (see Chapter 2), we determine that the mean excess function is asymptotically linearly increasing

$$e(x) \in \mathrm{RV}_\infty(1). \tag{6.56}$$

6.2.3 Representations and Characterizations

Several characterizations of the Singh–Maddala distribution are available in the statistical and econometric literature. Most of them are probably best understood by relating them to characterizations of the exponential distribution, the most widely known and used distribution on $\mathbb{R}_+$. The exponential distribution is remarkably well behaved, in that its c.d.f., quantile function, m.g.f., mean excess function, etc. are all available in simple closed form. Not surprisingly, it has generated a substantial characterizations literature. For a more extensive discussion of exponential characterizations, see Galambos and Kotz (1978) for results until the late 1970s and Johnson, Kotz, and Balakrishnan (1994, Section 19.8) for selected results that were obtained thereafter.

For any monotonic function $h(\cdot)$, the characterization of X is equivalent to that of $h(X)$. Thus, if a specific distribution can be associated with the exponential distribution, a host of characterization results become available. The Singh–Maddala distribution is related to the exponential distribution via

$$X \sim \mathrm{SM}(a, b, q) \Longleftrightarrow \log\left[1 + \left(\frac{X}{b}\right)^a\right] \sim \mathrm{Exp}(q), \tag{6.57}$$

where Exp(q) denotes an exponential distribution with scale parameter q. Hence,

$$h(x) = \log\left[1 + \left(\frac{x}{b}\right)^a\right], \quad a > 0,$$

is the required monotonic transformation, and many characterizations of Singh–Maddala distributions are now available by applying the transformation $h^{-1}(\cdot)$ on exponential variables.

The lack of memory property is perhaps the most popular and intuitively transparent characterization of the exponential distribution. One of its many equivalent expressions is given in terms of the functional equation

$$P(X > x + y) = P(X > x)P(X > y). \tag{6.58}$$

El-Saidi, Singh, and Bartolucci (1990) showed that the functional equation

$$P\left[1 + \left(\frac{X}{b}\right)^a > xy\right] = P\left[1 + \left(\frac{X}{b}\right)^a > x\right] \cdot P\left[1 + \left(\frac{X}{b}\right)^a > y\right],$$

where $a, b > 0$ and $x, y > 1$, characterizes the Singh–Maddala distribution (which they call a generalized log-logistic distribution) among all continuous distributions supported on $[0, \infty)$. Ghitany (1996) pointed out that this result, and others to be discussed below, derive essentially from the lack of memory property (6.58) of the exponential distribution. For the remainder of this section we shall ignore the scale parameter b and consider $h(x) = \log(1 + x^a)$.

Further Singh–Maddala characterizations are in terms of order statistics. Suppose that $X_1, \ldots, X_n$ are a random sample from an absolutely continuous distribution with c.d.f. F and let $X_{1:n}, \ldots, X_{n:n}$ denote their order statistics. Al-Hussaini (1991) showed that $X_i \sim \mathrm{SM}(a, 1, q)$, $i = 1, \ldots, n$, if and only if, for some $a > 0$,

$$Z_1 = 1 + X_{1:n}^a,\ Z_2 = \frac{1 + X_{2:n}^a}{1 + X_{1:n}^a},\ Z_2 = \frac{1 + X_{3:n}^a}{1 + X_{2:n}^a}, \ldots, Z_2 = \frac{1 + X_{n:n}^a}{1 + X_{n-1:n}^a}$$

are independent. Since the exponential distribution is characterized by the independence of successive spacings, this result can also be traced back to the exponential case via the transformation $h(\cdot)$ defined above [Ghitany (1996), who was however unaware of Al-Hussaini's (1991) result].

Another group of characterizations is related to the mean excess (or mean residual life) function. Dimaki and Xekalaki (1996) showed that the Singh–Maddala distribution is characterized by the property

$$E[\log(1+X^a)|X>y]=\log(1+y^a)+c, \tag{6.59}$$

for all nonnegative values of y and for some $a>0$, $c>0$ among the continuous distributions with the support on $[0,\infty)$ and such that $E|h(X)|<\infty$. This is essentially a restatement of the condition $E(X|X>y)=y+E(X)$, for all nonnegative y, which is known to characterize the exponential distribution (Shanbhag, 1970).

Fakhry (1996) provided three related results. First, for $h(x)=\log(1+x^a)$ the recurrence relation

$$E[h^k(X)|X>y]=h^k(y)+\frac{1}{q}kE[h^{k-1}(X)|X>y] \tag{6.60}$$

characterizes the SM$(a, 1, q)$ distribution. Under the additional condition that $Eh^2(X)<\infty$, the same is true if, for all y and some fixed c,

$$\text{var}[h(X)|X>y]=c^2. \tag{6.61}$$

Third, he observed that the condition

$$E[h^k(X_{i:n})]=E[h^k(X_{i-1:n})]+\frac{k}{q(n-i+1)}E[h^{k-1}(X_{i:n})] \tag{6.62}$$

also characterizes the SM$(a, 1, q)$ distribution. Again, all three results are connected to characterizations of the exponential distribution via the transformation h. Characterization (6.60) is a higher-order moment version of (6.59). The exponential characterization associated with (6.61) is due to Azlarov, Dzamirzaev, and Sultanova (1972) and Laurent (1974), whereas the characterization (6.62) via relationships between the moments of order statistics is reduced to an exponential characterization provided by Lin (1988).

Khan and Khan (1987) presented further characterizations based on conditional expectations of order statistics. If $a=q(n-i)$ is independent of x,

$$E[X^k_{i+1:n}|X_{i:n}=x]=\frac{a}{a-1}\left[x^k+\frac{1}{a}\right] \tag{6.63}$$

characterizes the Singh–Maddala distribution (referred to as the Burr XII by Khan and Khan) among all continuous distributions with support on $[0, \infty)$ and $F(0) = 0$. In view of $E[X_{n:n}^k | X_{n-1:n} = x] = E[X^k | X > x]$ and $E[X_{1:n}^k | X_{2:n} = y] = E[X^k | X \leq y]$, this can be expressed alternatively as

$$E[X^k | X \geq x] = \frac{q}{q-1}\left[x^k + \frac{1}{q}\right].$$

Also, the distribution is characterized by

$$E[X_{1:n}^k | X_{2:n} = x] = \frac{1}{q-1} - \frac{q}{q-1}\frac{1-F(x)}{F(x)}x^k. \tag{6.64}$$

Both results can be restated in terms of $E[X^k | X_{1:n} = x]$ or $E[X^k | X_{n:n} = x]$.

Specializing from the mixture representation of the GB2 distribution (6.21), the Singh–Maddala distribution can be considered a compound Weibull distribution whose scale parameter follows an inverse generalized gamma distribution (Takahasi, 1965; Dubey, 1968)

$$\text{Wei}(a, \theta) \bigwedge_{\theta} \text{InvGG}(a, b, q) = \text{SM}(a, b, q). \tag{6.65}$$

6.2.4 Lorenz Curve and Inequality Measures

As we have already mentioned, the quantile function of the Singh–Maddala distribution is available in closed form. Consequently, its (normalized) integral, the Lorenz curve, is also of a comparatively simple form, namely,

$$L(u) = I_y\left(1 + \frac{1}{a}, q - \frac{1}{a}\right), \quad 0 \leq u \leq 1, \tag{6.66}$$

where $y = 1 - (1-u)^{1/q}$. The subclass where $q = (a+1)/a$ is even simpler analytically and yields

$$L(u) = [1 - (1-u)^{a/(a+1)}]^{(a+1)/a}, \quad 0 < u < 1, \tag{6.67}$$

a Lorenz curve that is symmetric (in the sense of Section 2.1.1). It is interesting that this is also a subclass of the Rasche et al. (1980) family of Lorenz curves; cf. Section 2.1. Thus, the Rasche et al. model and the Singh–Maddala distribution share the same underlying structure.

For $X_i \sim \text{SM}(a_i, b_i, q_i)$, $i = 1, 2$, the necessary and sufficient conditions for $X_1 \geq_L X_2$ are

$$a_1 \leq a_2 \quad \text{and} \quad a_1 q_1 \leq a_2 q_2. \tag{6.68}$$

[See (6.24) and (6.25).] This result is due to Wilfling and Krämer (1993).

From (6.53) we obtain the generalized Gini coefficients (Kleiber and Kotz, 2002)

$$G_n = 1 - \frac{E(X_{1:n})}{E(X)} = 1 - \frac{\Gamma(nq - 1/a)\Gamma(q)}{\Gamma(nq)\Gamma(q - 1/a)}$$

for $n = 2, 3, 4, \ldots$, where $n = 2$ yields the ordinary Gini coefficient (Cronin, 1979; McDonald and Ransom, 1979a)

$$G = 1 - \frac{\Gamma(q)\Gamma(2q - 1/a)}{\Gamma(q - 1/a)\Gamma(2q)}. \tag{6.69}$$

The Theil measure is (McDonald, 1981)

$$T(X) = \frac{1}{a}\left[2\psi\left(\frac{1}{a}\right) - \psi(q+1) - \psi\left(q - \frac{1}{a}\right)\right] - \log(q) - \log B\left(1 + \frac{1}{a}, q - \frac{1}{a}\right),$$

and the Pietra index can be written, using (6.23) and the representation $P = F(\mu) - F_{(1)}(\mu)$ (Butler and McDonald, 1989),

$$P(X) = F_{\mathrm{SM}}(\mu; a, b, q) - F_{\mathrm{GB2}}\left(\mu; a, b, 1 + \frac{1}{a}, q - \frac{1}{a}\right),$$

where μ is the first moment. Finally, the variance of logarithms is (Schmittlein, 1983)

$$\mathrm{VL}(X) = \mathrm{var}(\log X) = \frac{1}{a^2}[\psi'(q) + \psi'(1)].$$

Unlike in the lognormal case, these measures are not very attractive analytically.

Klonner (2000) presented necessary as well as sufficient conditions for first-order stochastic dominance (FSD) within the Singh–Maddala family. The conditions $a_1 \geq a_2$, $a_1 q_1 \leq a_2 q_2$, and $b_1 \geq b_2$ are sufficient for $X_1 \geq_{\mathrm{FSD}} X_2$, whereas the conditions $a_1 \geq a_2$ and $a_1 q_1 \leq a_2 q_2$ are necessary. It is instructive to compare these conditions to those for the Lorenz ordering (6.68): Although $a_1 q_1 \leq a_2 q_2$ is also a condition for $X_1 \geq_L X_2$, the second condition $a_1 \geq a_2$ appears in reversed form in (6.68). The reason is that FSD describes "size," whereas the Lorenz ordering describes variability. Namely, $a_1 \leq a_2$ and $a_1 q_1 \leq a_2 q_2$ mean that X_1 is associated with both a heavier left and a heavier right tail and thus more variable than X_2. On the other hand, $a_1 \geq a_2$ and $a_1 q_1 \leq a_2 q_2$ mean that X_1 is associated with a lighter left and a heavier right tail and thus stochastically larger than X_2. See Chapter 2 for details of the argument in connection with the Lorenz ordering.

Zenga ordering within the Singh–Maddala family was studied by Polisicchio (1990). It emerges that $a_1 \leq a_2$ implies $X_1 \geq_Z X_2$, for a fixed q, and similarly that $q_1 \leq q_2$ implies $X_1 \geq_Z X_2$, for a fixed a. Under these conditions we know from

(6.68) that $X_1 \geq_L X_2$; however, a complete characterization of the Zenga order among Singh–Maddala distributions appears to be unavailable at present.

6.2.5 Estimation

Singh and Maddala (1976) estimated parameters by using a regression method minimizing

$$\sum_{i=1}^{n}\left\{\log[1-F(x_i)]+q\log\left[1+\left(\frac{x_i}{b}\right)^a\right]\right\}^2, \tag{6.70}$$

that is, a nonlinear least-squares regression in the Pareto diagram. Stoppa (1995) discussed a further regression method utilizing the elasticity $d\log F(x)/d\log x$ of the distribution. The resulting estimators could be used, for example, as starting values in ML estimation.

The log-likelihood for a complete random sample of size n equals

$$\log L = n\log q + n\log a + (a-1)\sum_{i=1}^{n}\log x_i - na\log b$$
$$-(1+q)\sum_{i=1}^{n}\log\left[1+\left(\frac{x_i}{b}\right)^a\right],$$

yielding the likelihood equations

$$\frac{n}{a}+\sum_{i=1}^{n}\log\left(\frac{x_i}{b}\right) = (1+q)\sum_{i=1}^{n}\log\left(\frac{x_i}{b}\right)\left[\left(\frac{b}{x_i}\right)^a+1\right]^{-1}, \tag{6.71}$$

$$n = (1+q)\sum_{i=1}^{n}\left[1+\left(\frac{b}{x_i}\right)^a\right]^{-1}, \tag{6.72}$$

$$\frac{n}{q} = \sum_{i=1}^{n}\log\left[1+\left(\frac{x_i}{b}\right)^a\right]. \tag{6.73}$$

The algorithmic aspects of this optimization problem are discussed in Mielke and Johnson (1974), Wingo (1983), and Watkins (1999).

Specializing from the information matrix for the GB2 distribution (6.30), we obtain for $\theta = (a, b, q)^\top$

$$I(\theta) = \left[-E\left(\frac{\partial^2\log L}{\partial\theta_i\partial\theta_j}\right)_{i,j}\right] =: \begin{pmatrix} I_{11} & I_{12} & I_{13} \\ I_{21} & I_{22} & I_{23} \\ I_{31} & I_{32} & I_{33} \end{pmatrix}, \tag{6.74}$$

where

$$I_{11} = \frac{1}{a^2(2+q)}\{q[(\psi(q) - \psi(1) - 1)^2 + \psi'(q) + \psi'(1)] + 2[\psi(q) - \psi(1)]\}, \tag{6.75}$$

$$I_{21} = I_{12} = \frac{1 - q + q[\psi(q) - \psi(1)]}{b(2+q)}, \tag{6.76}$$

$$I_{22} = \frac{a^2 q}{b^2(2+q)}, \tag{6.77}$$

$$I_{23} = I_{32} = -\frac{a}{b(1+q)}, \tag{6.78}$$

$$I_{31} = I_{13} = -\frac{\psi(q) - \psi(1) - 1}{a(1+q)}, \tag{6.79}$$

$$I_{33} = \frac{1}{q^2}. \tag{6.80}$$

[For I_{33} we used the identity $\psi'(x) - \psi'(x+1) = x^{-2}$.]

Schmittlein (1983) provided an explicit expression for the inverse of the Fisher information (using a different parameterization) as well as the information matrix when the data are grouped and/or type I censored. Comparing these formulae with the expressions above permits an evaluation of the information loss due to the effect of grouping and/or censoring. Asymptotic variances for functionals of the distribution can be obtained by the delta method. Since the resulting expressions for the asymptotic variances of, for example, the Gini, Pietra, and variance of logarithms measures of inequality are not very attractive analytically, we omit the corresponding formulae and refer the interested reader to Schmittlein (1983).

As an alternative to ML estimation, the maximum product of spacings (MPS) estimation was considered by Shah and Gokhale (1993). This method obtains estimates of a vector-valued parameter θ by maximizing

$$H = \frac{1}{n+1}\sum_{i=1}^{n+1} \log\{F(x_i, \theta) - F(x_{i-1}, \theta)\},$$

$i = 1, 2, \ldots, n+1$, with $x_0 = -\infty$ and $x_{n+1} = \infty$. From a simulation study employing ten parameter combinations and nine sample sizes ranging from $n = 10$ to $n = 150$, Shah and Gokhale concluded that MPS is superior to ML estimation, in the sense of smaller MSE, at least for small samples.

6.2.6 Empirical Results

Incomes and Wealth

Singh and Maddala (1976) compared their model to Salem and Mount's (1974) results for the gamma distribution using 1960–1972 U.S. family incomes and concluded that their model provides a better fit than either the gamma or lognormal functions. However, Cronin (1979), in a comment on the Singh and Maddala paper, observed that the implied Gini indices for the Singh–Maddala model almost always fall outside the Gastwirth (1972) bounds (calculated by Salem and Mount, 1974) for their data. He concluded that Singh and Maddala's claim that the "Burr distribution fits the data better [than the gamma distribution] would now appear to be questionable" (p. 774).

When compared to lognormal, gamma, and beta type I fittings for U.S. family incomes for 1960 and 1969 through 1975, the Singh–Maddala distribution generally outperforms all these distributions, with only the beta type I being slightly better in a few cases (McDonald and Ransom, 1979a).

Dagum (1983) fit a Singh–Maddala distribution to 1978 U.S. family incomes for which it outperforms the (two-parameter) lognormal and gamma distributions by wide margins. However, the (four- and three-parameter) Dagum type III and type I distributions fit even better.

In McDonald (1984) the distribution ranks second out of 11 considered models—being inferior only to the GB2 distribution—when fitted to 1970, 1975, and 1980 U.S. family incomes.

For Japanese incomes for 1963–1971, the distribution outperforms the Fisk, beta, gamma, lognormal, and Pareto (II) distribution. However, a likelihood ratio test reveals that the full flexibility of the Singh–Maddala distribution is not required and that the two-parameter Fisk distribution already provides an adequate fit (Suruga, 1982). The distribution was also fitted to various strata from the 1975 Japanese Income Redistribution Survey by Atoda, Suruga, and Tachibanaki (1988). Four occupational classes as well as primary and redistributed incomes were considered, and five different estimation techniques were applied. In a later study using the same data set, Tachibanaki, Suruga, and Atoda (1997) considered ML estimators on the basis of individual observations. Here the Singh–Maddala model is almost always the best out of six different functions [including (generalized) gamma and Weibull distributions] in terms of several fit criteria. However, when the AIC is employed for model selection, the Singh–Maddala distribution turns out to be essentially overparameterized for one stratum, with a two-parameter log-logistic special case providing an adequate fit.

Henniger and Schmitz (1989) employed the Singh–Maddala distribution when fitting five parametric models to data from the UK Family Expenditure Survey for 1968–1983. However, for the whole population all parametric models are rejected; for subgroups the Singh–Maddala distribution performs better than any other parametric model considered and appears to be adequate for their data, in terms of goodness-of-fit tests.

Majumder and Chakravarty (1990) considered the Singh–Maddala distribution when modeling U.S. income data from 1960, 1969, 1980, and 1983. The distribution is among the best three-parameter models. In a reassessment of Majumder and Chakravarty's findings, McDonald and Mantrala (1993, 1995) fit the Singh–Maddala distribution to 1970, 1980, and 1990 U.S. family incomes, using two different fitting methods and alternative groupings of the data. Here the distribution is outperformed by the more flexible GB2 as well as the Dagum distribution.

McDonald and Xu (1995) studied 1985 U.S. family incomes; out of 11 distributions of the beta and gamma type, the Singh–Maddala ranks fourth in terms of likelihood and several other goodness-of-fit criteria.

In an application to 1984–1993 German household incomes, the Singh–Maddala distribution emerges as one of two suitable models (Brachmann, Stich, and Trede, 1996), being comparable to the more general GB2 distribution. Two-parameter models such as the gamma or Weibull distributions are not appropriate for these data.

Bordley, McDonald, and Mantrala (1996) fit the Singh–Maddala distribution to U.S. family incomes for 1970, 1975, 1980, 1985, and 1990. The distribution ranks fourth out of 15 considered models of the beta and gamma type, being outperformed only by the GB2, GB, and Dagum type I distributions and improving on three- and four-parameter models such as the generalized gamma and GB1 distributions. For 1985 the relative ranking of the Singh–Maddala and Dagum distributions depends on the criterion selected; in terms of likelihood and χ^2, the Singh–Maddala does slightly better.

Bell, Klonner, and Moos (1999) fit the Singh–Maddala distribution to the per capita consumption expenditure data of rural Indian households for 28 survey periods, stretching from 1954–1955 to 1993–1994. They reported that for a total of 44.4% of all 378 possible pairs, a ranking in terms of first-order stochastic dominance is possible, and another 47.7% can be Lorenz-ordered.

Botargues and Petrecolla (1997) estimated the Singh–Maddala model for the income distribution in the Buenos Aires region, for the years 1990–1996. However, the Dagum distributions (of various types) perform better on these data.

Actuarial Losses

In the actuarial literature, Hogg and Klugman (1983) fit a Singh–Maddala distribution (under the name of Burr distribution) to 35 observations on hurricane losses in the United States. Compared to the Weibull and lognormal models, the distribution appears to be overparameterized for these data.

Cummins et al. (1990), in their hybrid paper already mentioned above, fit the Singh–Maddala distribution (under the name of Burr XII) to aggregate fire losses (the data are provided in Cummins and Freifelder, 1978). The distribution performs quite well; however, less generously parameterized limiting forms of the GB2 such as the inverse exponential distribution seem to do even better. The same authors also considered data on the severity of fire losses and fit the GB2 to both grouped and individual observations. Here the Singh–Maddala is indistinguishable from the four-parameter GB2 distribution for individual observations and is therefore a distribution of choice for these data.

6.2.7 Extensions

A generalization of the Singh–Maddala distribution was recently proposed by Zandonatti (2001). Using Stoppa's (1990a,b) method leading to a power transformation of the c.d.f., the resulting density is given by

$$f(x) = \frac{aq\theta x^{a-1}}{b^a}\left[1+\left(\frac{x}{b}\right)^a\right]^{-q-1}\left\{1-\left[1+\left(\frac{x}{b}\right)^a\right]^{-q}\right\}^{\theta-1}, \quad x > 0. \tag{6.81}$$

Clearly, for $\theta = 1$ we arrive at the Singh–Maddala distribution.

More than a decade before the publication of Singh and Maddala's pioneering 1976 paper, a multivariate Singh–Maddala distribution was proposed by Takahasi (1965), under the name of multivariate Burr distribution. This distribution is defined by the joint survival function

$$\bar{F}(x_1, \ldots, x_k) = \left\{1 + \sum_{i=1}^{k}\left(\frac{x_i}{b_i}\right)^{a_i}\right\}^{-q}, \quad x_i > 0, i = 1, \ldots, k, \tag{6.82}$$

and is often referred to as the *Takahasi–Burr distribution*. For the components X_i of a random vector $(X_1, \ldots, X_k)$ following this distribution, we have the representation

$$X_i \overset{d}{=} b_i\left(\frac{Y_i}{Z}\right)^{1/a_i}, \quad i = 1, \ldots, k, \tag{6.83}$$

where the Y_i's are i.i.d. following a standard exponential distribution and Z has a gamma(q, 1) distribution. Note that Y_i^{1/a_i} follows a Weibull and Z^{1/a_i} a generalized gamma distribution; hence, (6.83) is a direct generalization of the univariate mixture representation (6.65). The Takahasi–Burr distribution possesses Singh–Maddala marginals as well as conditionals.

As is well known from the theory of copulas (e.g., Nelsen, 1998), a multivariate survival function can be decomposed in the form

$$\bar{F}(x_1, \ldots, x_k) = G[\bar{F}_1(x_1), \ldots, \bar{F}_k(x_k)],$$

where the $\bar{F}_i(x_i)$, $i = 1, \ldots, x_k$, are the marginal survival functions and G is the copula (a c.d.f. on $[0, 1]^k$ that captures the dependence structure of F). The dependence structure within the Takahasi–Burr family has been studied by Cook and Johnson (1981); the copula is

$$G(u_1, \ldots, u_k) = \left\{\sum_{i=1}^{k} u_i^{-1/q} - (k-1)\right\}^{-q}, \quad 0 < u_i < 1, i = 1, \ldots, k, \tag{6.84}$$

which is often called the Clayton (1978) copula. However, considering data on annual incomes for successive years, Kleiber and Trede (2003) found that this model does not provide a good fit. When combining an elliptical copula (associated with, e.g., a multivariate normal distribution) with Singh–Maddala marginal distributions, they obtained rather encouraging preliminary results.

6.3 DAGUM DISTRIBUTIONS

Although introduced as an income distribution only one year after the Singh–Maddala model, the Dagum distribution is less widely known. Presumably, this is due to the fact that Dagum's work was published in the French journal *Economie Appliquée*, whereas the Singh–Maddala paper appeared in the more widely read *Econometrica*. However, in recent years there are indications that the Dagum distribution is, in fact, a more appropriate choice in many applications.

6.3.1 Definition and Motivation

The Dagum distribution is a GB2 distribution with the shape parameter $q = 1$; hence, its density is

$$f(x) = \frac{apx^{ap-1}}{b^{ap}[1 + (x/b)^a]^{p+1}}, \quad x > 0, \tag{6.85}$$

where $a, b, p > 0$.

Like the Singh–Maddala distribution considered in the previous section, the Dagum distribution was rediscovered many times in various fields of science. Apparently, it occurred for the first time in Burr (1942) as the third example of solutions to his differential equation defining the Burr system of distributions. Thus, it is known as the Burr III distribution. As mentioned above, the Dagum distribution is closely related to the Singh–Maddala distribution, specifically

$$X \sim \mathrm{D}(a, b, p) \Longleftrightarrow \frac{1}{X} \sim \mathrm{SM}\left(a, \frac{1}{b}, p\right). \tag{6.86}$$

This relation permits us to translate several results pertaining to the Singh–Maddala family to corresponding results for the Dagum distributions.

The Singh–Maddala is the Burr XII, or simply the Burr distribution, so it is not surprising that the Dagum distribution is also called the *inverse Burr distribution*, notably in the actuarial literature (e.g., Klugman, Panjer, and Willmot, 1998). Like the Singh–Maddala, the Dagum distribution can be considered a generalized log-logistic distribution. The special case where $a = p$ is sometimes called the inverse paralogistic distribution (Klugman, Panjer, and Willmot, 1998). Prior to its use as an income distribution, the Dagum family was proposed as a model for precipitation amounts in the meteorological literature (Mielke, 1973), where it is called the (three-parameter) *kappa distribution*.

Mielke and Johnson (1974) nested it within the GB2 and called it the beta-K distribution. In a parallel development—aware of Mielke (1973) but presumably unaware of Dagum (1977)—Fattorini and Lemmi (1979) proposed the distribution as an income distribution. [See also Lemmi (1987).] Nonetheless, the distribution is usually called the Dagum distribution in the income distribution literature, and we shall follow this convention below.

Dagum (1977) derived his model from the empirical observation that the income elasticity of the c.d.f. of income is a decreasing and bounded function of F. Starting from the differential equation,

$$\eta(x, F) = \frac{d\log F(x)}{d\log x} = ap\{1 - [F(x)]^{1/p}\}, \quad x \geq 0, \tag{6.87}$$

subject to $p > 0$ and $ap > 0$, one obtains the density (6.85).

Fattorini and Lemmi (1979) independently arrived at the Dagum distribution as the equilibrium distribution of a continuous-time stochastic process under certain assumptions on its infinitesimal mean and variance (see also Dagum and Lemmi, 1989).

6.3.2 Moments and Other Basic Properties

Like the c.d.f. of the Singh–Maddala distribution discussed in the previous section, the c.d.f. of the Dagum distribution is available in closed form, namely,

$$F(x) = \left[1 + \left(\frac{x}{b}\right)^{-a}\right]^{-p}, \quad x > 0. \tag{6.88}$$

This is also true of the quantile function

$$F^{-1}(u) = b[u^{-1/p} - 1]^{-1/a}, \quad \text{for } 0 < u < 1. \tag{6.89}$$

As was the case with the Singh–Maddala distribution discussed in the previous section, the Dagum family was considered in several equivalent parameterizations. Mielke (1973) and later Fattorini and Lemmi (1979) used $(\alpha, \beta, \theta) = (1/p, bp^{1/a}, ap)$, whereas Dagum (1977) employed $(\beta, \delta, \lambda) = (p, a, b^a)$.

From (6.12), the kth moment exists for $-ap < k < a$; it equals

$$E(X^k) = \frac{b^k B(p + k/a, 1 - k/a)}{B(p, 1)} = \frac{b^k \Gamma(p + k/a)\Gamma(1 - k/a)}{\Gamma(p)}. \tag{6.90}$$

[In view of (6.86), this result can alternatively be obtained upon replacing q with p and a with $-a$ in (6.46).]

Specifically,

$$E(X) = \frac{b\Gamma(p + 1/a)\Gamma(1 - 1/a)}{\Gamma(p)} \tag{6.91}$$

and

$$\operatorname{var}(X) = \frac{b^2\{\Gamma(p)\Gamma(p + 2/a)\Gamma(1 - 2/a) - \Gamma^2(p + 1/a)\Gamma^2(1 - 1/a)\}}{\Gamma^2(p)}. \tag{6.92}$$

Hence, the coefficient of variation is

$$\mathrm{CV} = \sqrt{\frac{\Gamma(p)\Gamma(p + 2/a)\Gamma(1 - 2/a)}{\Gamma^2(p + 1/a)\Gamma^2(1 - 1/a)} - 1} \tag{6.93}$$

and the shape factors are

$$\sqrt{\beta_1} = \frac{\Gamma^2(p)\lambda_3 - 3\Gamma(p)\lambda_2\lambda_1 + 2\lambda_1^3}{[\Gamma(p)\lambda_2 - \lambda_1^2]^{3/2}} \tag{6.94}$$

and

$$\beta_2 = \frac{\Gamma^3(p)\lambda_4 - 4\Gamma^2(p)\lambda_3\lambda_1 + 6\Gamma(p)\lambda_2\lambda_1^2 - 3\lambda_1^4}{[\Gamma(p)\lambda_2 - \lambda_1^2]^2}, \tag{6.95}$$

where we have set $\lambda_i = \Gamma(1 - i/a)\Gamma(p + i/a)$, $i = 1, 2, 3, 4$. These expressions are not easily interpreted.

From a moment-ratio diagram—a graphical display of $(\sqrt{\beta_1}, \beta_2)$—of the Dagum and the closely related Singh–Maddala distributions (Rodriguez, 1983; Tadikamalla, 1980) it may be inferred that both distributions allow for various degrees of positive skewness and leptokurtosis, and even for a considerable degree of negative skewness, although this feature does not seem to be of particular interest in our context. Tadikamalla (1980, p. 342) observed "that although the Burr III [=Dagum] distribution covers all of the region ... as covered by the Burr XII [=Singh–Maddala] distribution and more, much attention has not been paid to this distribution." Kleiber (1996) noted that the same has happened in the econometrics literature.

The mode of this distribution is at

$$x_{\text{mode}} = b\left(\frac{ap - 1}{a + 1}\right)^{1/a}, \quad \text{if } ap > 1, \tag{6.96}$$

and at zero otherwise. This built-in flexibility is an attractive feature in that the model can approximate income distributions, which are usually unimodal, and wealth distributions, which are zeromodal.

From (6.89) the median is (Dagum, 1977)

$$x_m = b[2^{1/p} - 1]^{-1/a}.$$

Moments of the order statistics can be obtained in an analogous manner to the Singh–Maddala case. In view of the relation (6.86) presented above and the corresponding result (6.53) for Singh–Maddala minima, we have the closure property

$$X \sim D(a, b, p) \Longrightarrow X_{n:n} \sim D(a, b, np) \tag{6.97}$$

and thus, using (6.90), we obtain

$$E(X_{n:n}) = \frac{b^k \Gamma(np + 1/a)\Gamma(1 - 1/a)}{\Gamma(np)}. \tag{6.98}$$

As in the Singh–Maddala case, the moments of other order statistics can be obtained using recurrence relations. This will be necessary for the computation of generalized Gini coefficients where expectations of sample minima are required. Domma (1997) provided some further distributional properties of the sample median and the sample range.

To the best of our knowledge, the hazard rate and mean excess function of the Dagum distribution have not been investigated in the statistical literature. Nonetheless, from the general properties of regularly varying functions (see Chapter 2) we can infer that the hazard rate is decreasing for large x, specifically $r(x) \in \mathrm{RV}_\infty(-1)$, and similarly that the mean excess function is increasing, $e(x) \in \mathrm{RV}_\infty(1)$.

6.3.3 Representations and Characterizations

There are comparatively few explicit characterizations of the Dagum (Burr III) distribution in the statistical literature. However, in view of the close relationship with the Singh–Maddala distribution, all characterizations presented for that distribution translate easily into characterizations of the Dagum distribution.

For example, El-Saidi, Singh, and Bartolucci (1990) showed that the functional equation

$$P\left[1 + \left(\frac{b}{X}\right)^a > xy\right] = P\left[1 + \left(\frac{b}{X}\right)^a > x\right] \cdot P\left[1 + \left(\frac{b}{X}\right)^a > y\right],$$

where $a, b > 0$ and $x, y > 1$, characterizes the Dagum (which they called a generalized log-logistic) distribution among all continuous distributions supported on $[0, \infty)$. This follows directly from the corresponding characterization of the

Singh–Maddala distribution considered above via the relation (6.86). Ghitany (1996) observed that this characterization can be considered a restatement of the well-known characterization of the exponential distribution in terms of its lack of memory property, as described in connection with the Singh–Maddala distribution.

Utilizing the mixture representation of the GB2 distribution (6.21), the Dagum distribution can be considered a compound generalized gamma distribution whose scale parameter follows an inverse Weibull distribution

$$\mathrm{GG}(a, \theta, p) \bigwedge_{\theta} \mathrm{InvWei}(a, b) = D(a, b, p). \tag{6.99}$$

6.3.4 Lorenz Curve and Inequality Measures

Since the quantile function of the Dagum distribution is available in closed form, its (normalized) integral, the Lorenz curve, is also of a comparatively simple form, namely (Dagum, 1977),

$$L(u) = I_z\left(p + \frac{1}{a}, 1 - \frac{1}{a}\right), \quad 0 \leq u \leq 1, \tag{6.100}$$

where $z = u^{1/p}$.

A subclass of the Dagum distributions, defined by

$$F(x) = \left[1 + \left(\frac{x}{b}\right)^{-a}\right]^{-1+1/a}, \quad x > 0, \tag{6.101}$$

where $a > 1$, exhibits symmetric Lorenz curves (in the sense of Chapter 2). Interestingly, this was noted by Champernowne (1956, p. 182) long before the distribution was proposed as an income distribution. However, Champernowne did not develop the model further.

From (6.24) and (6.25) the necessary and sufficient conditions for Lorenz dominance are

$$a_1 p_1 \leq a_2 p_2 \quad \text{and} \quad a_1 \leq a_2. \tag{6.102}$$

This was derived by Kleiber (1996) from the corresponding result for the Singh–Maddala distribution using (6.86); for a different approach see Kleiber (1999b, 2000a). [Dancelli (1986) had shown somewhat earlier that (income) inequality is decreasing to zero for both $a \to \infty$ and $p \to \infty$ and increasing to 1 for $a \to 1$ and $p \to 0$, respectively, keeping the other parameter fixed.]

Klonner (2000) presented necessary as well as sufficient conditions for first-order stochastic dominance within the Dagum family. The conditions $a_1 \geq a_2$, $a_1 p_1 \leq a_2 p_2$, and $b_1 \geq b_2$ are sufficient for $X_2 \geq_{\mathrm{FSD}} X_1$, whereas the conditions $a_1 \geq a_2$ and $a_1 p_1 \leq a_2 p_2$ are necessary. (See the corresponding conclusions for the Singh–Maddala distribution for an interpretation of these results.)

Zenga ordering among the Dagum distributions was studied by Polisicchio (1990). Similar to the Singh–Maddala case, it turns out that $a_1 \leq a_2$ implies $X_1 \geq_Z X_2$, for a fixed p, and analogously that $p_1 \leq p_2$ implies $X_1 \geq_Z X_2$, for a fixed a. Under these conditions, we know from (6.68) that the distributions are also Lorenz-ordered, specifically $X_1 \geq_L X_2$. However, a complete characterization of the Zenga order within the family of Dagum distributions seems to be currently unavailable.

The Gini coefficient is (Dagum, 1977)

$$G = \frac{\Gamma(p)\Gamma(2p + 1/a)}{\Gamma(2p)\Gamma(p + 1/a)} - 1. \tag{6.103}$$

The generalized Gini coefficients can be obtained as follows: Combining the well-known recurrence relation (Arnold and Balakrishnan, 1989, p. 7)

$$E(X_{k:n}) = \sum_{j=1}^{n} (-1)^{j-i} \binom{n}{j} \binom{j-1}{i-1} E(X_{j:j})$$

and the expression for the expectations of Dagum maxima (6.98) yields

$$G_n = \frac{\Gamma(p)}{\Gamma(p + 1/a)} \sum_{j=1}^{n} (-1)^{j-1} \binom{n}{j} \frac{\Gamma(jp + 1/a)}{\Gamma(jp)}. \tag{6.104}$$

The Zenga index ξ_2 is (Latorre, 1988)

$$\begin{aligned} \xi_2 &= 1 - \exp\left\{ E(\log X) - \frac{E(X \log X)}{E(X)} \right\} \\ &= 1 - \exp\left\{ \frac{1}{a} \left[\psi(p) + \psi\left(1 - \frac{1}{a}\right) - \psi\left(p + \frac{1}{a}\right) - \psi(1) \right] \right\}. \end{aligned}$$

6.3.5 Estimation

Dagum (1977) discussed five methods for estimating the model parameters and recommended a nonlinear least-squares method minimizing

$$\sum_{i=1}^{n} \left\{ F_n(x_i) - \left[1 + \left(\frac{x_i}{b} \right)^{-a} \right]^{-p} \right\}, \tag{6.105}$$

a minimum distance technique based on the c.d.f. A further regression-type estimator utilizing the elasticity (6.87) was considered by Stoppa (1995).

The log-likelihood for a complete random sample of size n is

$$\log L = n\log a + n\log p + (ap-1)\sum_{i=1}^{n}\log x_i - nap\log b - (p+1) \times \sum_{i=1}^{n}\log\left[1+\left(\frac{x_i}{b}\right)^a\right], \tag{6.106}$$

yielding the likelihood equations

$$\frac{n}{a} + p\sum_{i=1}^{n}\log\left(\frac{x_i}{b}\right) = (p+1)\sum_{i=1}^{n}\frac{\log(x_i/b)}{1+(b/x_i)^a}, \tag{6.107}$$

$$np = (p+1)\sum_{i=1}^{n}\frac{1}{1+(b/x_i)^a}, \tag{6.108}$$

$$\frac{n}{p} + a\sum_{i=1}^{n}\log\left(\frac{x_i}{b}\right) = \sum_{i=1}^{n}\log\left[1+\left(\frac{x_i}{b}\right)^a\right]. \tag{6.109}$$

However, likelihood estimation in this family is not without problems: Considering the distribution of $\log X$, a generalized logistic distribution, Zelterman (1987) showed that there is a path in the parameter space along which the likelihood becomes unbounded. This implies that the global maximizer of the likelihood does not define a consistent estimator of the parameters. Fortunately, there nonetheless exists a sequence of local maxima that yields consistent estimators (Abberger and Heiler, 2000).

Apparently unaware of these problems, Domański and Jedrzejczak (1998) provided a simulation study for the performance of MLEs for samples of size $n =$ 1,000(1,000)10,000. It emerges that estimates of the shape parameters a, p can be considered as unbiased for samples of sizes 2,000–3,000, and as approximately normally distributed and efficient for $n \geq 7{,}000$. Reliable estimation of the scale parameter seems to require even larger samples. Estimators appear to be unbiased for $n \geq 4{,}000$, but even for $n = 10{,}000$ there are considerable departures from normality.

Analogously to the Singh–Maddala distribution, we can obtain the Fisher information matrix

$$I(\theta) = \left[-E\left(\frac{\partial^2 \log L}{\partial\theta_i \partial\theta_j}\right)_{i,j}\right] =: \begin{pmatrix} I_{11} & I_{12} & I_{13} \\ I_{21} & I_{22} & I_{23} \\ I_{31} & I_{32} & I_{33} \end{pmatrix}, \tag{6.110}$$

where $\theta = (a, b, p)^\top$, from the information matrix of the GB2 distribution (6.30). This yields

$$I_{11} = \frac{1}{a^2(2+p)} \{p[(\psi(p) - \psi(1) - 1)^2 + \psi'(p) + \psi'(1)] + 2[\psi(p) - \psi(1)]\}, \tag{6.111}$$

$$I_{21} = I_{12} = \frac{p - 1 - p[\psi(p) - \psi(1)]}{b(2+p)}, \tag{6.112}$$

$$I_{22} = \frac{a^2 p}{b^2(2+p)}, \tag{6.113}$$

$$I_{23} = I_{32} = \frac{a}{b(1+p)}, \tag{6.114}$$

$$I_{31} = I_{13} = \frac{\psi(2) - \psi(p)}{a(1+p)}, \tag{6.115}$$

$$I_{33} = \frac{1}{p^2}. \tag{6.116}$$

We note that there are at least two earlier derivations of the Fisher information in the statistical literature: a detailed one using Dagum's parameterization due to Latorre (1988) and a second one due to Zelterman (1987). As mentioned above, the latter article considers the distribution of $\log X$, a generalized logistic distribution, using the parameterization $(\theta, \sigma, \alpha) = (\log b, 1/a, p)$. Latorre (1988) also provided asymptotic standard errors for the Gini and Zenga coefficients derived from MLEs for the Dagum model.

However, an inspection of the scores (6.107–6.109) reveals that $\sup_x \|\partial L/\partial\theta\| = \infty$; thus, the scores function is unbounded in the Dagum case. This implies that the MLE is rather sensitive to isolated observations located sufficiently far away from the majority of the data. There appears therefore to be some interest in more robust procedures. For a robust approach to the estimation of the Dagum model parameters using an optimal B-robust estimator (OBRE), see Victoria-Feser (1995). (The basic ideas underlying this estimator are outlined in Section 3.6 in connection with the Pareto distribution.)

6.3.6 Extensions

Dagum (1977, 1980a) introduced two further variants of his distribution; thus, we shall refer to the previously discussed standard version as the Dagum type I distribution in what follows. The Dagum type II distribution has the c.d.f.

$$F(x) = \alpha + (1-\alpha)\left[1 + \left(\frac{x}{b}\right)^{-a}\right]^{-p}, \quad x \geq 0, \tag{6.117}$$

where $a, b, p > 0$ and $\alpha \in (0, 1)$. Clearly, this is a mixture of a point mass at the origin with a Dagum (type I) distribution over the positive halfline. Thus, the kth moment exists for $0 < k < a$.

The type II distribution was proposed as a model for income distributions with null and negative incomes, but more particularly to fit wealth distributions, which frequently present a large number of economic units with null gross assets and with null and negative net assets.

There is also a Dagum type III distribution, defined via

$$F(x) = \alpha + (1-\alpha)\left[1 + \left(\frac{x}{b}\right)^{-a}\right]^{-p}, \tag{6.118}$$

where again $a, b, p > 0$ but $\alpha < 0$. Consequently, the support of this variant is $[x_0, \infty)$, $x_0 > 0$, where $x_0 = \{b[(1-1/a)^{1/p} - 1]\}^{-1/a}$ is determined implicitly from the constraint $F(x) \geq 0$. Clearly, for the Dagum (III) distribution the kth moment exists for $k < a$.

Both the Dagum type II and type III can be derived from the differential equation

$$\eta(x, F) = \frac{d[\log F(x) - \alpha]}{d\log x} = ap\left\{1 - \left[\frac{F(x)-\alpha}{1-\alpha}\right]^{1/p}\right\}, \quad x \geq 0,$$

subject to $p > 0$ and $ap > 0$. This is a generalization of the differential equation (6.87) considered above.

Investigating the relation between the functional and personal distribution of income, Dagum (1999) obtained the following bivariate c.d.f. specifying the joint distribution of human capital and wealth:

$$F(x_1, x_2) = (1 + b_1x_1^{-a_1} + b_2x_2^{-a_2} + b_3x_1^{-a_1}x_2^{-a_2})^{-p}, \quad x_i > 0, i = 1, 2. \tag{6.119}$$

If $b_3 = b_1b_2$,

$$F(x_1, x_2) = (1 + b_1x_1^{-a_1})^{-p}(1 + b_2x_2^{-a_2})^{-p}.$$

Hence, the marginals are independent. As far as we are aware, there are no empirical applications of this multivariate Dagum distribution.

However, there is a recent application of a bivariate Dagum distribution in the actuarial literature. In the section dealing with the Singh–Maddala distribution, we presented a bivariate income distribution defined in terms of a copula. This is also the approach of Klugman and Parsa (1999), who combined Frank's (1979) copula

$$G(u_1, u_2) = -\frac{1}{\alpha}\log\left\{1 + \frac{(e^{-\alpha u_1} - 1)(e^{-\alpha u_2} - 1)}{e^{-\alpha} - 1}\right\},$$

$$0 < u_i < 1, i = 1, 2, \alpha \neq 1, \tag{6.120}$$

with two Dagum marginal distributions F_i, $i = 1, 2$, and applied the resulting model

$$F(x_1, x_2) = G[F_1(x_1), F_2(x_2)],$$

to the joint distribution of loss and allocated loss adjustment expense on a single claim. They obtained $\hat{\alpha} = 3.07$, indicating moderate positive dependence.

6.3.7 Empirical Results

Although the Dagum distribution was virtually unknown in major English language economics and econometrics journals until well into the 1990s, there were several early applications to income and wealth data, most of which appeared in French, Italian, and Latin American publications.

Incomes and Wealth

Dagum (1977, 1980a) applied his type II distribution to U.S. family incomes of 1960 and 1969, for which the model outperforms the Singh–Maddala, gamma, and lognormal functions, and in his *Encyclopedia of Statistical Sciences* entry of 1983 he fit Dagum types I and III as well as Singh–Maddala, gamma, and lognormal distributions to 1978 U.S. family incomes. Here the Dagum types III and I rank first and second; both outperform the (two-parameter) lognormal and gamma distributions by wide margins.

Espinguet and Terraza (1983) employed the Dagum type II distribution when modeling French wages, stratified by occupation for 1970–1978. The distribution is superior to a Box–Cox-transformed logistic, the Singh–Maddala and three-parameter lognormal and Weibull distributions, as well as a four-parameter beta type I model.

Dagum and Lemmi (1989) fit the Dagum type I–III distributions to Italian income data from the Banca d'Italia sample surveys of 1977, 1980, and 1984, for which the fit is in general quite satisfactory. The data were disaggregated by sex, region, and source of income.

Majumder and Chakravarty (1990) considered the Dagum type I distribution when modeling U.S. income data for 1960, 1969, 1980, and 1983. The distribution improves upon all other two- and three-parameter models. In a reassessment of Majumder and Chakravarty's findings, McDonald and Mantrala (1993, 1995) fit the Dagum type I distribution to 1970, 1980, and 1990 U.S. family incomes using two different fitting methods and alternative groupings of the data; for the 1970 and 1990 data the distribution performs almost as well as the four-parameter GB2 distribution, confirming Majumder and Chakravarty's conclusions.

McDonald and Xu (1995) studied 1985 U.S. family incomes. Out of 11 distributions considered, the Dagum type I ranks third in terms of likelihood and several other criteria, being outperformed only by the four-parameter GB2 and a five-parameter generalized beta distribution [see (6.142) below].

Victoria-Feser (1995, 2000) applied a Dagum type I distribution to incomes of households in receipt of social benefits using the 1979 UK Family Expenditure Survey (FES) and to incomes from the 1985 UK FES. She employed the MLE as well as an optimal B-robust estimator (OBRE) and concluded that the latter provides a better fit for the bulk of the 1979 data. The Dagum distribution is also preferred over the gamma distribution here. For the 1985 data, however, the differences between the two estimators are insignificant and the Dagum distribution does not do appreciably improve on the gamma.

Bantilan et al. (1995) modeled incomes from the Family Income and Expenditure Surveys (FIES) in the Philippines for 1957, 1961, 1965, 1971, 1985, and 1988 using a Dagum type I distribution. They noted that the model fits the data rather well, particularly in the tails.

Bordley, McDonald, and Mantrala (1996) fit the Dagum type I distribution to U.S. family incomes for 1970, 1975, 1980, 1985, and 1990. For all data sets it turns out to be the best three-parameter model, being inferior only to the GB2 distribution and an observationally equivalent generalization and outperforming three- and four-parameter models such as the generalized gamma and GB1 distributions by wide margins.

Botargues and Petrecolla (1997, 1999a,b) estimated Dagum type I–III models for income distribution in the Buenos Aires region, for all years from 1990–1997. They found that the Dagum models outperform lognormal and Singh–Maddala distributions, sometimes by wide margins.

Actuarial Losses

In the actuarial literature, Cummins et al. (1990) fit the Dagum type I distribution (under the name of Burr III) to aggregate fire losses from Cummins and Freifelder (1978). The distribution performs rather well; however, its full flexibility is not required and a one-parameter limiting case, the inverse exponential, is fully adequate. The same paper also considered data on the severity of fire losses and fit the GB2 and its subfamilies to both grouped and individual observations. Although the fit of the Dagum model is very good, that of the Singh–Maddala distribution is slightly better.

6.4 FISK (LOG-LOGISTIC) AND LOMAX DISTRIBUTIONS

In this section we collect some results on the one- and two-parameter subfamilies of the GB2 distribution. Although these models may not be sufficiently flexible in the present context, they have also been considered in various applications and some results pertaining to them—but not to their generalizations—are available. Since the moments and other basic properties have been presented in a more general form in the previous sections, we shall be brief and mention only those results that have not been extended to more general distributions.

6.4.1 Fisk Distribution

The first of these distributions is the Fisk distribution with c.d.f.

$$F(x) = 1 - \left[1 + \left(\frac{x}{b}\right)^a\right]^{-1} = \left[1 + \left(\frac{x}{b}\right)^{-a}\right]^{-1}, \quad x > 0, \qquad (6.121)$$

and p.d.f.

$$f(x) = \frac{ax^{a-1}}{b^a[1+(x/b)^a]^2}, \quad x > 0, \tag{6.122}$$

where a, $b > 0$, a Singh–Maddala distribution with $q = 1$. Alternatively, it can be considered a Dagum distribution with $p = 1$. This model is also a special case of the three-parameter Champernowne distribution to be discussed in Chapter 7 and was actually briefly considered by Champernowne (1952). However, in view of the more extensive treatment by Fisk (1961a,b), it is usually called the Fisk distribution in the income distribution literature. Some authors, for example, Dagum (1975) and Shoukri, Mian, and Tracy (1988), refer to the Fisk distribution as the log-logistic distribution, whereas Arnold (1983) calls it a Pareto (III) distribution and includes an additional location parameter. The term log-logistic may be explained by noting that the distribution of $\log X$ is logistic with scale parameter a and location parameter $\log b$.

A useful property of the Fisk distribution is that it allows for nonmonotonic hazard rates, specifically

$$h(x) = \frac{ax^{a-1}}{b^a[1+(x/b)^a]}, \quad x > 0, \tag{6.123}$$

which is decreasing for $a \leq 1$ and unimodal with the mode at $x = b(a-1)^{1/a}$ otherwise. Among the distributions discussed in the present chapter, the Fisk distribution is the simplest model with this property. In contrast, more popular two-parameter distributions such as the Weibull only allow for monotonic hazard rates.

Dagum (1975) considered a mixture of this distribution with a point mass at the origin, a model that may be viewed as a predecessor of the Dagum type II distribution considered in the previous section.

Interestingly, the distributions of the order statistics from a Fisk distribution have been encountered earlier in this chapter: For a Fisk(a,b) parent distribution, the p.d.f. of $X_{i:n}$ is

$$\begin{aligned} f_{i:n}(x) &= i\binom{n}{i}F(x)^{i-1}[1-F(x)]^{n-i}f(x) \\ &= \frac{n!ax^{ai-1}}{(i-1)!(n-i)!b^{ai}[1+(x/b)^a]^{n+1}}. \end{aligned} \tag{6.124}$$

This can be recognized as the p.d.f. of a GB2 distribution, specifically, $X_{i:n} \sim \text{GB2}(a, b, i, n-i+1)$ (Arnold, 1983, p. 60). [Much earlier, Shah and Dave (1963) already presented percentile points of log-logistic order statistics for $n = 1(1)10$, $1 \leq k \leq n$, when $a = b = 1$.]

Since b is a scale and a is the only shape parameter, we cannot expect much flexibility in connection with inequality measurement. For the Fisk distributions the Lorenz order is linear, specifically for $X_i \sim \text{Fisk}(a_i, b_i)$, $i = 1, 2$, we get from (6.24)

$$X_1 \geq_L X_2 \Longleftrightarrow a_1 \leq a_2, \tag{6.125}$$

provided that $a_i > 1, i = 1, 2$. It should also be noted that the expression for the Gini coefficient is even simpler than for the classical Pareto distribution; it is just

$$G = \frac{1}{a}. \tag{6.126}$$

The Fisk distribution is characterized by

$$P\left[1 + \left(\frac{X}{b}\right)^a > xy\right] = P\left[1 + \left(\frac{X}{b}\right)^a > x\right] \cdot P\left[1 + \left(\frac{X}{b}\right)^a > y\right],$$

where $a, b > 0$ and $x, y > 1$, among all continuous distributions supported on $[0, \infty)$ (Shoukri, Mian, and Tracy, 1988). As discussed above in Sections 6.2 and 6.3, this was generalized to the Singh–Maddala and Dagum distributions by El-Saidi, Singh, and Bartolucci (1990) and is intimately related to the lack of memory property of the exponential distribution.

Following the earlier work of Arnold and Laguna (1977), Arnold, Robertson, and Yeh (1986) provided a characterization of the Fisk distribution (which they call a Pareto (III) distribution) in terms of geometric minimization. Suppose N_p is a geometric random variable independent of $X_i \sim \text{Fisk}(a, b)$, $i = 1, 2, \ldots,$ with $P(N_p = i) = p(1-p)^{i-1}$, $i = 1, 2, \ldots,$ for some $p \in (0, 1)$. Then

$$U_p = \min_{i \leq N_p} X_i \sim \text{Fisk}\left(bp^{1/a}, \frac{1}{a}\right)$$

and

$$V_p = \max_{i \leq N_p} X_i \sim \text{Fisk}\left(bp^{-1/a}, \frac{1}{a}\right),$$

that is,

$$p^{-1/a} U_p \sim p^{1/a} V_p \sim X_1. \tag{6.127}$$

Under the regularity condition $\lim_{x \downarrow 0^+} x^{-\alpha} F(x) = b^{-a}$, it follows that if any one of the statements in (6.127) holds for a fixed $p \in (0, 1)$ then $X_i \sim \text{Fisk}(a, b)$. [This mechanism has been generalized in several directions by Pakes (1983), among other things relaxing the assumption that N_p be geometrically distributed. However, in general it is no longer possible to obtain the resulting income distribution in closed form.]

Regarding estimation, it is worth noting that the Fisher information of the Fisk model is diagonal; thus, the distribution is characterized by orthogonal parameters. This is a direct consequence of the fact that the distribution of $Y = \log X$ is a location-scale family, $af_Y[(y - \log b)a]$, with an f_Y that is symmetric about the origin (clearly, f_Y is logistic). See Lehmann and Casella (1998), Example 2.6.5. For Fisk distributions there is therefore no loss of asymptotic efficiency in estimating a or b when the other parameter is unknown. This property does not extend to the three- and four-parameter distributions discussed in this chapter.

Shoukri, Mian, and Tracy (1988) considered probability-weighted moments estimation of the Fisk model parameters. The probability-weighted moments (PWMs) are defined as

$$W_{l,j,k} = E\{X^l\{F(X)\}^j[1 - F(X)]^k\},$$

where l, j, k are real numbers. For $l = 1$ and $k = 0$, $W_r = W_{1,r,0} = E[XF(X)^r]$ will denote the PWMs of order r. For the Fisk distribution,

$$W_r = \frac{b\Gamma(r + 1 + 1/a)\Gamma(1 - 1/a)}{\Gamma(r + 2)}.$$

In particular,

$$W_0 = b\Gamma\left(1 + \frac{1}{a}\right)\Gamma\left(1 - \frac{1}{a}\right) = \frac{b\pi}{a}\left(\sin\frac{\pi}{a}\right)^{-1} \tag{6.128}$$

and

$$W_1 = \frac{1 + a}{a} W_0. \tag{6.129}$$

Given a complete random sample of size n, the estimation of W_r is most conveniently based on the order statistics. The statistic

$$\hat{W}_r = \frac{1}{n}\sum_{j=1}^{n} x_{j:n} \prod_{i=1}^{r} \frac{j - i}{n - i}$$

is an unbiased estimator of W_r (Landwehr, Matalas, and Wallis, 1979). The PWM estimators are now solutions of (6.128) and (6.129) when the W_r are replaced by their estimators $\hat{W}_r$. Thus,

$$a^* = \frac{\hat{W}_0}{2\hat{W}_1 - \hat{W}_0}$$

and

$$b^* = \frac{\hat{W}_0^2 \sin(\pi/a^*)}{\pi(2\hat{W}_1 - \hat{W}_0)}$$

are the PWM estimators of the parameters a and b, respectively. Shoukri, Mian, and Tracy presented the asymptotic covariance matrix of the estimators that can be derived from the general properties of statistics representable as linear functions of order statistics. They also showed that for the parameter a the PWM estimator is asymptotically less biased than the ML estimator for $a \geq 4$, and that for $a > 7$ each parameter estimator has asymptotic efficiency of more than 90% relative to the MLE. A small simulation study comparing PWM and ML estimators for samples of size $n = 15$ and $n = 25$ shows that the PWMs compare favorably with the MLEs: The PWN estimators seem to be less biased and almost consistently have smaller variances. In addition, for the Fisk distribution the PWM estimators are fast and straightforward to compute and always yield feasible values for the estimated parameters. However, for shape parameters $a \leq 6$ the MLE is generally more efficient, and this seems to be the relevant range in the present context.

Chen (1997) derived exact confidence intervals and tests for the Fisk shape parameter a. Observing that the distribution of the ratio $\xi = M_A/M_G$ of the arithmetic mean and the geometric mean of

$$\left(\frac{X_{1:n}}{b}\right)^a, \left(\frac{X_{2:n}}{b}\right)^a, \ldots, \left(\frac{X_{k:n}}{b}\right)^a$$

is parameter-free with a strictly increasing c.d.f., he obtained percentile points for $3 \leq k \leq n \leq 30$ by a Monte Carlo simulation. These tables can also be used to perform tests on the shape parameter a. In view of the small sample sizes considered, these results will perhaps be more useful in the actuarial field rather than in the income distribution area.

We conclude this section by mentioning that Zandonatti (2001) recently suggested a "generalized" Fisk distribution employing the procedure leading to Stoppa's generalized Pareto distribution. However, since a power transformation of the c.d.f. leads to the c.d.f. of $X_{1:n}$ (or rather a generalization of its distribution with noninteger n) and the order statistics of a Fisk parent follow a GB2 distribution [see (6.124)], this approach does not lead to a "new" distribution.

6.4.2 Lomax (Pareto II) Distribution

A further two-parameter special case of the GB2 distribution, the Lomax distribution—more precisely, the first Lomax distribution, since Lomax (1954) introduced two distributions—has the c.d.f.

$$F(x) = 1 - \left[1 + \left(\frac{x}{b}\right)\right]^{-q}, \quad x > 0, \tag{6.130}$$

and the density

$$f(x) = \frac{q}{b}\left(1 + \frac{x}{b}\right)^{-q-1}, \quad x > 0. \tag{6.131}$$

Hence, it is a Singh–Maddala distribution with $a = 1$. The Lomax distribution is perhaps more widely known as the Pareto (II) distribution—this term is used, for example, by Arnold (1983)—and is related to the classical Pareto distribution via $X \sim \text{Lomax}(b, q) \Longleftrightarrow X + b \sim \text{Par}(b, q)$. It is also a Pearson type VI distribution. Lomax (1954) considered it a suitable model for business failure data.

There is not as much variety in the possible basic shapes of the Pareto (II) distribution—after all, it is just a shifted classical Pareto distribution—as with other two-parameter models such as the gamma, Weibull, or Fisk, all of which allow for zeromodal as well as unimodal densities. Its hazard rate is given by

$$r(x) = \frac{q}{b + x}, \quad x > 0, \tag{6.132}$$

which is a strictly decreasing function for all admissible values of the parameters.

Nair and Hitha (1990) presented a characterization of the Lomax distribution [under the name of the Pareto (II) distribution] in terms of the "equilibrium distribution" defined via the p.d.f. $f_Z(x) = [1/E(X)]\{1 - F_X(x)\}$ (this distribution is of special significance in renewal theory). They showed that the condition $e_X(x) = pe_Z(x)$ (where e denotes the mean excess function) for all $x > 0$, for some $0 < p < 1$, characterizes the Lomax distribution. The distribution is also characterized by the condition of proportionality of the corresponding hazard rates, that is, by the condition $h_X(x) = ph_Z(x)$ for all $x > 0$, for some $p > 1$.

The distribution can furthermore be characterized within the framework of a model of underreported incomes (Revankar, Hartley, and Pagano, 1974); see Section 3.4 for a detailed discussion. This result essentially exploits the linearity of the mean excess function

$$e(x) = \frac{x + b}{q - 1}, \quad x > 0. \tag{6.133}$$

As in the case of the Fisk distribution, the Lorenz order is linear for Lomax random variables, specifically for $X_i \sim \text{Lomax}(b_i, q_i)$, $i = 1, 2$,

$$q_1 \leq q_2 \Longleftrightarrow X_1 \geq_L X_2, \tag{6.134}$$

provided $q_i > 1$. The Gini coefficient is also of a very simple form

$$G = \frac{q}{2q-1}. \tag{6.135}$$

Note that $1/2 \leq G \leq 1$, which casts some doubt on the usefulness of this model to approximate income distributions of, for example, some European Union countries, for which such extreme inequality is not observed.

Finally, we present some results pertaining to estimation.

For the scale parameter $b = 1$, the statistic $T = \sum_{i=1}^{n} \log(1 + x_i)$ is sufficient and complete for q^{-1}, and a minimum variance unbiased (MVU) estimator of q is given by $(n-1)/T$ (Patel, 1973). The c.d.f. $F(x) = 1 - (1+x)^{-q}$, $x \geq 0$, can also be estimated in an unbiased way; the estimator is

$$\hat{F}(x) = \begin{cases} 1 - \left[1 - \dfrac{\log(1+x)}{T}\right]^{n-1}, & T > \log(1+x) \\ 1, & T \leq \log(1+x). \end{cases},$$

For further unbiased estimators of functions of Lomax parameters, see Voinov and Nikulin (1993, pp. 435–436), who refer to these results as results for the Burr—meaning Burr XII, that is, Singh–Maddala—distribution that involves an additional shape parameter a. However, since all results given therein require a to be known, we prefer to consider them as results pertaining to the Lomax subfamily. Algorithmic aspects of ML estimation in the Lomax distribution (under the name of a Pareto distribution) are discussed by Wingo (1979), who used a numerical method for the univariate global optimization of functions expressible as the sum of a concave and a convex function.

A generalization of the Lomax distribution was recently suggested by Zandonatti (2001). Following Stoppa's (1990b,c) approach leading to a generalized Pareto (I) distribution (see Section 3.8), he arrived at a distribution with density

$$f(x) = \frac{q\theta}{b}\left(1 + \frac{x}{b}\right)^{-q-1}\left[1 - \left(1 + \frac{x}{b}\right)^{-q}\right]^{\theta-1}, \quad x > 0. \tag{6.136}$$

6.4.3 Empirical Applications

Incomes and Wealth

Fisk (1961a) considered weekly earnings in agriculture in England and Wales for 1955–1956 and U.S. income distribution for 1954 (by occupational categories). He concluded that the distribution may prove useful when income distributions that are homogeneous in at least one characteristic (here occupation) are examined.

Using nonparametric bounds on the Gini coefficient developed by Gastwirth (1972), Gastwirth and Smith (1972) found that the implied Gini indices derived from a Fisk distribution fall outside these bounds for U.S. individual adjusted gross incomes for 1955–1969 and concluded that Fisk distributions are inappropriate for modeling these data.

Arnold and Laguna (1977) fit the Fisk distribution, with an extra location parameter, to income data for 17 metropolitan areas in Peru for the period 1971–1972 and concluded that their results are reasonably consonant with a (shifted) Fisk distribution. It is of historical interest to note that Vilfredo Pareto in his *Cours d' économie politique* (1897) also used Peruvian income data (for the year 1800).

Harrison (1979) used the Fisk distribution (under the name of sech^2 distribution; see Chapter 7 for an explanation of this terminology) for the gross weekly earnings of full-time male workers aged 21 and over, in Great Britain, collected in April 1972, for seven occupational groups. The distribution performs about as well as a lognormal distribution when the data are disaggregated, but considerably better in the upper tail for the aggregate data.

McDonald (1984) fit the Fisk distribution to 1970, 1975, and 1980 U.S. family incomes. However, the distribution does not do well: It is outperformed by the (G)B1, (G)B2, Singh–Maddala, and generalized gamma distributions, usually by wide margins, and even the gamma and Weibull distributions (having the same number of parameters) are preferable.

For the Japanese incomes (in grouped form) for 1963–1971, the Fisk distribution does only slightly worse than the Singh–Maddala and outperforms the beta, gamma, lognormal, and Pareto (II) distributions (Suruga, 1982). In fact, a likelihood ratio test reveals that there are no significant differences between the Fisk and Singh–Maddala distributions for these data; hence, the simpler Fisk distribution is entirely adequate. The distribution was also fitted to several strata from the 1975 Japanese Income Redistribution Survey by Atoda, Suruga, and Tachibanaki (1988). Four occupational classes as well as primary and redistributed incomes were considered, and five different estimation techniques applied. In a later study using the same data set, Tachibanaki, Suruga, and Atoda (1997) applied ML estimation on the basis of individual observations. In both studies, it became clear that the data require a more flexible model such as the Singh–Maddala distribution. Among the two-parameter functions, gamma and Weibull distributions both fit better.

Henniger and Schmitz (1989) considered the Fisk distribution when comparing five parametric models for the UK Family Expenditure Survey for 1968–1983 to nonparametric fittings. Although for the entire population all parametric models are rejected, the Fisk distribution performs reasonably well for some subgroups.

Actuarial Losses

In the actuarial literature, Benckert and Jung (1974) employed the Pareto (II) distribution to model the distribution of fire insurance claims in Sweden for the period 1958–1969. They found that for one class of buildings (wooden houses) the distribution provides a good fit, with estimates of the tail index q less than 1 in all cases.

Hogg and Klugman (1983) fit the Lomax distribution (under the name of Pareto distribution) to data on malpractice losses, for which it is preferable over the beta II, Singh–Maddala, lognormal, and Weibull distributions. The data require a very heavy-tailed model with a parameter q slightly below 1.

Cummins et al. (1990) applied the distribution to two sets of fire liability data but the performance of the Lomax distribution is not impressive. It ranks only 12th and 13th out of 16 distributions of the gamma and beta type.

In a recent investigation studying losses from catastrophic events in the United States, Burnecki, Kukla, and Weron (2000) employed a Pareto type II distribution and obtained a tail index q in the vicinity of 2.7.

In summary, it would seem that most data on size distributions require a more flexible distribution than the Fisk or Lomax distributions. Specifically, an additional shape parameter appears to be appropriate.

6.5 (GENERALIZED) BETA DISTRIBUTION OF THE FIRST KIND

Occasionally, distributions supported on a bounded domain have been considered for the modeling of size phenomena, notably the distribution of income. The most flexible of these are the generalized beta distribution of the first kind (hereafter referred to as GB1) and its special case, the standard beta distribution. We refer to Johnson, Kotz, and Balakrishnan (1995, Chapter 25) for the basic properties of this well-known distribution, and we shall briefly mention below some aspects pertaining to size phenomena. The GB1 was introduced by McDonald (1984) as an income distribution, whereas the B1 was used for the same purpose more than a decade earlier by Thurow (1970).

6.5.1 Definition and Properties

The GB1 is defined in terms of its density

$$f(x) = \frac{ax^{ap-1}[1-(x/b)^a]^{q-1}}{b^{ap}B(p,q)}, \quad 0 \le x \le b, \tag{6.137}$$

where all four parameters a, b, p, q are positive. Here b is a scale and a, p, q are shape parameters. When $a = 1$, we get

$$f(x) = \frac{x^{p-1}[1-x/b]^{q-1}}{b^{p}B(p,q)}, \quad 0 \le x \le b, \tag{6.138}$$

the three-parameter beta distribution.

The GB1 is related to the GB2 distribution via the relation

$$X \sim \mathrm{GB2}(a, b, p, q) \Longrightarrow \left(\frac{X^a}{1+X^a}\right)^{1/a} \sim \mathrm{GB1}(a, b, p, q).$$

This generalizes a well-known relationship between the B1 and B2 distributions.

The c.d.f.'s of the GB1 and B1 distributions cannot be expressed in terms of elementary functions. However, in view of (6.8), they are available in terms of Gauss's hypergeometric function ${}_2F_1$, in the form (McDonald, 1984)

$$F(x) = \frac{(x/b)^a}{pB(p, q)} {}_2F_1\left[p, 1-q; p+1; \left(\frac{x}{b}\right)^a\right]. \tag{6.139}$$

In analogy with the GB2 case discussed in Section 6.1, they can also be written as an incomplete beta function ratio

$$F(x) = I_z(p, q), \quad \text{where } z = \left(\frac{x}{b}\right)^a, \tag{6.140}$$

in the GB1 case (of course, $a = 1$ yields the c.d.f. of the three-parameter B1 distribution).

The moments of the GB1 exist for $-ap < k < \infty$; they are

$$E(X^k) = \frac{b^k B(p + k/a, q)}{B(p, q)} = \frac{b^k \Gamma(p + k/a)\Gamma(p+q)}{\Gamma(p+q+k/a)\Gamma(p)}. \tag{6.141}$$

An analysis of the hazard rate is more involved than for either the generalized gamma or the GB2 distribution. Monotonically decreasing, monotonically increasing, $\bigcap$ as well as $\bigcup$-shapes are possible; see McDonald and Richards (1987) for a discussion of these possibilities.

The Lorenz ordering within the GB1 family was studied by Wilfling (1996c), who provided four sets of sufficient conditions. Noting that the GB1 density is regularly varying at the origin with index $-ap - 1$, it can be deduced along the lines of Kleiber (1999b, 2000a) that $a_1 p_1 \leq a_2 p_2$ is a necessary condition for $X_1 \geq_L X_2$. A complete characterization of the Lorenz ordering within this family appears to be unavailable at present.

However, Sarabia, Castillo, and Slottje (2002) provided Lorenz ordering results for nonnested income distributions. These include the following: That for $X \sim \mathrm{GB2}(a, b, p, q)$ and $Y \sim \mathrm{GB1}(a, b, p, q)$, one has $X \geq_L Y$, and for $X \sim \mathrm{GB1}(a, b, p, q)$ and Y following a generalized gamma distribution (cf. Chapter 5), that is, $Y \sim \mathrm{GG}(\tilde{a}, \tilde{b}, \tilde{p})$ with $a \geq \tilde{a}$ and $ap \geq \tilde{a}\tilde{p}$, it follows that $Y \geq_L X$.

McDonald (1984) provided the Gini coefficient of the GB1 as a somewhat lengthy expression involving the generalized hypergeometric function ${}_4F_3$. For the

B1 subfamily the expression is less cumbersome and equals (McDonald and Ransom, 1979a)

$$G = \frac{2B(p+q, 1/2)B(p+1/2, 1/2)}{\pi B(q, 1/2)} = \frac{\Gamma(p+q)\Gamma(p+1/2)\Gamma(q+1/2)}{\sqrt{\pi}\Gamma(p+q+1/2)\Gamma(p+1)\Gamma(q)}.$$

The Pietra index and the (first) Theil coefficient of the B1 distribution are (McDonald, 1981; Pham-Gia and Turkkan, 1992)

$$P = \frac{[p/(p+q)]^p}{p(p+1)B(p, q)} {}_2F_1\left(p, 1-q, p+2; \frac{p}{p+q}\right)$$

and

$$T_1(X) = \psi(p+1) - \psi(p+q+1) - \log\left[\frac{p}{p+q}\right],$$

respectively.

A distribution encompassing both the GB1 and GB2 families was proposed by McDonald and Xu (1995). This five-parameter generalized beta distribution has the p.d.f.

$$f(x) = \frac{|a|x^{ap-1}[1+(1-c)(x/b)^a]^{q-1}}{b^{ap}B(p, q)[1+c(x/b)^a]^{p+q}}, \tag{6.142}$$

where $0 < x^a < b^a/(1-c)$, and all parameters a, b, p, q are positive. Again, b is a scale and a, p, q are shape parameters. The new parameter $c \in [0, 1]$ permits a smooth transition between the special cases $c = 0$, the GB1 distribution, and $c = 1$, the GB2 distribution. The moments of this generalized distribution exist for all k if $c < 1$ or for $k < aq$ if $c = 1$; they are

$$E(X^k) = \frac{b^k B(p+k/a, q)}{B(p, q)} {}_2F_1\left(p+\frac{k}{a}, \frac{k}{a}; p+q+\frac{k}{a}; c\right).$$

However, when fitting this model to 1985 family incomes, McDonald and Xu found that the GB2 subfamily is selected (in terms of likelihood and several other criteria). Thus, it appears that the five-parameter GB distribution does not provide additional flexibility, at least in our context. The authors of this book remain skeptical about the usefulness and meaning of five-parameter distributions and consider the distribution described by (6.142) to be just a curious theoretical generalization.

A further application of beta type I income distribution models is given in Pham-Gia and Turkkan (1997), who derived the density of income $X = X_1 + X_2$, where X_1 is true income and X_2 an independent (additive) reporting error under the assumption that X_i, $i = 1, 2$, follow $\text{B1}(p_i, q_i)$ distributions with general support $[0, \theta_i]$. The density of X can be expressed in terms of an Appell function (a bivariate hypergeometric function).

6.5.2 Empirical Applications

In an influential paper, Thurow (1970) applied the B1 distribution to U.S. Census Bureau constant dollar (1959) income distributions for households (families and unrelated individuals) for every year from 1949–1966, stratified by race. [The estimated parameters are not given in Thurow (1970) but in McDonald (1984).] He also studied the impact of various macroeconomic factors on the parameters of the distribution via regression techniques. In particular, his results raise questions as to whether economic growth is associated with a more egalitarian distribution and also suggest that inflation may lead to a more equal distribution for whites. However, McDonald (1984) expressed some doubts concerning Thurow's results. He pointed out that none of the estimated densities is $\bigcap$-shaped (as is to be expected) and that the implied Gini coefficients differ from census estimates by about 30%, concluding that theses differences highlight estimation problems.

McDonald and Ransom (1979a) employed the B1 distribution for approximating U.S. family incomes for 1960 and 1969 through 1975. When utilizing three different estimators, it turns out that the distribution is preferable to the gamma and lognormal distributions but inferior to the Singh–Maddala, which has the same number of parameters.

McDonald (1984) estimated both the B1 and GB1 distribution for 1970, 1975, and 1980 U.S. family incomes. The performance of the GB1 is comparable to the generalized gamma and B2 distribution—both of which have the same number of parameters—but inferior to the GB2 or Singh–Maddala distributions.

The B1 distribution was also fitted to Japanese income data, in grouped form, from the 1975 Income Redistribution Survey by Atoda, Suruga, and Tachibanaki (1988). Four occupational classes as well as primary and redistributed incomes were considered, and five different estimation techniques applied. The distribution provides a considerably better fit than the (generalized) gamma and lognormal distributions, but the Singh–Maddala is superior.

Bordley, McDonald, and Mantrala (1996) fit the (G)B1 and GB distributions to U.S. family incomes for 1970, 1975, 1980, 1985, and 1990. The GB distribution is observationally equivalent to the GB2 distribution for four of the data sets; for the 1970 data it provides a slight improvement. The GB1 comes in fifth out of 15 beta- and gamma-type distributions, being observationally equivalent to the generalized gamma distribution for all these years. However, the three-parameter Dagum type I and Singh–Maddala distributions perform considerably better.

Brachmann, Stich, and Trede (1996) estimated both the B1 and GB1 distributions for household income data from the German Socio-Economic Panel (SOEP) for 1984–1993. They noted that the ML estimation of the GB1 proved to be rather difficult since the gradient of the log likelihood in the parameter b was rather small. Also, both models tend to underestimate the mean for these data.

All these studies appear to suggest that there are several distributions supported on an unbounded domain which provide a considerably better fit than a GB1 or B1 distribution.

CHAPTER SEVEN

Miscellaneous Size Distributions

In this chapter we shall study a number of size distributions that may not be in the mainstream of current research, but are definitely of historical interest as well as containing potential applications. We have tried to unify the results scattered in the literature, sometimes in the most unexpected sources.

7.1 BENINI DISTRIBUTION

As was discussed in detail in Chapter 3 (the reader may wish to consult Section 3.2), Vilfredo Pareto (1896, 1897)—the father of the statistical-probabilistic theory of income distributions—announced in his classical works *La courbe de la répartition de la richesse* and *Cours d'économie politique* the remarkable discovery that the survival function of an income distribution is approximately linear in a double-logarithmic plot. He provided empirical verifications of his law for a multitude of data, showing that the relationship is valid for any (geographical) location, any time period (for which the data are available), and any economic level of a country or region. An alert and energetic Italian statistician and demographer, Rodolfo Benini (a short biography of Benini is presented in Appendix A) was able to confirm almost immediately in 1897 that the Pareto law indeed holds for incomes as well as various other economic variables. While analyzing additional different data sources, Benini subsequently discovered in 1905–1906 that for the distribution of legacies a quadratic (rather than linear) function

$$\log \bar{F}(x) = a_0 - a_1 \log x - a_2(\log x)^2 \tag{7.1}$$

provides a better fit. This leads to a distribution with the c.d.f.

$$F(x) = 1 - \exp\{-\alpha(\log x - \log x_0) - \beta(\log x - \log x_0)^2\}, \tag{7.2}$$

Statistical Size Distributions in Economics and Actuarial Sciences, By Christian Kleiber and Samuel Kotz.
ISBN 0-471-15064-9

where $x \geq x_0 > 0$ and α, $\beta > 0$. For parsimony, Benini (1905) considered only the case where $\alpha = 0$, that is,

$$F(x) = 1 - \exp\{-\beta(\log x - \log x_0)^2\} \tag{7.3}$$

$$= 1 - \left(\frac{x}{x_0}\right)^{-\beta(\log x - \log x_0)}, \quad x \geq x_0 > 0.$$

Subsequently, this idea was used by several other Italian economists and statisticians, including Bresciani Turroni (1914) and Mortara (1917, 1949). They however introduced higher-order terms. Independently, a well-known Austrian statistician Winkler (1950) some 45 years later also suggested that, in the Pareto diagram, a higher-order polynomial in $\log x$ may provide an even better fit to empirical income distributions than the Pareto distribution and he fit a quadratic—that is, the original Benini distribution (7.1)—to the U.S. income distribution of 1919.

Also independently, but somewhat later, in the actuarial literature, the distribution with c.d.f. (7.1) was proposed as a model for the size-of-loss distribution. In Head (1968) it appears as a nameless distribution resulting in a better fit than the Pareto distribution for several empirical fire loss-severity distributions. Ramachandran (1969) also found this model to be preferable to the Pareto distribution when dealing with UK fire losses. DuMouchel and Olshen (1975) called it an *approximate lognormal distribution*, whereas Shpilberg (1977) referred to it as the *quasi-lognormal distribution*. We shall refer to all the variants (i.e., $\alpha = 0$ and $\alpha \neq 0$) of this multidiscovered distribution as the *Benini distribution*.

It is perhaps worthwhile to examine briefly the actuarial motivation for (7.2). Ramachandran (1969) and Shpilberg (1977) searched for a model for the probability distribution of an individual fire loss amount and started from the hazard rate of fire duration, assuming that, for a homogeneous group of risks, the fire loss increases exponentially with the duration of the fire. They observed that an exponential distribution of fire duration leads to a Pareto distribution as the distribution of the fire loss amount (see Section 3.2) and claimed that a model which reflects a gradual decrease in the probability of survival of the fire (as implied by an increasing failure rate distribution) would be more plausible empirically. The simplest functional form for the hazard rate with the required properties is the linear function

$$r(t) = \alpha + \beta t, \quad (\alpha, \beta > 0), \tag{7.4}$$

where t is the duration of the fire, resulting in a fire duration distribution with the c.d.f.

$$F(t) = 1 - \exp\left(-\alpha t - \frac{1}{2}\beta t^2\right).$$

This distribution was previously discussed by Flehinger and Lewis (1959) in a reliability context. Under the assumption that $x/x_0 \propto e^{kt}$, $k > 0$, that is, the ratio of the fire loss amount x to the minimum discernible loss x_0 increases exponentially with the fire duration, one directly obtains the Benini c.d.f. in the form (7.1).

The density of the three-parameter Benini distribution (7.2) is

$$f(x) = \exp\left\{-\alpha\log\left(\frac{x}{x_0}\right) - \beta\left[\log\left(\frac{x}{x_0}\right)\right]^2\right\}\left\{\frac{\alpha}{x} + \frac{2\beta\log(x/x_0)}{x}\right\}, \quad x_0 \le x, \quad (7.5)$$

whereas in the two-parameter case (7.3) we get

$$f(x) = 2\beta x^{-1} \cdot \exp\left\{-\beta\left[\log\left(\frac{x}{x_0}\right)\right]^2\right\} \cdot \log\left(\frac{x}{x_0}\right), \quad x_0 \le x. \qquad (7.6)$$

Here x_0 is a scale and α and β are shape parameters. For $\alpha = 0$, that is, Benini's original form, it moreover follows from (7.3) that the quantile function is available in a closed form, namely,

$$F^{-1}(u) = x_0 \exp\sqrt{-\frac{1}{\beta}\log(1-u)}, \quad \text{for } 0 < u < 1. \qquad (7.7)$$

This is an attractive feature for simulation purposes.

Figure 7.1 depicts several two-parameter Benini densities.

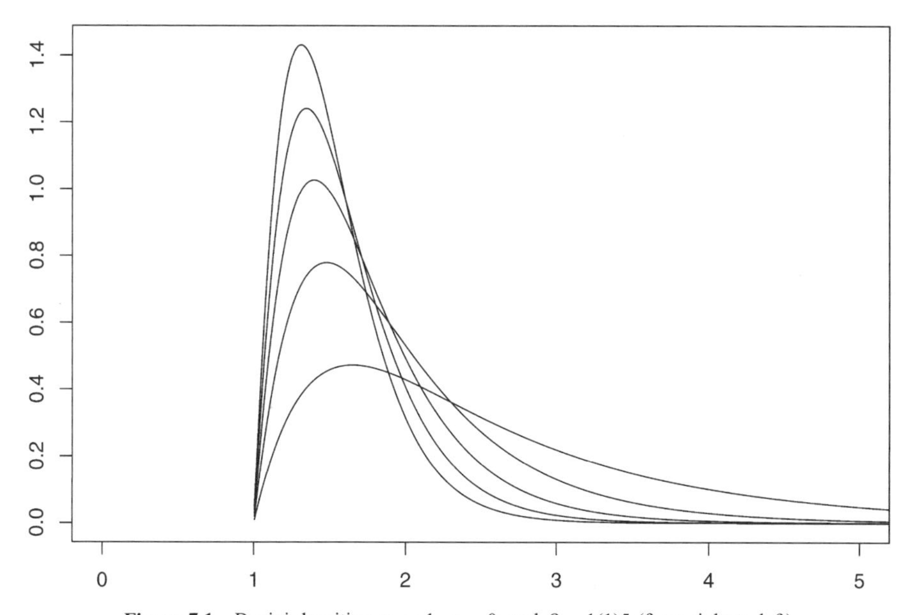

Figure 7.1 Benini densities: $x_0 = 1$, $\alpha = 0$, and $\beta = 1(1)5$ (from right to left).

For $\alpha = 0$ and $x_0 = 1$, the distribution can also be recognized as a log-Rayleigh distribution [namely, if Y follows a Rayleigh distribution, with c.d.f. $F(y) = 1 - \exp(-y^2)$, $y > 0$, then $X := \exp(Y)$ has the distribution with c.d.f. (7.3), where $\beta = 1$ and $x_0 = 1$]. Hence, the log-Weibull family, with c.d.f. $F(x) = 1 - \exp\{-(\log x)^p\}$, can be viewed as a natural generalization of the Benini distribution. Indeed, Benini (1905, p. 231) discussed rather briefly this model and reported that, for his data, when $p = 2.15$, the fit is superior to that using model (7.3).

The Benini distribution is a member of Dagum's generating system of income distributions (Dagum 1980a, 1983, 1990a); see Section 2.4.

An attractive feature of the three-parameter Benini distributions is that this family contains the Pareto distribution as a special case (for $\beta = 0$). This led DuMouchel and Olshen (1975) to suggest a method of testing Pareto vs. Benini distributions. Their test is based on the idea that, if X follows a Par(x_0, α) distribution, then $\log(X/x_0)$ is an exponentially distributed variable and hence the mean and standard deviation of $\log(X/x_0)$ coincide. Therefore, an appropriate test of H_0: $\beta = 0$ vs. H_1: $\beta > 0$ depends on the ratio of the sample standard deviation to the sample mean. Their test statistic is

$$\sqrt{\frac{n}{8}}\left(1 - \frac{s^2}{\bar{y}^2}\right), \tag{7.8}$$

where n is the sample size, $\bar{y}$ the sample mean, and s^2 the sample variance of the logarithms of the observations. This test is derived by using the Neyman's $C(\alpha)$ principle and is therefore locally asymptotically most powerful. Under H_0, the test statistic is asymptotically standard normal and the test rejects the null hypothesis for large values of (7.8).

Since the Benini distribution is quite close to the lognormal distribution, the DuMouchel-Olshen test may be viewed as a test of the Pareto distribution vs. the lognormal, that is, it enables one to choose at least approximately between the two classical size distributions.

Benini (1905, 1906) estimated the parameters of his distribution via regression in the Pareto diagram using Cauchy's method that is seldom used nowadays (see, e.g., Linnik, 1961); later authors (Winkler, 1950; Head, 1968) preferred ordinary least-squares.

7.2 DAVIS DISTRIBUTION

Harold T. Davis (1892–1974), in his 1941 monographs *The Theory of Econometrics* and *Analysis of Economic Time Series*, proposed another income size distribution, which is a generalization of the so-called *Planck's law of radiation* from statistical physics.

Davis was one of the pioneers of econometrics in the United States; he helped to found the professional journal *Econometrica* and served as its Associate Editor for

26 years. He also worked on the staff of the Cowles Commission in its early days in Colorado Springs. See Farebrother (1999) for a recent account of Davis's life.

In an attempt to derive an expression that would represent not merely the upper tail of the distribution of income, Davis required an appropriate model with the following properties:

- $f(x_0) = 0$, for some $x_0 > 0$, that may be interpreted as the subsistence income. In Davis's words, x_0 represents the *wolf point*, since below this point "the wolf, which lurks so close to the doors of those in the neighborhood of the modal income [here he assumes that the distribution is highly skewed with a mode close to the lower bound of the support, *our addition*], actually enters the house" (1941a, p. 405).
- A modal income exists.
- For large x the distribution approaches a Pareto distribution

$$f(x) \backsim A(x - x_0)^{-\alpha-1}.$$

Davis then postulated that f is of the form

$$f_D(z) = \frac{C}{z^{\nu+1}} \frac{1}{e^{1/z} - 1}, \quad \nu > 1,$$

where $z = x - x_0$. This is, for $x_0 = 0$ and $\nu = 4$, the distribution of $1/V$, where V follows the *Planck distribution* mentioned above. Clearly, $f_D(0) = 0$ and the density exhibits an interior mode. It remains to determine the normalizing constant C. From, for example, Prudnikov et al. (1986, Formula 2.3.14.6), we have

$$\int_0^\infty f_D(z)\, dz = C \cdot \Gamma(\nu)\zeta(\nu), \tag{7.9}$$

where $\zeta(\nu) = \sum_{j=1}^\infty j^{-\nu}$, $\nu > 1$, is the Riemann zeta function. If we add scale and location parameters, the Davis density becomes

$$f(x) = \frac{b^\nu}{\Gamma(\nu)\zeta(\nu)} (x - x_0)^{-\nu-1} \frac{1}{\exp\{b(x - x_0)^{-1}\} - 1}, \quad 0 < x_0 \leq x, \tag{7.10}$$

where $x_0, b > 0$ and $\nu > 1$. Utilizing the well-known series expansion

$$\frac{x}{e^x - 1} = \sum_{n=0}^\infty \frac{B_n}{n!} x^n,$$

where B_n are the Bernoulli numbers, it is not difficult to see that the parameter ν is related to Pareto's α as $\nu = \alpha + 1$.

For x close to x_0, the density is approximately of the form

$$f(x) \simeq C \cdot (x - x_0)^{-\nu-1} \exp\{-b(x - x_0)^{-1}\}, \quad 0 < x_0 \leq x,$$

and therefore resembles the Vinci (inverse gamma) distribution of Section 5.2.

From (7.9) we directly obtain the moments

$$E[(X - x_0)^k] = \frac{b^k \Gamma(\nu - k)\zeta(\nu - k)}{\Gamma(\nu)\zeta(\nu)}, \tag{7.11}$$

provided $k + 1 < \nu$.

It is quite remarkable that the Davis distribution—not easily related to other continuous univariate distributions—is a member of D'Addario's (1949) generating system of income distributions; cf. Section 2.4. [See also Dagum (1990a, 1996).]

Davis (1941a,b) fit his model to the distribution of income among personal-income recipients in the United States in 1918 and obtained a value of ν in the vicinity of 2.7. His method of estimation is a two-stage procedure: After estimating Pareto's parameter α by least-squares in the Pareto diagram, yielding $\hat{\nu} = \hat{\alpha} + 1$, he essentially determined b from the estimating equation (7.11) for $k = 1$.

The distribution was later used by Champernowne (1952), who reconsidered Davis's data but found the model not to fit as well as his own three-parameter distribution. We shall next study the Champernowne distribution.

7.3 CHAMPERNOWNE DISTRIBUTION

Champernowne (1937, 1952) considered the distribution of log-income $Y = \log X$, also termed "income power," as the starting point and assumed that it has a density function of the form

$$f(y) = \frac{n}{\cosh[\alpha(y - y_0)] + \lambda}, \quad -\infty < y < \infty, \tag{7.12}$$

where α, λ, y_0, n are positive parameters, n being the normalizing constant and hence a function of the others. (The reader will hopefully consult the biography of Champernowne presented in the appendix to learn more about this colorful person.)

The function given by (7.12) defines a symmetrical distribution, with median y_0, whose tails are somewhat heavier than those of the normal distribution. It is included in a large family of generalized logistic distributions due to Perks (1932), a British actuary. In view of the relation $\cosh y = (e^y + e^{-y})/2$, (7.12) may be rewritten as

$$f(y) = \frac{n}{1/2e^{\alpha(y-y_0)} + \lambda + 1/2e^{-\alpha(y-y_0)}}.$$

This explains why (7.12) defines a generalization of the logistic distribution. For $y_0 = 0$, $\alpha = 1$, $n = 1/2$, and $\lambda = 1$ we obtain

$$f(y) = \frac{1}{e^y + 2 + e^{-y}} = \frac{e^y}{(1 + e^y)^2},$$

the density of the standard logistic distribution, in view of $\operatorname{sech} y = (\cosh y)^{-1}$ also known as the *sech square distribution*.

If we set $\log x_0 = y_0$, the density function of the income $X = \exp Y$ is given by

$$f(x) = \frac{n}{x[1/2(x/x_0)^{-\alpha} + \lambda + 1/2(x/x_0)^{\alpha}]}, \quad x > 0. \tag{7.13}$$

By construction, x_0 is the median value of income. The form of the c.d.f. depends on the value of λ. There are three variants: $|\lambda| < 1$, $\lambda = 1$, and $\lambda > 1$, which are discussed in Champernowne's publications at some length.

The simplest case occurs for $\lambda = 1$, only briefly mentioned by Champernowne (1952). However, it was discussed in greater detail by Fisk (1961a,b) and is therefore often referred to as the *Fisk distribution* (see Section 6.4). Its density is

$$f(x) = \frac{\alpha}{2x\{\cosh[\alpha \log(x/x_0)] + 1\}} = \frac{\alpha x^{\alpha-1}}{x_0^{\alpha}[1 + (x/x_0)^{\alpha}]^2}, \quad x > 0. \tag{7.14}$$

Here x_0 is a scale parameter and $\alpha > 0$ is a shape parameter. The parameter α is the celebrated Pareto's alpha. This special case of the Champernowne distribution is also a special case of the Dagum type I distribution (for $p = 1$) and of the Singh–Maddala distribution (for $q = 1$). Consequently, all the distributional properties of this model were presented in Chapter 6. (In the notation of Chapter 6, α equals a and x_0 equals b.) The c.d.f. of (7.14) is

$$F(x) = 1 - \frac{1}{1 + (x/x_0)^{\alpha}} = \left\{1 + \left(\frac{x}{x_0}\right)^{-\alpha}\right\}^{-1}, \quad x > 0. \tag{7.15}$$

For $|\lambda| < 1$, the Champernowne density can be written as

$$f(x) = \frac{\alpha \sin\theta}{2\theta x\{\cosh[\alpha \log(x/x_0)] + \cos\theta\}}, \quad x > 0, \tag{7.16}$$

or alternatively,

$$f(x) = \frac{\alpha \sin\theta}{\theta x[(x/x_0)^{-\alpha} + 2\cos\theta + (x/x_0)^{\alpha}]}, \tag{7.17}$$

where $\cos\theta = \lambda$ and $-\pi < \theta < \pi$. (Actually, one must confine oneself to $0 < \theta < \pi$ because otherwise the model may not be identifiable.) The latter expression can be rewritten as

$$f(x) = \frac{\alpha \sin\theta (x/x_0)^{\alpha-1}}{\theta x_0[1 + 2\cos\theta (x/x_0)^{\alpha} + (x/x_0)^{2\alpha}]}$$

$$= \frac{\alpha \sin\theta (x/x_0)^{\alpha-1}}{\theta x_0\{[\cos\theta + (x/x_0)^{\alpha}]^2 + \sin^2\theta\}},$$

which integrates to the c.d.f.

$$F(x) = 1 - \frac{1}{\theta}\arctan\left[\frac{\sin\theta}{\cos\theta + (x/x_0)^{\alpha}}\right], \quad x > 0. \tag{7.18}$$

Parameters α and x_0 play the same role as in the case where $\lambda = 1$. Unfortunately, the new parameter θ evades simple interpretation; Champernowne (1952) noted that it may be regarded as a parameter for adjusting the kurtosis of the distribution of log income. Harrison (1974) observed that, for $\theta \to \pi$, the distribution approaches a point mass concentrated at x_0. For $\theta \to 0$, the distribution becomes the one with $\lambda = 1$, that is, the Fisk distribution. The role of the parameter θ is illustrated in Figure 7.2.

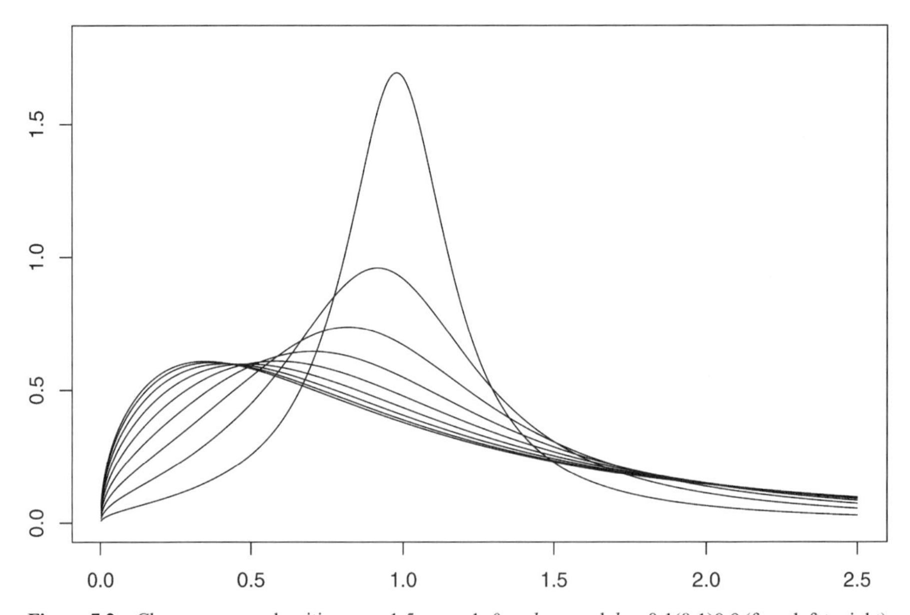

Figure 7.2 Champernowne densities: $\alpha = 1.5$, $x_0 = 1$, $\theta = d \cdot \pi$, and $d = 0.1(0.1)0.9$ (from left to right).

A stochastic model leading to the Champernowne distribution as the equilibrium distribution was briefly discussed by Ord (1975).

Equation (7.18) implies that the quantile function is available in closed form; it is

$$F^{-1}(u) = x_0 \left\{ \frac{\sin\theta}{\tan[\theta(1-u)]} - \cos\theta \right\}^{1/\alpha}, \quad 0 < u < 1. \tag{7.19}$$

As always, this is an attractive feature for simulation purposes.

The mode of this form of the Champernowne distribution occurs at

$$x_{\text{mode}} = x_0 \left\{ \frac{\sqrt{\alpha^2 - \sin^2\theta} - \cos\theta}{\alpha + 1} \right\}^{1/\alpha}. \tag{7.20}$$

Finally, for $\lambda > 1$ the density becomes

$$f(x) = \frac{\alpha \sinh \eta}{\eta x[(x/x_0)^\alpha + 2\cosh\eta + (x/x_0)^{-\alpha}]}, \quad x > 0, \tag{7.21}$$

and the corresponding c.d.f. is

$$F(x) = 1 - \frac{1}{2\eta}\log\left(\frac{x^\alpha + e^\eta x_0^\alpha}{x^\alpha + e^{-\eta}x_0^\alpha}\right), \tag{7.22}$$

where $\cosh\eta = \lambda$. The mode of this form of the Champernowne distribution occurs at

$$x_{\text{mode}} = x_0 \left\{ \frac{\sqrt{\alpha^2 - \sinh^2\eta} - \cosh\eta}{\alpha + 1} \right\}^{1/\alpha}. \tag{7.23}$$

In view of the earlier remarks, all three forms of the Champernowne distribution may be regarded as "generalized log-logistic distributions." However, Champernowne (1952) reported that the majority of income distributions which he graduated throughout his investigations give values of λ numerically less than 1. Most authors (e.g., Harrison, 1974; Kloek and van Dijk, 1977, 1978; Campano, 1987) who have used the Champernowne distribution have confined themselves to the case $|\lambda| < 1$. We shall follow their convention below. (The case $\lambda = 1$, of course, is dealt with in Chapter 6.)

During the course of preparing the section on the Champernowne distribution, the authors—to their delight—were able to discover in the widely scattered literature a number of somewhat unexpected sources in which the distribution (7.16) is discussed, albeit under different names. (We are not sure whether Champernowne—who continued to develop and modify his original distributions for some 35 years—was aware of these new forms and descriptions of his model.)

This fact (similar to the situation with the Benini distribution; see Section 7.1) is another indication of the wide diversity of sources for derivations and uses of statistical distributions, especially in the last 50 years. [The Champernowne distribution is possibly not the mainstream distribution utilized in standard income studies; it was revisited after some 30 years in a paper by Campano (1987), who found it to be appropriate to more recent income data.]

Our observations pertain to the appearance of the distribution in the applied and theoretical probabilistic literature and to its strong relation to stable distributions of various kinds, which are of increasing importance in financial applications.

1. The three-parameter Champernowne distribution (7.18) with $\alpha = 1$ is nothing but a truncated Cauchy distribution, that is, the distribution with the density

$$f(x) = \frac{1}{\pi\sigma[1 + \{(x - \vartheta)/\sigma\}^2]}, \qquad -\infty < x < \infty, \tag{7.24}$$

truncated from below at zero [for the Champernowne distribution (7.18), $\vartheta = -\cos\theta$ and $\sigma = \sin\theta$]. This fact was overlooked even in the comprehensive two-volume compendium *Continuous Univariate Distributions* by Johnson, Kotz, and Balakrishnan (1994, 1995), who discussed these distributions in different volumes without connecting them.

2. More unexpectedly, another special case of the Champernowne distribution, namely, the case where $\alpha \in (0, 2]$ and $\theta = \pi\alpha/2$, is discussed in Section 3.3 of the classical book on stable distributions by Zolotarev (1986). There it occurs in connection with the distribution of functions (ratios) of independent strictly stable variables with the same characteristic exponent α. Zolotarev also provided an alternative integral representation of this special case of the Champernowne c.d.f.

$$F_\alpha(x) = \frac{1}{2} + \frac{1}{\alpha\pi^2}\int_0^\infty \log\left(\frac{1 + |x + y|^\alpha}{1 + |x - y|^\alpha}\right)\frac{dy}{y}. \tag{7.25}$$

He further pointed out that $F_\alpha(\cdot)$ is a distribution function even for $\alpha > 2$ and moreover, for any integer $n \geq 1$, $F_{2n}(\cdot)$ is a mixture of Cauchy distributions with a linearly transformed argument.

3. The Champernowne distribution is also associated with the so-called Linnik (1953) distribution (a generalization of the Laplace distribution) that possesses a simple characteristic function

$$\frac{1}{1 + \sigma^\alpha |t|^\alpha}, \qquad \alpha \in (0, 2], \tag{7.26}$$

but does not have a simple expression for the density and c.d.f. Specifically, it appears in a mixture representation of the Linnik distribution discussed by Kozubowski (1998).

These interesting theoretical properties in the spirit of modern developments in the theory of statistical distributions could lead to a better understanding of the hidden structure of the Champernowne distribution and perhaps provide an avenue for further generalizations and discoveries. Needless to say, the original motivation and model leading to this distribution are quite removed from its genesis related to the stable and Linnik laws. See Kleiber (2003b) for further details and applications of the above interrelations.

In the case where $|\lambda| < 1$, the moments $E(X^k)$ of (7.17) exist only for $-\alpha < k < \alpha$. They are [Champernowne (1952), for $k = 1$, and Fisk (1961a), for a general k]

$$E(X^k) = x_0^k \frac{\pi}{\theta} \frac{\sin k\theta/\alpha}{\sin k\pi/\alpha}. \tag{7.27}$$

The coefficient of variation is therefore given by

$$\mathrm{CV} = \sqrt{\frac{\theta}{\pi} \cot\left(\frac{\theta}{\alpha}\right) \cot\left(\frac{\pi}{\alpha}\right) - 1}. \tag{7.28}$$

Harrison (1974) presented the Atkinson measures of inequality for the distribution (7.16) that are given by

$$A_\epsilon = 1 - \left(\frac{\pi}{\theta}\right)^{\epsilon/(1-\epsilon)} \left\{ \frac{\sin[(1-\epsilon)\theta/\alpha]}{\sin[(1-\epsilon)\pi/\alpha]} \right\}^{1/(1-\epsilon)} \frac{\sin(\pi/\alpha)}{\sin(\theta/\alpha)}. \tag{7.29}$$

Here ϵ is the inequality aversion parameter. In the case where $\epsilon = 1$, the last formula is not applicable. Instead, one obtains the simple expression

$$A_1 = 1 - \frac{x_0}{E(X)} = 1 - \frac{\sin(\pi/\alpha)}{\sin(\theta/\alpha)}, \tag{7.30}$$

an inequality measure that was previously proposed by Champernowne (1952). Harrison (1974) showed graphically that the measure A_ϵ is, for a fixed ϵ, monotonically decreasing in θ, for $0 < \theta < \pi$, approaching complete equality of incomes for $\theta \to \pi$, that is, for a point mass concentrated at x_0.

All three forms of the Champernowne distribution discussed above possess the property of symmetry for the distribution of the income power. There is, however, an asymmetric (on a log-scale) four-parameter generalization of the case where $|\lambda| < 1$, given by the c.d.f. (Champernowne, 1952)

$$F(x) = \begin{cases} 1 - \dfrac{1}{(1+\sigma)\theta}\left[(\sigma-1)\theta + 2\arctan\left\{\dfrac{\sin\theta}{\cos\theta + (x/x_0)^{\sigma\alpha}}\right\}\right], & \text{for } 0 \le x \le x_0, \\ 1 - \dfrac{2\sigma}{(1+\sigma)\theta}\arctan\left\{\dfrac{\sin\theta}{\cos\theta + (x/x_0)^{\alpha}}\right\}, & \text{for } x_0 < x, \end{cases} \tag{7.31}$$

where the new parameter σ may be viewed as a skewness parameter, since when $\sigma > 1$, the curve exhibits positive skewness, and for $\sigma < 1$, the curve is negatively skewed. The case where $\sigma = 1$ yields the distribution (7.18). The p.d.f. of (7.31) is given by

$$f(x) = \frac{2\sigma\alpha \sin\theta (x/x_0)^{\psi}}{x(1+\sigma)\theta[1 + 2\cos\theta(x/x_0)^{\psi} + (x/x_0)^{\alpha\psi}]}, \tag{7.32}$$

where $\psi = \sigma\alpha$ for $x \le x_0$ and $\psi = \alpha$ for $x > x_0$. [We note that (7.31) is sometimes referred to as "Champernowne's five-parameter formula" (see, e.g., Campano, 1987), although (7.31) defines a four-parameter distribution. This is presumably due to Champernowne's unusual notation (7.12), where n is a function of the remaining three parameters y_0, α, and λ.]

As far as the estimation is concerned, only methods for grouped data have been discussed in the literature. Rather than employing the computationally complex maximum likelihood method—note the precomputer era of his work!—Champernowne (1952) attributed special importance to methods which yield "solutions that agree with the observed distribution in such economically important matters as the total number of incomes, the average income and the position and slope of the Pareto line for high incomes" (pp. 597–598). Most of the methods he considered start with the total number of incomes and some form of average income (mean, median, mode), proceed with an estimation of Pareto's alpha by a regression technique, and finally obtain θ via interpolation.

He fit the three-parameter distribution to 1929 U.S. family incomes, 1938 Japanese incomes (previously also considered by Hayakawa, 1951), 1930 Norwegian incomes, and 1938–1939 UK incomes, and the four-parameter model to 1918 (previously considered by Davis, 1941a,b) and 1947 U.S. incomes. The fit of the 1918 data is comparable to that provided by Davis's model (see Section 7.2). In addition, it emerges that for 1933 Bohemian incomes the two-parameter Champernowne (i.e., Fisk) distribution provides as good a fit as several three-parameter models.

Harrison (1974) suggested a minimum-distance estimator, determining parameters simultaneously using an iterative generalized least-squares approach. Analyzing British data for the years 1954–1955, 1959–1960 and 1969–1970, he contrasted it with Champernowne's methods and found his estimator to be more reliable.

When fitted to the distribution of 1969 fiscal incomes in three regions of the Netherlands, it outperforms gamma, generalized gamma Weibull, and log-logistic distributions (Bartels, 1977). A Box–Cox-transformed Champernowne distribution does even better on these data.

Kloek and van Dijk (1977) analyzed the distribution of Australian family incomes during 1966–1968. Using minimum χ^2 as well as maximum likelihood estimates, they noted that, compared to the lognormal, gamma, log Pearson type IV, and log-t distributions, the Champernowne function "...fits rather well, but...the parameters cannot easily be interpreted" (p. 446). Employing *Cox tests* (Cox, 1961), they found that "...neither the log-t [when] compared with the Champernowne nor the log Pearson IV compared with the Champernowne can be rejected, and vice versa." In other words, the distributions are statistically on the same footing.

Kloek and van Dijk (1978) utilized the four-parameter Champernowne distribution when analyzing Dutch earnings data for 1973. Estimating parameters by the minimum χ^2 method, they found that the distribution outperforms the two-parameter lognormal and two-parameter gamma distributions and is comparable to the three-parameter generalized gamma and the three-parameter log-t distributions.

Fattorini and Lemmi (1979) fit the three-parameter Champernowne distribution to Italian, U.S., and Swedish data for the years 1967–1976, 1954–1957, 1960–1962, and 1965–1966, respectively. They found that it almost always outperforms the lognormal distribution, but their kappa 3 (i.e., Dagum type I) distribution is even better, in terms of the sum of squared errors and the Kolmogorov distance. Interestingly enough, their estimates of λ turn out to be numerically greater than 1 for Italian and Swedish data, whereas Champernowne considered this case to be rather exceptional.

More recently, Campano (1987), a statistician with the United Nations in Geneva, reconsidered Champernowne's own results for the four-parameter version (7.32) for U.S. income data of 1947. He also compared Champernowne and Dagum type I fittings for 1969 U.S. family incomes (previously investigated by Dagum, 1977) and found the Champernowne distribution to fit slightly better. (However, this version of the Champernowne distribution has one additional parameter compared to the Dagum type I.)

7.4 BENKTANDER DISTRIBUTIONS

Starting from the observation that empirical mean excess functions point to distributions that are intermediate between the Pareto and exponential distributions, the Swedish actuary Gunnar Benktander (1970) discussed two new loss models. Whereas for the exponential distribution the mean excess function is given by

$$e(x) = \lambda, \quad x \geq 0, \tag{7.33}$$

λ being the exponential scale parameter, in the Pareto case we have

$$e(x) = \frac{x}{\alpha - 1}, \quad x \geq x_0. \tag{7.34}$$

However, empirically one observes mean excess functions that are increasing but at a decreasing rate. Two intermediate versions are therefore defined by

$$e_I(x) = \frac{x}{a + 2b\log x}, \quad a > 0, b \geq 0, \tag{7.35}$$

and

$$e_{II}(x) = \frac{x^{1-b}}{a}, \quad a > 0, 0 < b \leq 1. \tag{7.36}$$

It follows from Watson and Wells (1961) that the first form is of the lognormal type, whereas the second asymptotically resembles the mean excess function of a Weibull distribution (Beirlant and Teugels, 1992).

Using the relation between the mean excess function and the c.d.f.,

$$F(x) = 1 - \frac{e(x_0)}{e(x)} \exp\left\{-\int_{x_0}^{x} \frac{1}{e(t)}\, dt\right\}, \quad x \geq x_0,$$

we obtain after some calculations

$$F_I(x) = 1 - x^{-(1+a+b\log x)}\left\{1 + \frac{2b}{a}\log x\right\}, \quad 1 \leq x, \tag{7.37}$$

where $a > 0$, $b \geq 0$, and

$$F_{II}(x) = 1 - x^{-(1-b)} \exp\left\{\frac{a}{b} - \frac{ax^b}{b}\right\}, \quad 1 \leq x, \tag{7.38}$$

where $a > 0$, $0 < b \leq 1$. Equation (7.37) defines the Benktander type I and (7.38) defines the Benktander type II distribution. We note that although the distributions are commonly known under these names (see Beirlant, Teugels, and Vynckier, 1996; Embrechts, Klüppelberg, and Mikosch, 1997), they seem to have appeared for the first time in Benktander and Segerdahl (1960).

The densities are given by

$$f_I(x) = x^{-2-a-b\log x}\left\{\left(1 + \frac{2b\log x}{a}\right)(1 + a + 2b\log x) - \frac{2b}{a}\right\}, \quad 1 \leq x, \tag{7.39}$$

and

$$f_{II}(x) = \exp\left\{\frac{a}{b} - \frac{ax^b}{b}\right\} x^{-2+b}\{(1-b) + ax^b\}, \quad 1 \le x, \tag{7.40}$$

respectively.

Using the relation between the mean excess function and the hazard rate,

$$r(x) = \frac{1 + e'(x)}{e(x)},$$

the hazard rates are easily obtained as

$$r_I(x) = \frac{a + 1 + 2b\log x}{x} - \frac{2b}{x(a + 2b\log x)}, \quad 1 \le x, \tag{7.41}$$

and

$$r_{II}(x) = \frac{a}{x^{1-b}} + \frac{1-b}{x}, \quad 1 \le x, \tag{7.42}$$

from which it is again obvious that the Pareto distribution is a limiting case (for $b = 0$ in both cases).

It follows directly from the expressions for the mean excess functions that the means of both Benktander distributions are equal to

$$E(X) = 1 + \frac{1}{a}$$

and are independent of b.

Klüppelberg (1988) has shown that both distributions, as well as the associated integrated tail distributions

$$F_{\text{int}}(x) = \frac{1}{E(X)} \int_0^x \bar{F}(y)\, dy, \quad x \ge 0, \tag{7.43}$$

belong to the class of subexponential distributions, that is, they satisfy the condition

$$\lim_{x \to \infty} \frac{\bar{F}^{*2}(x)}{\bar{F}(x)} = 2,$$

where F^{*2} stands for the convolution product $F * F$. This property is of importance in risk theory (and other fields) since it allows for a convenient treatment of problems related to convolutions of heavy-tailed distributions. It is interesting that the Benktander I survival function is proportional to the Benini density (see

Section 7.1), that is, the distributions are related as in (7.43) with F_{int} being the Benini c.d.f. Klüppelberg's results therefore imply that the Benini distribution is also subexponential.

To the best of our knowledge, the classical methods of estimation have not been used in connection with these two models. Since the Benktander distributions are close to the lognormal and Weibull distributions, respectively, the authors are inclined to consider the more tractable lognormal and Weibull distributions as preferable in applications, at least if the mean excess function itself is not of central importance.

APPENDIX A

Biographies

The theory of economic size distributions is closely associated with two quite different personalities whose contributions had substantial impact.

These are the Swiss-Italian Vilfredo Pareto (1848–1923), a brilliant scholar but somewhat eccentric person, and Max Otto Lorenz (1876–1959), a very stable American statistician and economist whose paper in 1905 generated a flood of investigations, mostly after his death.

Although Pareto's name is currently widely associated with concepts to which he contributed only indirectly (e.g., the so-called Pareto chart and the Pareto set of designs in engineering statistics), his contributions to income distributions are fundamental. Pareto had two distinguished careers in his lifetime: economics up to about 1907 and sociology until his death in 1923. His initially very liberal political views rapidly changed and he eventually became sympathetic to the cause of fascism—albeit in its very early stages—a fact that is often obscured by his biographers.

For most of his career, Max Otto Lorenz served as a bureaucrat in Washington, D.C., mainly at the Interstate Commerce Commission for some 35 years. He is well known professionally for a single innovative paper that was only indirectly related to his Ph.D. thesis in Economics at the University of Wisconsin.

Other personalities whose biographies are presented in this chapter include Rodolfo Benini, Corrado Gini, Luigi Amoroso, Raffaele D'Addario, Robert Gibrat, and David Champernowne. Pareto's biography is by far the longest due to an abundance of biographical material, the references cited here represent just a small portion of them. Note the predominance of Italian scholars. For cultural and historical reasons, it seems that there was greater interest in the quantitative analysis of income

Statistical Size Distributions in Economics and Actuarial Sciences, By Christian Kleiber and Samuel Kotz.
ISBN 0-471-15064-9

distributions in Italy in the twentieth century than in Germany, France, and Great Britain. A possible impetus is the influence of Pareto and Gini.

A.1 VILFREDO FEDERICO DOMASO PARETO, MARCHESE DI PARIGI

Born: July 15, 1848, Paris, France.

Died: August 19, 1923, Céligny (near Geneva), Switzerland.

Vilfredo Pareto was the only son of Genoese nobleman Raffaele Pareto—a follower of the Italian revolutionary Mazzini, exiled in 1836 from his native city because of his political views—and French calvinist Marie Méténier, a woman of "modest social origins." He had two sisters Aurelia and Cristina. The title of Marquis was bestowed on Pareto's great-great-great-grandfather in 1729 (according to other sources, on his grandfather) and, after his father's death in 1882, it passed to Pareto himself. He never used the title, insisting that since it was not earned, it held little meaning for him.

It should be noted at the outset that Pareto's life and work are surrounded by legends and shrouded in contradictions, and there is even conflicting information about his Christian names. In about a dozen biographies of Pareto, his Christian name is invariably given as Vilfredo Federico Domaso. It was therefore with great interest that we were able to discover based on a clue provided in Eisermann (1987) that his dissertation submitted to the School of Engineering Applications in Turin in 1869 (entitled *Principj fondamentali della teoria della elasticità de'corpi solidi e richerche sulla integrazione delle equazioni differenziali che ne definiscono l'equilibrio*) explicitly states his name on the front page as Fritz Wilerid, whereas the signature on the last page of the paper (where Pareto thanks his professors Genocchi and Rosellini for developing his love of mathematics) is correctly given as Fritz Wilfrid Pareto. The thesis is reprinted on pp. 591–639 of his *Scritti teorici* (Milan, 1952). In view of the fluidity of his father's situation at the time of Pareto's birth, this official German name adds an additional level of mystery to Pareto's background. Turin was in 1869 an exclusively Italian city with no trace of a German or Swiss connection (if anything, it had remained under French rule during an earlier period). Although we were unable to examine Pareto's birth records, the name as given in the thesis should be viewed as Pareto's original name, which was apparently "Italianized" after the family returned to Italy. It remains a mystery as to why Pareto's parents may have chosen a German name for their son.

It was only in 1858 after 20 years in exile and 10 years after Vilfredo Pareto's birth that political amnesty enabled the family to return to Italy. Because his father earned a reasonably comfortable living as a successful hydrological engineer, Pareto was reared in an upper-middle-class environment, enjoying the many advantages that accrued to people of his class during that age. He received a quality education in both France and Italy. His secondary studies were mainly mathematical and classical, but his superior intellect transcended the boundaries of applied sciences into the realm of pure general concepts, realizing that all exact sciences are fundamentally one.

After the family's return to Italy, Raffaele Pareto briefly moved in 1859 to Casale Monferrato, where he taught various subjects at the Leardi Technical School. Its principal at that time was the mathematician F. Pio Rosellini, whose course in mathematical physics Vilfredo Pareto later attended. At the same time, young Pareto studied Italian (not spoken at home) and also Latin and Greek. Studying the classics and Greek, he cultivated a knowledge of literature and history. In late 1862 the family moved to Florence (the capital of Italy at that time), and in 1864 at age 16, Vilfredo Pareto obtained his high-school diploma and that same year he passed the entrance exams to the University of Turin. In his initial studies at Turin, Pareto concentrated on courses in mathematical studies, and in 1866 he published his first paper on the application of asymmetric designs. He received his degree in mathematics and physical sciences in 1867 and immediately entered the engineering school at the University of Turin (the school was established in 1859), from which he graduated in January 1870 after writing a thesis entitled *Theory of Elasticities of Solid Bodies* (1869). Pareto then started work as an engineer at the Joint Stock Railway Company of Florence (Val d'Arno Lines) but soon realized the limited career prospects associated with such a profession. He left this job in 1873, accepting the post of Deputy Head of the Iron Works Company at San Giovanni in Valdarno.

In 1872 he met the mayor of Florence, Ubaldino Peruzzi, and began to attend the literary salon held by Peruzzi's wife, where he became acquainted with politicians, artists, and writers. Pareto started writing articles on commerce, the state of Italian industry, and railway problems. In 1874 he was nominated as a member of the Accademia di Georgofili of Florence and continued his writings in support of free trade, advocating against state intervention in economic activities. He fought against protectionism, custom duties, and state subsidies to industry. He was one of the founders of the Adam Smith Society that upheld the doctrine of economic liberalism.

In May 1875 Pareto became the technical director of the Iron Works company, and in 1877 he was elected to the Town Council of San Giovanni, Valdarno. During 1876–1880 Pareto traveled extensively in Belgium, France, Switzerland, England, and Germany. His famous comment "we are given to believe that the English are hard working people—it is but an illusion" may serve as an indication of his anti-British feelings, which later became quite clear in his fiery confrontation with F. Y. Edgeworth and K. Pearson.

In 1880 he was appointed the director-general of the Iron Works Company, whose name he changed to Ferriere Italiane (Italian Ironworks Company), and in 1883 he was awarded the knighthood of the order of the Crown of Italy. He continued to publish fiercely liberal free-trade articles and his relationship with the government branches public administration deteriorated as a result. He became increasingly isolated.

Pareto married in December 1889. His first spouse Alessandrina (Dina) Bakunin, of Russian descent, apparently loved an active social life, which stood in direct conflict with Pareto's own preference of privacy and solitude. Later after 12 years of marriage, Dina abandoned her husband, running away with a young servant (a socialist according to another version).

In the meantime, the financial situation of Ferriere Italiane became extremely precarious, and in July 1890 the company's board of directors requested Pareto's resignation. He left the company and was able (with his bride) to retire to Villa Rosa in Frisolc and live comfortably on consulting fees.

A chance meeting—on a train according to some sources—with a famous and influential Italian economist Maffeo Panteleoni in 1890 had a significant impact on Pareto's future career. They became good friends, and Panteleoni advised Pareto to study the works of the political and mathematical economist Leon Walras, whom he met in September 1890 in Switzerland. Although he had at first been rather indifferent to Walras's work, on rereading it, Pareto was much impressed by the theory of general economic equilibrium. Walras was considering resigning his professorship at Lausanne University and was pleased to find somebody capable of understanding the scope and importance of his work, he suggested to Pareto that he might become his successor. In 1893 Walras did retire, at the age of 58 due to ill health, and Pareto, who was then 45, succeeded him. During the final decades of the nineteenth century Pareto acquired a deep interest in the political life of his country and expressed his views on a variety of topics in public lectures and in articles for various journals, and in direct political activity. Steadfast in his support of the economic theory of free enterprise and free trade, he never ceased arguing that these concepts were vital necessities for the development of Italy. Vociferous and polemical in his advocacy of these ideas, and sharp in his denunciation of his opponents (who happened to be in power in Italy at that time), Pareto's public lectures were sufficiently controversial that they were sometimes raided and closed down by the police, and occasionally brought threats of violence from hired thugs.

Pareto's professorial appointment at the University of Lausanne marked the beginning of his scientific career, during the course of which he produced books of remarkable quality: the *Cours d'économie politique* (1896–1897); the *Systèmes socialistes* (1902–1903); the *Manuale di economia politica*, which appeared in 1906 and was published in French, with various improvements, as *Le manuel d'économie politique* in 1909; and the *Trattato di sociologica generale*, which appeared in Italian in 1916 and in French in 1917–1919. His original appointment was as an associate professor in April 1893; exactly one year later he became a full professor. In 1896 he was elected a senior member of the law faculty at the University of Lausanne.

In October 1891 the Parisian *Revue des deux mondes* published Pareto's article "L'Italie économique," which became renowned for its condemnation of the Italian government's economic policies. The publication unleashed a storm of protest and controversy in Italy and abroad. Pareto, however, carried on undaunted, producing articles and giving lectures attacking Italian economic policy. In 1893—the year of his appointment at the University of Lausanne—he wrote a refutation of Marx' theory of value that was published as the introduction to an anthology of passages from Karl Marx's *Das Kapital*. Although widespread dissatisfaction was expressed particularly by Marx's "comrade in arms" Friedrich Engels, Pareto's activism did not harm his relations with socialists and he continued to write for left-wing publications and adopt radical ideas.

Pareto's two main contributions to economics, which also contain his developments in income distributions, are *Cours d'économie politique* (1896–1897) written during his tenure at the University of Lausanne and *Manuale di economia politica* (1906) written after his semiretirement in 1900. His massive contribution to sociology was the four-volume *The Mind and Society: A Treatise on General Sociology* (1916) that he completed during the last decade of his active life. Both the *Cours* and the first two chapters of the *Manual* contain a remarkable discussion of the methodology of economics, which shows Pareto to have been widely read in the literature on the philosophy of science, perhaps better than any other economist of the period. The uneven and poorly organized *Cours* contained, among its many historical and statistical illustrations, the so-called *Pareto's law* of income distribution which states in a nutshell that the slope of the line connecting the percentage of income received above a certain level to the percentage of relevant income-receivers is a constant, thus demonstrating that the distribution of income in all countries and in all ages conforms to an invariant pattern.

The *Manual* is famous for at least three ideas: the (unsuccessful) attempts to banish the term *utility* and replace it with *ophelimité*, a word coined by Pareto to denote the power of goods to satisfy demands; the clear distinction between cardinal and ordinal utility, and the demonstration, via Edgeworth's indifference curves, that ordinal utility or the mere ranking of preferences is sufficient to deduce all the important propositions of demand theory; and the apparently innocent, noncontroversial definition of an economic optimum as that configuration of prices which commands unanimous approval—any other configuration might make some individuals better off but only by making at least one person worse off—the far-reaching concept of *Pareto optimality*, coupled with the (not altogether successful) attempt to show that a perfectly competitive economy in fact achieves a Pareto optimum, and vice versa.

Along with Walras and the American Irving Fisher of Yale University, Pareto may be regarded as one of the three founders of modern economic science. The three, although different in their personalities and backgrounds, have much in common. Walras and Fisher, like the early Pareto, were ardent champions of normative ideas, and in all three the scientist struggled with the crusader. They were either ignored or hated, especially by their compatriots, but the reputations of all three have been steadily rising ever since.

As to Pareto's law and curve, which is the topic of main interest to the readers of this book, it should be stressed that he developed here a fundamental yardstick.

Unfortunately, Pareto failed to make a proper distinction between the law in the rational sense and the concept of an empirical law although he was conscious that his law was an empirical one. Looking back on Pareto's discovery from the vantage point of a hundred years, we may think it being somewhat odd that he was so determined to contrast his law of the distribution of income with "the law of probability," by which, in effect, he meant a model of the lottery type. As pointed out by the leading British economist and statistician F. Y. Edgeworth as early as 1896 [the year in which the initial contribution of Pareto, *La courbe de la répartition de la richesse* (Lausanne, 1896), on this topic was published],

> The reference to asymmetrical statistical curves leads me on to supplement my remarks on this subject in this *Journal* [of the Royal Statistical Society] for September, 1895, by calling attention to a remarkable formula for the frequency of incomes which has been lately given by the eminent mathematical economist Professor Pareto in his *La courbe de la répartition de la richesse* (Lausanne 1896). Designating each amount of incomes as x, and the number of incomes equal or superior to x as N he finds the following simple relation between the logarithms of these quantities
>
> $$\log N = \log A - \alpha \log x;$$
>
> where α is a constant which proves to be much the same for different countries, whence
>
> $$N = \frac{A}{x^\alpha}.$$
>
> This law is of considerable economic interest. The approximate identity of the law as ascertained for different countries, points to the dependence of the distribution of income upon constant causes not to be easily set aside by hasty reformers. (pp. 529–539)

Edgeworth compared Pearson's formula that is of the form

$$N = \frac{A}{(x+a)^\alpha} 10^{-\beta x},$$

α and β being constant, with Pareto's formula cited above by emphasizing that N in Pearson's formula denotes only the valuations at a certain x. He, however, pointed out that the difference between the two curves occurs in form rather than in essence. He then added the following rather puzzling comment:

> I do not forget that there are certain theoretical arguments in favour of the Pearsonian formula, and I have allowed a certain weight to them in the critical paper above referred to. I have allowed great weight to the authority of Professor Karl Pearson. Where opinions on a matter of this sort differ, the presumption is certainly in favour of the author who has made the greatest advance in the science of Probabilities which has been made since the era of Poisson.

This comment justifiably infuriated Pareto, who responded to it in a letter to his friend and mentor Pantaleoni, accusing Edgeworth of saying "all he could against my work; nevertheless since he did not understand it, he directed against me a superficial criticism, highly dishonest." He also blasted Edgeworth in his two papers in *Giornale degli economisti*. Edgeworth's response in his 1899 paper "On the representation of statistics by mathematical formulae", also reproduced in his article on Pareto's law published in *Palgrave* (second edition) 1926, was as follows:

> An article in the *Journal of the (Royal) Statistical Society* (1896) in which the discontinuity of incomes at the lower extremity of the curve was pointed out called forth from Pareto a somewhat acrimonious explanation ... which is of interest as throwing light not only on the character of the curve but also on that of its discoverer.

Nevertheless in the same *Palgrave* entry, Edgeworth presented a rather positive (albeit somewhat restrained) assessment of Pareto's contributions to economics. (It should be mentioned—to emphasize the magnitude of the animosity—that in another letter to Pantaleoni, Pareto referred to Edgeworth as a "real Jesuit [who] could only show how to solve the equations of exchange. These are the only ones he knows.")

The interpretation and significance of the change in the value of the constant α in Pareto's formula

$$\log N = \log A - \alpha \log x$$

(which measures the slope of the income curve) was a source of another minor controversy. In his original publication Pareto stated that as α increases, so does the inequality of incomes. Arthur Bowley (1869–1957), the well-known and influential British statistician and economist, in his early writings adhered to Pareto's interpretation, but later on he declared that Pareto was wrong. Benini (see the next biographical sketch) and some other Italian statisticians did not accept Pareto's interpretation either. There were some who characterized Pareto's assertion as a "curious slip" or even as a printer's error. Bresciani Turroni—an eminent economist—in a 1939 paper published in the Egyptian journal *Revue Al Qanoun Wal Iqtisad* asserted that Pareto's statement is a logical consequence of his concept of inequality. According to it, incomes tend toward inequality when "poor become rich," ignoring the fact that the increase in the proportion of higher incomes is not an indication of less inequality of individual incomes.

We are not going to discuss Pareto's contributions to sociology which he began writing in 1912 in detail. Rather, we will refer the interested reader, among other sources, to the slim volume by F. Borkenau in the Modern Sociologists series entitled *Pareto* (1936) for this purpose. It would seem that Pareto's most important sociological writings dealt with his theory of the "circulation of elites." He wrote on the sociology of the political process, in which history consists essentially of a succession of elites, whereby those with superior ability in the prevailing lower strata at any time challenge, and eventually overcome, the existing elite in the topmost stratum and replace them as the ruling minority. In Pareto's view, this pattern is repeated over and over again.

In 1899 on the death of an uncle (his father's brother, an ambassador to the Ottoman Empire), Pareto inherited a substantial fortune that allowed him to resign his professorship in Lausanne (which included teaching duties) and to move to the town of Céligny, in the canton of Geneva, Switzerland. From 1900 until his death in 1923, he led a rather reclusive life, wholly devoted to his studies and writings. Pareto remained at his country villa, seldom venturing into nearby cities and entertaining only a few close friends, thus becoming "the lone thinker of Céligny."

However, Pareto did not completely sever his connections with the University of Lausanne and—according to a personal communication from Professor F. Mornati there—from 1900 until June 4, 1909 (the day of his last lecture in economics) he was still teaching but less and less. He was instrumental in reforming the teaching of social studies at the university in 1910. According to some sources, Pareto officially retired in 1911, but in May 1916 he gave a final series of lectures on sociology.

As was earlier mentioned, Pareto suffered a personal misfortune in 1901 when his wife left him; luckily in February 1902 (when he was 53 years old) he met a 22-year-old French woman Jeanne Régis, who became his devoted companion until his death in 1923. The first edition of his *Trattato di sociologica* (1916) was dedicated to "Signora Jane Regis." He was finally able to obtain a formal divorce from his first wife in 1922 (after becoming a citizen of the Free State of Fiume), and he married Jeanne two months before his death in June 1923. He was afflicted with a heart ailment toward the end of his life and his last years were marked by considerable ill health. Throughout his life, Pareto also suffered from insomnia, which was perhaps one reason for his outstanding productivity.

Borkenau (1936) asserted that there are certain analogies between Pareto's sociology and Hitler's *Mein Kampf.* He pointed out the similarity between Pareto's theory of elites and Hitler's theory of race. He also quoted two of Pareto's statements that resonate in Hitler's work and his convictions. The first deals with the principles of mass propaganda:

> Speaking briefly but without scientific exactness, ideas must be transformed into passions in order to influence society, or in other words, the derivations must be transformed into residues.

The second concerns with the dictum of repeating your statements over and over again, not necessarily to prove them:

> Repetition works principally upon the feelings, proofs upon reason, and then, at best, modify the derivations, but have little influence upon feelings.

Pareto sometimes had this tendency to generalize a certain single aspect prevalent in his own time and to apply it to the theory of mankind—thus substantially reducing its scientific value.

During the period 1900–1919 in his "semiretirement," Pareto wrote his most important works, which present a continuous discussion of the problems of liberalism. In his *Cours d'économie politique* (1896–1897) and *Manuel d'economie politique* (Italian version 1906, French 1909), he tried to prove mathematically the absolute superiority of free trade over any other economic system. On the other hand, he later developed the reasons why this solution "does not obtain in practice as a rule," which led him to his study of sociology. He turned to sociology rather late in life, but his previously cited *Treatise on General Sociology* and two smaller volumes, *The Rise and Fall of the Elites* and (in 1921) *The Transformation of Democracies*, are acclaimed masterpieces in the field. In the latter work he described the dissolution of the Italian democratic state. This work was written only a few months before the final victory of fascism in Italy in October 1922. In 1898 witnessing the police repression that followed the May riots in Milan, Pareto probably made a conscious decision to reject liberalism at that time. As indicated above he later gravitated toward fascism, but somewhat hesitatingly. Possibly, Pareto's intellectual detachment made it quite impossible for him to be an ardent adherent of any political movement.

In the first years of Mussolini's rule, the policies executed were along the lines advocated by Pareto, such as replacing state management by private enterprise, diminishing taxes on property, and favoring industrial development. When political events compelled King Victor Emmanuel III to appoint Mussolini as the Prime Minister of Italy in October 1922, Pareto was able to rise from his sick bed and utter a triumphant, "I told you so." Yet unlike his friend and mentor Pantaleoni, he never joined the Fascist Party. In his autobiography Mussolini provided a few comments on Pareto's lectures in Lausanne: "I looked forward to every one ... for here was a teacher who was outlining the fundamental economic philosophy of the future." He even attended a course on applied economics at Lausanne during the first term of the academic year 1903–1904 that was officially taught by Pareto (but mostly delivered by Pareto's successor Boninsegni). The Mussolini government extended several honors on Pareto. He was designated as a delegate to the Disarmament Conference in Geneva (but excused himself on account of poor health); he was nominated as a senator of the Italian Kingdom in 1923 but (according to G. Busino) refused to submit the required papers for ratification. Other sources (Cirillo, 1979) claim that he accepted the nomination. Pareto also contributed to the Duce's personal periodical *Gerarchia*. Some of Pareto's biographers claim that after a lifetime of disappointments with politics, he welcomed fascism as "a sort of messianic revolution," and in turn the new fascist government regraded with sympathy the man whose sociological theories seemed to favor their doctrine.

One thing is certain: In the last years of his life, Pareto was a bitter man. He was disappointed by the treatment he received from his "native" country of Italy when he left in 1893 for the professorship at Lausanne. He was asked to teach a course in sociology at the University of Bologna for a few weeks in 1906, but much later, when an offer of professorship finally came, he was already too ill to accept it.

In the six years preceding his death, Pareto's main joy were his Angora cats, said to number at least 18 at one time. They were given precedence over anyone else, and were often fed before his (very occasional) visitors. Near the end of his life, Pareto developed an affinity for birds and squirrels. His home in Céligny was called Villa Angora and boasted the best wine cellar in Switzerland. He alienated many of his friends by his irony and sarcasm. It was in Pareto's nature to find faults with nearly all political regimes and many human beings.

All his life, Pareto was confident in his own intellect and could be stubborn in his views, disdainful of those with divergent opinions and often careless of other people's feelings. On the other hand, in his early years especially up to and including 1898, he was generous to "underdogs" and offered money, shelter, and counsel to political exiles and strongly protested against antisemitic incidents in Algeria. He was (like his father) conservative in his personal tastes and inclinations, but in some respects—because he was a free thinker—he combined this trait with an ardent belief (at least in his younger years) in personal liberty. In one of his last essays written shortly before his death, he declared "... most people are incapable of having a will, not to mention a will which has a positive effect on the nation, whose fate always seems to be in the hands of a minority which imposes itself with faith helped by force, and often by means of the dictatorship of a group, an aristocracy, a party, a few men, or one man."

Pareto always shunned scientific honors so common among Italian scientists: He became a member of only one academy—the Academy of Science of Turin—the city where he received his academic education. His physical appearance in later years reminded one of Michelangelo's "Moses"; he felt himself to be profoundly Italian and considered the fact that he was born in France an "unfortunate circumstance."

The almost unprecedented scholarly activity surrounding Pareto's work including Italian, Swiss, British, and American scholars who dedicated their lives to the clarification and popularization of his ideas serve as testimony to his great talent and originality and his impact on economics and sociology. Pareto's influence, particularly as propagated by his student Enrico Barone (1858–1924), on the famous interwar controversy on the economic merit of a socialist (collectivist) economy and the very feasibility of central planning was also substantial. As was already mentioned in connection with the Pareto distribution in Chapter 3, there exists a Walras–Pareto Web site maintained by the University of Lausanne. There is also a French journal entitled *Cahiers Vilfredo Pareto/Revue européenne d'histoire des sciences sociales* (now entitled *Revue européenne des sciences sociales et Cahiers Vilfredo Pareto*), founded in 1963.

His widow, who died in 1948, had a rough time after Pareto's death. In 1936 she had to surrender the copyright to her husband's work in exchange for a pension, which was slow in coming and forced her to sell Villa Angora. She was not entitled to a Swiss state pension or a widow's pension from the University of Lausanne.

REFERENCES

Allais, M. (1968). Vilfredo Pareto: Contributions to economics. In: D. D. Sills (ed.): *International Encyclopedia of the Social Sciences*, New York: Macmillan, Vol. 11, pp. 399–411.

Biaudet, J.-C. (1975). *Vilfredo Pareto et Lausanne.* Convegno Internazionale Vilfredo Pareto, Rome: Accademia Nazionale dei Lincei, pp. 75–108.

Blaug, M. (ed.) (1992). *Vilfredo Pareto (1848–1923).* Aldershot: Edward Elgar. (Contains 23 papers on Pareto's works and philosophy.)

Borkenau, F. (1936). *Pareto*. New York: John Wiley.

Bousquet, G. H. (1928). *Vilfredo Pareto, sa vie et son oeuvre*. Paris: Payot.

Bousquet, G. H. (1960). *Pareto (1848–1923), le savant et l'homme*. Lausanne: Payot.

Busino, G. (1977). *Vilfredo Pareto e l'Industria del Ferro nel Valdarno*. Banca Commerciale Italiana Milano.

Busino, G. (1987). Pareto Vilfredo. In: *The New Palgrave*, Vol. 3, pp. 799–804.

Busino, G. (1989). *L'Italia di Vilfredo Pareto*. 2 Vols. Banca Commerciale Italiana Milano. (These volumes include an extensive collection of Pareto's correspondence.)

Cirillo, R. (1979). *The Economics of Vilfredo Pareto*. London: Frank Cass Ltd.

Eisermann, G. (1987). *Vilfredo Pareto*. Tübingen: J. C. B. Mohr (Paul Siebeck).

Homans, G. C., and Curtis, C. P. (1956). *An Introduction to Pareto*. New York: Knopf.

Kirman, A. K. (1987). Pareto as an economist. In: *The New Palgrave*, Vol. 3, pp. 804–808.

Mornati, F. (2000). Pareto observateur du libéralism économique Suisse et vaudois fin de siècle. *Schweizerische Zeitschrift für Geschichte*, 50, 403–420.

Mornati, F. (2001). Personal correspondence.

Pareto, V. (1896). *La courbe de la répartition de la richesse*. Lausanne.

Pareto, V. (1897). *Cours d'économie politique*. Lausanne: Rouge.

Pareto, V. (1907). *Manuale d'economia politica*. Milan: Società Editrice Libraria.

Pareto, V. (1952). *Scritti teorici*. Milan: Malfasi Editore.

Pareto, V. (1964–1970). *Oeuvres complètes de Vilfredo Pareto*. 14 Vols. Edited by G. Busino. Geneva: Librairie Droz.

Parsons, T. (1933). Pareto, Vilfredo. *Encyclopedia of the Social Sciences*, Vol. XI, New York: Macmillan, pp. 576–578.

Parsons, T. (1968). Vilfredo Pareto: Contributions to sociology. In: D. D. Sills (ed.): *International Encyclopedia of the Social Sciences*, New York: Macmillan, Vol. 11, pp. 411–416.

Powers, C. H. (1987). *Vilfredo Pareto*. In: J. H. Turner (ed.): *Masters of Social Theory*, Vol. 5, Newbury Park, CA: Sage Publications.

Powers, C. H. (2001). Personal correspondence.

Roy, R. (1949). Pareto statisticien: La distribution des revenus. *Revue d'économie politique*, 59, 555–577.

Schumpeter, J. A. (1951). *Ten Great Economists from Marx to Keynes*. Oxford: Oxford University Press.

Thornton, F. J. (2000). Vilfredo Pareto: A concise overview of his life, work and philosophy. Available at http://counterrevolution.net/resources/misc/pareto.html (A biased account by a right-wing group that capitalizes on reactionary views of Pareto during the last decade of his life, ignoring—on the whole—his progressive ideas and active participation in liberal causes.)

A.2 RODOLFO BENINI*

Born: June 11, 1862, Cremona, Italy.
Died: February 12, 1956, Rome, Italy.

At the beginning of his university career, Rodolfo Benini taught a history of commerce course at the University of Bari (1889–1895). Then he became a professor of economics at the University of Perugia (1896), a professor of statistics at the University of Pavia (1897–1907) and at the Bocconi University of Milan (1905–1909). Finally, he moved to the University of Rome (currently La Sapienza

*The following biography is a slightly edited version of text provided to the authors by Professor Giovanni M. Giorgi of the "La Sapienza" University in Rome.

University), where he was the first professor of statistics (1908–1928) and then professor of political economy (1928–1935). He was appointed professor emeritus at the University of Rome in 1935. Benini held various positions at the national and international levels, including president of the Higher Council of Statistics and of the Commission of Statistics and Law at the Ministry of Justice. He also represented the Italian government at the Geneva Conference in 1921 and served as the president of the Commission of Statistics at the General Assembly for the World Agriculture Census in 1926. In addition, he taught social sciences to the Crown Prince in 1922–1923 when the monarchy still existed in Italy.

Benini was a member of various societies and scientific bodies: the Lincei Academy, the Academy of Italy, the International Statistical Institute, and the American Statistical Association. He had a wide range of scientific interests, but his scientific output can be classified into the following main groups: statistics, demography and economics, social sciences, as well as studies on Dante's *Divine Comedy*.

With regard to statistics, Benini used economic and demographic empirical studies with the aim of constructing tools of general validity. In this context, we may include the study of the relations between the distribution of particular economic phenomena and the distribution of more general phenomena such as income or patrimony. His pioneering book *Principles of Statistics* (1906) contains his main contributions and successfully combines statistical methodology into a unified theory. In this way, Benini tried to give statistics an autonomous role in relation to economics, demography, and social sciences, with which it often became confused at the end of nineteenth and the beginning of the twentieth century. Among Benini's major original contributions we may mention, for example, the attraction indices, the extension to the patrimonies of the Paretian laws of income (Benini, 1897) discussed in Section 7.1 of this volume, and a probabilistic study of factors determining the proportion of the sexes in twins.

Despite his varied scientific interests, Benini was mainly a statistician, "the first complete Italian statistician" as C. Gini defined him. He considered the statistician to be a scholar who would not stop at the formal aspect but proceed to study phenomena in greater depth. In his opinion formulae should be adapted to reality and not vice versa.

As far as demography is concerned, which is subdivided into qualitative and quantitative population theory, Benini viewed it as a statistical science of human categories. In this field he completed an exemplary work on the classification of knowledge in his book *Principles of Demography* (1901). He also held a statistical vision of political economy that allowed him to keep in touch with demographic-social reality on the basis of a "statistical demand" for the critical review of reality. Benini was the first to use multiple regression for the empirical analysis of the curve of demand (Benini, 1907) not as a simple application of a particular statistical tool, but within the framework of a program that he called "inductive economics" used in order to "reduce economic sciences to a type of experimental science" (Benini, 1908). His view of economics is reflected in the organization of the final version of his *Lectures on Political Economy* (1936). Benini's thoughts were conditioned so much by statistics that he even used it, albeit indirectly, in his studies of Dante's *Divine Comedy* (1948).

REFERENCES

Benini's complete bibliography (137 scientific works) appears in *Bibliografie con brevi cenni biografici* (1959), Bologna: Biblioteca di Statistica, Cappelli Editore, pp. 45–51.

Benini, R. (1897). Di alcune curve descritte da fenomeni economici aventi relazioni colla curva del reddito o con quella del patrimonio. *Giornale degli Economisti*, 14, 177–214.

Benini, R. (1901). *Principii di demografia*. Florence: Barbera Editore.

Benini, R. (1906). *Principi di statistica metodologica*. Turin: Unione Tipografico-Editrice.

Benini, R. (1907). Sull'uso delle formule empiriche nella economia applicata. *Giornale degli Economisti*, 35 (supplement 2), 1053–1063.

Benini, R. (1908). Una possibile creazione del metodo statistico: L'Economia politica induttiva. *Giornale degli Economisti*, 36 (Series 2), 11–34.

Benini, R. (1936). *Lezioni di economia politica*. Bologna: Zanichelli.

Benini, R. (1948). Le interpretazioni della Cantica dell'Inferno alla luce della Statistica. *Rivista Italiana di Demografia e Statistica*, 2.

Boldrini, M. (1957). L'opera scientifica di Rodolfo Benini. *Giornale degli Economisti e Annali di Economia*, 16, 599–618.

Dall'Aglio, G. (1966). Benini, Rodolfo. In: *Dizionario degli Italiani*. Rome: Istituto della Enciclopedia Italiana fondata da Giovanni Treccani, Vol. 8, pp. 536–538.

A.3 MAX OTTO LORENZ

Born: September 19, 1876, Burlington, Iowa.
Died: July 1, 1959, Sunnyvale, California.

Max Otto Lorenz was the son of Carl Wilhelm Otto and Amalie Marie (Brautigam) Lorenz. His father was born in Essen, Germany, in 1841. In 1851 his (father's) parents came to America and after a short sojourn in New Jersey took up permanent residence in Burlington, Iowa. For a number of years Otto Lorenz maintained a wholesale and retail grocery store and afterwards became a successful businessman engaging in wholesale trade—cigars, teas, coffees, and spices. He was a sterling citizen who served several times on the city council and provided an excellent education to his three children. Max's brother Charles was a physicist in Cleveland. His sister also had her Ph.D. and served as the head of the Spanish department at a college in Wisconsin. She married Charles Wachsmuth—a distinguished paleontologist in Burlington—and assisted him in amassing his fossil collection, one of the most complete collections known in the world at the time.

Max O. Lorenz was educated in public schools in Burlington and graduated from the University of Iowa with a B.A. in 1899 and obtained his Ph.D. at the University of Wisconsin in 1906 for a thesis entitled, *Economic Theory of Railroad Rates*. He was a high-school teacher in Burlington, Iowa, during the years 1899–1901 and an instructor of economics while studying for his doctorate. In June 1905 he published in the *Publications of the American Statistical Association* (now the *Journal of the American Statistical Association*) his most influential paper "Methods of measuring the concentration of wealth," in which he proposed what is now known as the Lorenz curve as a basic tool for measuring inequalities in population or income and wealth, although it is applicable to any problem of a similar nature. As applied to the distribution of income or wealth, Lorenz suggested that one

> plot along one axis cumulated per cents of the population from poorest to richest, and along the other the per cent, of the total wealth held by these per cents, of the population.

He criticized Pareto's approach to measuring inequality or income concentration:

> Pareto, in his "Cours d'Economie Politique," does this, but in an erroneous way. He represents logarithms of class divisions in wealth along one axis, and the logarithms of the number of persons having more than each class division along the other. The error in this procedure lies in adhering to a fixed classification for two epochs. The number of persons having more than, say, $10 in each of the two periods of time is, as we have seen, of no significance in the question of degree of concentration when the per capita wealth of the community is growing.

An approach similar to that of Lorenz has been advocated independently by L. G. Chiozza Money in his classical book *Riches and Poverty* (1905), but even in the revised tenth edition (1911) of this work, Money still had not developed his idea any further. He continued to present his original "Lorenz curve" based on only three subdivisions: rich, comfortable, and poor.

The Lorenz curve received substantial attention in the literature (both economic and statistical) especially in the early 1970s. It remains a topic of active current research.

The Lorenz curve was initially used extensively by the Bureau of the Census and other U.S. government agencies from the early years of the twentieth century. It has been mistakenly assumed that it was developed by Lorenz during his association with the Bureau of the Census in 1910. After earning his Ph.D. degree, Lorenz served for several years as deputy commissioner of labor and industrial statistics for the state of Wisconsin. After a brief stint at the U.S. Bureau of the Census (1909–1910) and at the Bureau of Railway Economics in Washington, D.C. (1910–1911), he joined in 1911 the Interstate Commerce Commission (I.C.C.) in Washington, D.C., where he remained for 38 years until his retirement (to California) in 1944. From 1920 onward he was the director of the Bureau of Statistics and later the Bureau of Transport Economics and Statistics. His work with the I.C.C. consisted largely of planning, collecting, tabulating, and publishing data on rail, bus, and water transportation in the United States. These data were used for rate-making purposes and for improving operations.

He was married in Columbus, Ohio, in October 1911 to Nellie Florence (née Sheets), the daughter of a prominent Columbus lawyer; they had three sons during the course of their 48 years of marriage. Lorenz was the director of the so-called Eight-Hour Commission in 1916–1917, which was established to assess the effects of the eight-hour standard workday mandated by the U.S. Congress in September 1916 on production (it was the first U.S. legislation protecting workers' rights).

Lorenz was the co-author (with three other economists) of a popular textbook *Outlines in Economics* (4th edition, 1923). He was a member of the American Statistical Association, Western Economic Association, and Cosmos Club of Washington, D.C. His special interests included scientific developments, calendar reform, and Interlingua, a proposed international language.

REFERENCES

Lorenz, Max Otto. In: *American National Encyclopedia*, Vol. 47, p. 490.
Lorenz, Max Otto. In: *Who Was Who in America*, Vol. 5 (1969–1973). Chicago: Marquis.

A.4 CORRADO GINI*

Born: May 23, 1884, Motta di Livenza, Italy.
Died: March 13, 1965, Rome, Italy.

Corrado Gini is considered the leading Italian statistician of the twentieth century. He was born into an old family of landed gentry in northeast Italy and graduated with a degree in law in 1905 from the ancient and prestigious University of Bologna (founded in 1088). His thesis was on gender from a statistical point of view. During his studies at the university, which demonstrate the multidisciplinary interests characterizing his life, he took additional courses in mathematics and biology. In 1908 he was awarded the *libera docenza*, which qualified him to teach statistics at the university level; in 1909 he was appointed a temporary professor of statistics at the University of Cagliari and by 1910 he held the chair of statistics at the same university. In 1913 Gini moved to the University of Padua and in 1925 he was called to the University of Rome (today La Sapienza University), first to the chair of political and economic statistics and then to the chair of statistics from 1927–1955. During his long academic career, he also taught other subjects, such as economics, economic statistics, demography, biometry, and sociology. He was the founder and director of the Institutes of Statistics at the Universities of Padua and Rome, and it was at the University of Rome that in 1936 he founded the Faculty of Statistical, Demographic, and Actuarial Sciences, of which he was dean until 1954. In 1955 he was awarded the title professor emeritus. He was also awarded *honoris causa* degrees from various universities in Italy and abroad (e.g., in economics at the Catholic University of Milan, in sociology at the University of Geneva, in the sciences at Harvard University, and in social science at the University of Cordoba in Argentina). Gini devoted a lot of time to scientific publication; in fact, in 1920 he founded *Metron*, an international journal in statistics, and in 1934 *Genus*, a journal of the Italian Committee for the Study of Population Problems. He was director of

*The following biography is a slightly edited version of text provided to the authors by Professor Giovanni M. Giorgi of the "La Sapienza" University in Rome.

both until his death. From 1926–1943 he also directed the periodical *La Vita Economica Italiana*, which provided analyses of and information on various aspects of the Italian economy.

Gini belonged to many national and international societies and scientific bodies; he was honorary fellow of the Royal Statistical Society, president of the International Sociological Institute, president of the Italian Genetics and Eugenics Society, president of the International Federation of Eugenics Society in Latin-Language Countries, president of the Italian Sociological Society, honorary member of the International Statistical Institute, president of the Italian Statistical Society, and a national member of the Lincei Academy.

Gini held many public positions at both the national and international level, so many so that here we will only mention the most important and that is his leadership as the president of the Central Institute of Statistics from 1926–1932. During this period he attempted to centralize statistical services and was therefore in direct contact with central and peripheral state institutions, including Benito Mussolini who came to power in 1922 and seemed to have an interest in statistics as a government instrument (Leti, 1996). Although he possessed only limited economic resources and his positions often differed from those of state administrators, he was able to modernize the Italian statistical system. In connection with this period, it is necessary to underscore the fact that although Gini respected political power, he did not submit to its authority; on the contrary, he exploited it in an attempt to improve the quality of the official statistical service in Italy and to reach other scientific objectives.

Gini was a renaissance type of scholar with numerous and varied interests in the fields of statistics, economics, demography, sociology, and biology. He published a very large number of scientific writings (over 800, including articles, conference papers, books, etc.), of which it is impossible to mention even some briefly here (see Benedetti, 1984; Castellano, 1965; Federici, 1960; Forcina, 1982). As far as statistics is concerned, his principal contributions mainly concerned three topics: mean values, variability, and the association between statistical variates. An important group of papers concerning variability and concentration were collected in one volume (Gini, 1939), whereas his studies on fundamental issues of probability and statistics were reproduced in two posthumous volumes (Gini, 1968), for which the scientific material published had been chosen by Gini himself shortly before his death. He particularly wished to insert an inedited paper on the logical and psychological theory of probability that he had left in the drawer of his desk for more than 50 years, as if he had wanted to wait for his experience in life to confirm some of his ideas.

Despite the large number of seminal contributions in the most varied fields, Gini is especially remembered for his concentration ratio (Gini, 1914), a summary inequality index that is still of topical interest today (Giorgi, 1990, 1993, 1999). He also proposed another index (Gini, 1909) that is a direct measure of concentration and is more sensitive than Pareto's.

Gini was married and had two daughters, but according to one of his biographers (Giorgi, 1996), the numerous positions that he held at the university and elsewhere as well as the efforts dedicated to scientific production did not leave him much time for his family.

REFERENCES

Benedetti, C. (1984). A proposito del centenario della nascita di Corrado Gini. *Metron*, 42 (1–2), 3–19.

Castellano, V. (1965). Corrado Gini: A memoir. *Metron*, 24(1–4), 3–84. (Contains a full bibliography of Gini's 827 scientific works.)

Federici, N. (1960). L'opera di Corrado Gini nell'ambito delle scienze Sociali. In: *Studi in Onore di Corrado Gini*. Rome: Istituto di Statistica, Facoltà di Scienze Statistiche, Università di Roma, Vol. 2, pp. 3–34.

Federici, N. (2000). Gini Corrado. In: *Dizionario degli Italiani*. Rome: Istituto della Enciclopedia Italiana fondata da Giovanni Treccani, Vol. 55, pp. 18–21.

Forcina, A. (1982). Gini's contribution to the theory of inference. *International Statistical Review*, 50, 65–70.

Gini, C. (1909). Il diverso accrescimento delle classi sociali e la concentrazione della ricchezza. *Giornale degli Economisti*, 20, 27–83.

Gini, C. (1910). *Indici concentrazione e di dipendenza*. Turin: Biblioteca dell'Economista, UTET.

Gini, C. (1914). Sulla misura della concentrazione e della variabilità dei caratteri. *Atti del reale Istituto Veneto di Scienze, Lettere ed Arti*, a.a. 1913–1914, 73 (Part 2), 1203–1248.

Gini, C. (1939). *Memorie di metodologia statistica: Variabilità e concentrazione*. Milan: Giuffrè.

Gini, C. (1968). *Questioni fondamentali di probabilità e statistica*. 2 Vols. Rome: Biblioteca di Metron, Università di Roma.

Giorgi, G. M. (1990). Bibliographic portrait of the Gini concentration ratio. *Metron*, 48 (1–4), 183–221.

Giorgi, G. M. (1993). A fresh look at the topical interest of the Gini concentration ratio. *Metron*, 51 (1–2), 83–96.

Giorgi, G. M. (1996). Encounters with the Italian statistical school: A conversation with Carlo Benedetti. *Metron*, 54 (3–4), 3–23.

Giorgi, G. M. (1999). Income inequality measurement: The statistical approach. In: J. Silber (ed.): *Handbook of Income Inequality Measurement*. Boston: Kluwer Academic Publishers.

Leti, G. (1996). L'ISTAT e il Consiglio Superiore di Statistica dal 1926 al 1945. *Annali di Statistica*, Instituto Nazionale di Statistica, Rome, 8 (supplement 10).

Regazzini, E. (1997). Gini Corrado. In: N. L. Johnson and S. Kotz (eds.): *Leading Personalities in Statistical Sciences*. New York: Wiley. pp. 291–296.

A.5 LUIGI AMOROSO*

Born: May 23, 1886, Naples, Italy.

Died: October 28, 1965, Rome, Italy.

Luigi Amoroso's father Nicola played an important role in his son's moral, intellectual, and scientific development. He was an engineer and important technical official, first in the Southern Railway Company and then in the Italian State Railway.

*The following biography is a slightly edited version of text provided to the authors by Professor Giovanni M. Giorgi of the "La Sapienza" University in Rome.

Despite his heavy work load, in the evenings, after supper, the father would dedicate his time to research in pure mathematics, thereby setting an example and demonstrating the importance of study to his six sons, of whom Luigi was the eldest. In 1903 Luigi entered the *Normale* in Pisa where he studied mathematics. In 1905, when the family moved to Rome, he left the *Normale* and enrolled at Rome University from which he graduated with a degree in mathematics in 1907. He wrote his thesis on a difficult problem concerning the theory of holomorphic functions.

Amoroso's university career began in 1908 when he became assistant professor of analytic geometry at the University of Rome (today La Sapienza University). While studying mathematics, he also carried out research in the field of administration and economics, and in recognition of his eclecticism, he was awarded the *libera docenza*, which qualified him to teach at the university level in economics (1910) and mathematical physics (1913). In 1914 he held the chair in financial mathematics at the University of Bari and in 1921 at the University of Naples. In 1926 Amoroso was called to the University of Rome to hold the chair in economics at the Faculty of Political Science, of which he served as dean between 1950 and 1961. During the same period he was also Director of the Institute of Economic, Financial and Statistical Studies at the same Faculty. In 1962 he was awarded the title of professor emeritus.

Amoroso held additional important positions, such as government advisor to the Bank of Naples, vice commissioner and later counsellor of the National Insurance Institute, managing director of the company Le Assicurazioni d'Italia, and advisor to the Banca Nazionale del Lavoro. He was member of the Higher Council for Statistics, the National Research Council, and the Higher Council for Public Education. He was also a member of various societies and scientific bodies, including the Lincei Academy, the Econometric Society, and the International Statistical Institute. Amoroso had a very rigid lifestyle, inspired by his deep Catholic faith and a multitude of interests not only concerning pure and financial mathematics, economics, and statistics but also the humanities and philosophy. His various pursuits permitted a healthy and prolific eclecticism.

Amoroso's scientific publications began in 1909 with a paper on economic theory according to Pareto that was followed by a completely different type of note on the criteria of resolvability of the integral linear equation of the first kind. His last work appeared in 1961, when his bad health prevented him from dedicating time to research, as a volume on the natural laws of economics. In more than half a century of scientific research, the area of study that attracted him the most and to which he made the most important contributions was without doubt economic science. Most notable, are some books in which he skilfully synthesized his most important results; in particular, we refer the interested reader to his *Economic Mechanics* (1942), in which he collected his lectures given in 1940–1941 at the National Institute of Higher Mathematics. In the topics dealt with in this volume, Amoroso used sophisticated mathematical tools to reach his objectives and, in extreme synthesis, presented the main results concerning the generalization of Walras's and Pareto's equilibrium, which is static, through the dynamic consideration of the factors that intervene in the optimal solution of consumer and producer problems. In his *Lessons on Mathematical Economics* (1921), he lucidly outlined the methods, limits, and aims of mathematical economics, whereas in *The Natural Laws of Economics* (1961) he laid out a definitive description of the fundamental aspects of his favorite scientific problem, that is, the parallelism between mechanical and economic phenomena and the transfer of mechanical laws in the economic world. Finally, we must mention Amoroso's (1925) contribution on the analytical representation of income distribution, with which he proposed an equation that is an extension and generalization (see D'Addario, 1936) of those proposed by March and Vinci, suitable for describing, according to the value of the parameters, both unimodal and zeromodal distributions.

REFERENCES

Amoroso's full bibliography (150 scientific works) appears in *Luigi Amoroso: Discorsi commemorativi pronunciati dai Lincei Mauro Picone e Volrigo Travaglini* (1967), Rome: Accademia Nazionale dei Lincei.

Amoroso, L. (1909). La teoria dell'equilibrio economico secondo il Prof. Vilfredo Pareto. *Giornale degli Economisti*, 39 (Series 2), 353–367.

Amoroso, L. (1910). Sulla risolubità della equazione lineare di prima specie. In: *Rendiconti dell'Accademia dei Lincei*. Rome: Classe di scienze fisiche, matematiche e naturali, Accademia Nazionale dei Lincei.

Amoroso, L. (1921). *Lezioni di economia matematica*. Bologna: Zanichelli.

Amoroso, L. (1925). Ricerche intorno alla curva dei redditi. *Annali di matematica pura e applicata*, 2 (Series 4), 123–157.

Amoroso, L. (1942). *Meccanica Economica*. Città di Castello: Macrì Editore.

Amoroso, L. (1961). *Le leggi naturali dell'economia politica*. Turin: UTET.

D'Addario, R. (1936). Sulla curva dei redditi di Amoroso. In: *Annali dell'Istituto di Statistica dell'Università di Bari*, Vol. 10. Bari: Macrì Editore.

Giva, D. (1988). Amoroso Luigi. In: *Dizionario degli Italiani*, Rome: Istituto della Enciclopedia Italiana fondata da Giovanni Treccani, Vol. 34, pp. 113–116.

A.6 RAFFAELE D'ADDARIO*

Born: December 17, 1899, Grottaglie, Italy.

Died: September 1, 1974, Rome, Italy.

Before he was even eighteen years old, Raffaele D'Addario served in World War I, an unforgettable experience that remained with him for all of his life. After the war he enrolled at Bari University from which he graduated in 1924 with a degree in economics and business administration. Carlo Emilio Bonferroni supervised

*The following biography is a slightly edited version of text provided to the authors by Professor Giovanni M. Giorgi of the La Sapienza University in Rome.

D'Addario in the writing of his thesis and soon thereafter D'Addario became Bonferroni's assistant in mathematics. Later he resigned from the University of Bari and moved to Rome. Here he worked for ISTAT, the National Institute of Statistics, where he became head of the Office of Studies in 1929 and met Corrado Gini who at that time was president of the institute. However, D'Addario did not remain long at ISTAT. According to some biographers, it may have been due to incompatibility with the president; others maintain that it was because he met Luigi Amoroso who, as administrator of the company Le Assicurazioni d'Italia, appointed him as the company's technical consultant in October 1931. The experience that he gained first at ISTAT and then in the field of insurance gave D'Addario not only the opportunity to analyze concrete phenomena, but also provided stimuli for continuing theoretical-methodological studies within the university environment. He was awarded the *libera docenza*, which qualified him to teach statistics at the university level, in 1931, and in 1936 he was appointed to the chair of statistics at the Faculty of Economics and Business Administration at the University of Bari. During the academic year 1950–1951 he moved to Rome University (now La Sapienza University) to the Faculty of Political Science, of which he was dean from the academic year 1963–1964 until the time of his death.

D'Addario held important public positions at the National Council of Labour and Economics, the Higher Council for Public Education, and the Higher Council for Statistics. He was a member of various societies and scientific bodies, including the Lincei Academy, the International Statistical Institute, the International Institute of Public Finance, the International Union for the Scientific Study of the Population, the Econometric Society, the Italian Actuarial Institute, the Italian Society of Economists, and the Italian Society of Economics, Demography and Statistics.

D'Addario's scientific interests were rather wide and may be classified into the following main groups: the distribution of personal income and wealth, the economics and statistics of insurance, and the division of taxes. His most important studies belong to the first group. D'Addario was motivated to study income distribution following the results obtained by Pareto on the topic. After having ascertained (D'Addario, 1934) that of the three curves [called first (I), second (II), and third (III) approximation] by which the Paretian model is expressed (the literature at that time mainly dwelled on the I-type curve), he investigated the analytical properties of the II-type curve and proposed three methods for determining its parameters. Subsequently, he studied the subject in more detail and suggested a further method for calculating the parameters. D'Addario (1936, 1949, 1953) also showed that the functions proposed by various authors (e.g., Pareto, Kapteyn, March, Vinci, Amoroso, Davis, Mortara, and Benini, etc.) for the analytical representation of the income and patrimony distributions are particular cases of a more general function that formally synthesizes Boltzmann, Bose–Einstein, and Fermi–Dirac statistics.

As far as research in the field of insurance is concerned, we can briefly say that D'Addario dealt, among other things, with the "rational evaluations" of the "rates" and "reserves," and of the existence—restricted to the branch of accidents and civil liability—of a specific law for the distribution of damages. Furthermore, he analyzed insurance problems in relation to fluctuations in currency value.

Finally, D'Addario's studies on the division of taxes were mainly concerned with the issues relating to progressive taxation. In this context he proposed some methods for measuring the structural progressivity of taxation and a general method for establishing the scale of tax rates. He also investigated the influence of a proportional property tax on corresponding incomes as well as the influence of a proportional income tax on corresponding property.

REFERENCES

D'Addario, R. (1934). *Sulla misura della concentrazione dei redditi*. Rome: Poligrafico dello Stato.

D'Addario, R. (1936). Le trasformate Euleriane. In: *Annali dell'Istituto di Statistica dell'Università di Bari*, Vol. 8. Bari: Macrì Editore.

D'Addario, R. (1936). Sulla curva dei redditi di Amoroso. In: *Annali dell'Istituto di Statistica dell'Università di Bari*, Vol. 10. Bari: Macrì Editore.

D'Addario, R. (1939). La curva dei redditi: sulla determinazione numerica dei parametri della seconda equazione paretiana. In: *Annali dell'Istituto di Statistica dell'Università di Bari*, Vol. 12. Bari: Macrì Editore.

D'Addario, R. (1939). Un metodo per la rappresentazione analitica delle distribuzioni statistiche. In: *Annali dell'Istituto di Statistica dell'Università di Bari*, Vol. 16. Bari: Macrì Editore.

D'Addario, R. (1949). Ricerche sulla curva dei redditi. *Giornale degli Economisti e Annali di Economia*, 8, 91–114.

D'Addario, R. (1953). Su una funzione di ripartizione dei redditi del Mortara. *Giornale degli Economisti e Annali di Economia*, 12, 15–54.

D'Addario, R. (1980). *Scritti scelti*. Rome: Istituto di Studi Economici, Finanziari e Statistici, Facoltà di Scienze Politiche, Università di Roma. (Contains D'Addario's full scientific bibliography.)

Guerrieri, G. (1985). D'Addario Raffaele. In: *Dizionario degli Italiani*, Vol. 31, Rome: Istituto della Enciclopedia Italiana fondata da Giovanni Treccani, pp. 613–614.

A.7 ROBERT PIERRE LOUIS GIBRAT

Born: March 23, 1904, Lorient, France.

Died: May 13, 1980, Paris, France.

The son of a chief navy physician, Robert Pierre Louis Gibrat studied in Lorient, Rennes, and Brest (Normandy) and later on at the Lycée Saint-Louis in Paris. He entered the École Polytechnique in 1922 and specialized in mining engineering at the École des Mines. He started his professional career as a technical consultant for private firms. While working, he studied at the universities of science and law in Paris and Lyon and received a joint bachelor's degree in science and doctorate in law with honors. His doctoral thesis was on economic inequality. He married Yseult Viel in 1928 with whom he had three daughters. During the period from 1927–1931 he was employed as a professor at and served as an assistant director of the St. Etienne École des Mines.

Returning to Paris, he taught at the École des Mines and became Professor in 1936. It is at this point that there is a gap in Gibrat's biography. David E. R. Gay (1988) claims that he stayed with the École des Mines until 1968; the school records show that he left in 1940 but the library there possesses a copy of his industrial electricity course for 1943–1944. More importantly, it should be noted that his 1931 doctoral thesis at Lyon was a historical study of the lognormal law and its utilization for modeling revenue (income) distributions. In his thesis Gibrat confined himself to a study of one-parameter distributions of wealth and income. This research continued to preoccupy him until 1940. After that he pursued two further professional interests: the hydroelectric energy of rivers and tides and nuclear energy.

In the 1930s, after joining a nonconformist political group Ordre Nouveau, Gibrat became an active member of the avant-garde group X-Crise. Created after the Great Depression of 1929, it focused on solving economic and social problems, without the petty quarrels and dogmatic points of view characteristic of political parties. Gibrat published in the journal *X-Information*, which became the monthly bulletin of the *Centre Polytechnicien d'Études Économiques*.

The above mentioned 1931 doctoral dissertation entitled

> Les inégalités économiques. Applications: aux inégalités des richesses, a la concentration des enterprises, aux populations des villes, aux statistiques des familles, etc., d'une loi nouvelle, la loi de l'effet proportionnel*

remained Gibrat's main economic publication. In 1930 he contributed a lengthy article in the *Bulletin de la statistique generale de la France* and also later reprinted his dissertation under the title *Les inégalités économiques* as well as writing a note *La loi de l'effet proportionnel* for the Paris Academy of Sciences' *Comptes rendus*.

Like Pareto, Gibrat used the term law, but explicitly admitted that it is essentially statistical in the form of a mathematical model. He introduced his own "law," brushing aside the Pareto and Pearson empirical fittings. In the second part of his thesis he discussed a measure of inequality made possible by this law and defended its superiority over other approaches. He criticized Pearson's method of moments, claiming that "its automatism makes any control difficult." For this contribution he was elected a Fellow of the Econometric Society in 1948.

After 1940 Gibrat had a distinguished scientific and administrative career, although his behavior during the German occupation of France may be regarded in a negative light. In June 1940, he was appointed Director of Electricity for the Ministry of Industrial Production in the first puppet pro-German Laval government; he was named Secrétaire d'Etat aux Communications during the second Laval regime in April 1942. In November 1942 Gibrat happened to be in North Africa inspecting the Trans-Saharian

*"Economic inequalities. Applications: to the inequality of wealth, the concentration of industry, the population of cities, the statistics of families, etc., of a new law, the law of proportional effect."

railway when the American landing took place. It is not clear when he returned to France (after the Germans occupied the entire country), but it is known that he resigned from his post. After the liberation he spent a year in a Fresnes prison where he had the opportunity to develop (theoretically) a new type of hydraulic power station. In March 1946 he was sentenced to 10 years of "national unworthiness."

All of this did not prevent Gibrat from resuming his professional activities in 1945. He served as a consulting engineer for French Electric on tidal energy (1945–1968) (he had published his first article on the subject in the *Revue de l'industrie minerale* in 1944). He then held high-level positions in institutions involved in the development of French atomic policy. These posts included his role as general manager for atomic energy at the Groupement de l'Industrie Atomique (INDATOM) (1955–1974), general manager of the Société pour l'Industrie Atomique (SOCIA) (1960), and as president of the European Atomic Energy Community (EURATOM) Scientific Committee (1962). He was also a long-term consulting engineer for Central Thermique (1942–1980).

Gibrat was a member of the International Statistical Institute and served as president of the French Society of Electricians and the Civil Engineer Society of France (1966), the Statistical Society of Paris (1966), the French Statistical Society (1978), the Technical Committee for the Hydrotechnical Society of France and the French Meteorological Society (1969), the World Federation of Organizations of Engineers and the French Section of the American Nuclear Society (1972).

He was one of the first French members of the Econometric Society (founded in December 1930 in Cleveland, Ohio) and before World War II published three articles in *Econometrica*.

Gibrat was the author of numerous reports to the Academy of Sciences, some 100 professional articles, and two books (on economics and tidal energy). He was also a Knight of the Legion of Honor.

REFERENCES

Armatte, M. (1998). Robert Gibrat and the law of proportional effect. In: W. J. Samuels, S. Böhm, and P. Anderson (eds.): *European Economists of the Early 20th Century*. Cheltenham, UK: Edward Elgar, pp. 94–111.

Gay, D. E. R. (1988). Gibrat, Robert Pierre Louis. In: J. Eatwell (ed.): *The New Palgrave*. London: Macmillan, p. 521.

A.8 DAVID GAWEN CHAMPERNOWNE

Born: July 9, 1912, Oxford, England.

Died: August 19, 2000, Devon, England.

David Gawen Champernowne was the only child of Francis Champernowne, the bursar of Keble College at Oxford University. The family was originally from

South Devon, and one of his ancestor, Katherine Champernowne, was the mother of Sir Walter Ralegh. Champernowne was educated at the College of Winchester, excelling in mathematics. Later on as a scholar at King's College in Cambridge, he became very friendly with Alan Turing and together they designed and constructed a chess computer named "Turochamp." He was one of the few mathematicians able to keep up with Turing. Champernowne achieved special distinction when he published a scientific paper in the *Journal of the London Mathematical Society* (1933) that produced an example of a normal number 0.1234567891011. . . , nowadays known as Champernowne's constant. He then developed an interest in economics and obtained a First in the Economics Tripos. Using his mathematical skills, Champernowne was particularly gifted in the application of statistical techniques to economics. He was made an assistant lecturer at the London School of Economics in 1936 and in 1938 returned to Cambridge as a university lecturer in statistics.

At LSE Champernowne worked with William Beveridge, the eventual architect of the British welfare state. W. Beveridge and J. M. Keynes greatly influenced his approach to economics and statistics. During World War II he first (in 1940) worked in the statistical section of the Prime Minister's office. Then Prime Minister Winston Churchill asked his "think tank" to explore questions about home and enemy resources, and the unit's head, Professor F. A. Lindemann (later Lord Cherwell) became, in turn, extremely demanding of his staff. Champernowne did not get on well with him, and in 1941 he moved to the Ministry of Aircraft Production, staying there for the remainder of the war.

In 1944 Champernowne got involved with J. M. Keynes of the Cambridge economics faculty in establishing a separate Department of Applied Economics. In 1945 he became Director of the Oxford Institute of Statistics and a Fellow of Nuffield College, attaining a full professorship in 1948.

In 1948 Champernowne married Wilhelmina ("Mieke") Dullaert who had had the good grace to lose to "Turochamp" in their chess matches. They had two sons. That same year he also presented a pioneering paper at the Royal Statistical Society on the Bayesian approach to the time series analysis of autoregressive processes. A related paper presented in 1960 dealt with the spurious correlation problem, anticipating the work of Granger and Newbold (*Journal of Econometrics*, 1974) on this topic. However, Champernowne came to regret his change of university and began to look for a way to return to Cambridge. This he could achieve only through a highly unusual step, by assuming a lower academic rank, which he did when he became a reader in economics at Cambridge and a teaching fellow at Trinity from 1959–1970.

During the subsequent decade Champernowne combined his work on income distributions with investigations of uncertainty in economic analysis in a three-volume book: *Uncertainty and Estimation in Economics* (1969). In this magnum opus he attempted to integrate quantitative analysis with decision theory. This unique (but currently overlooked) study resulted in Champernowne's being given a personal chair at Cambridge in economics and statistics and his election as a Fellow of the British Academy in 1970. During 1971–1976 he was a co-editor of the *Economic Journal*. As mentioned above, his economic studies in the early 1930s were supervised by J. M. Keynes and he made a valuable contribution in his published review of Keynes's *General Theory* (1936) to the debate between Keynes's approach and upholders of the classical theory of employment, specifically whether money or real wages are the proper subject of bargaining between management and workers. He courageously (at the age of 23) took a controversial stand, respectfully disagreeing with his eminent supervisor and history proved him right.

From 1935, for over 50 years, Champernowne carried out research on income distributions, producing a rigorous stochastic model using the technique of the Markov chain (see Chapter 3 for a brief description of his model leading to the Pareto distribution). His 1937 work earned him a prize fellowship at King's College and was widely discussed and consulted by researchers worldwide. However, this work was only officially published some 36 years later (Champernowne, 1973). His last book *Economic Inequality and Income Distribution* he wrote jointly with a former student, Frank Cowell; it was published in 1998.

Champernowne retired in 1978 but remained an emeritus professor for almost 20 years. When he fell ill in his final years, he moved with Mieke to Budleigh Salterton in Devon, close to his family roots—the birthplace of Sir Walter Raleigh and to be near one of his sons.

REFERENCES

Champernowne, D. G. (1933). The construction of decimals normal in the scale of ten. *Journal of the London Mathematical Society*, 8, 254–260.

Champernowne, D. G. (1936). Unemployment, basic and monetary: The Classical analysis and the Keynesian. *Review of Economic Studies*, 3, 201–216.

Champernowne, D. G. (1969). *Uncertainty and Estimation in Economics*. 3 Vols. London: Oliver & Boyd.

Champernowne, D. G. (1973). *The Distribution of Income Between Persons*. Cambridge: Cambridge University Press.

Champernowne, D. G., and Cowell, F. A. (1998). *Economic Inequality and Income Distribution*. Cambridge: Cambridge University Press.

Harcourt, G. C. (2001). David Gawen Champernowne, 1912–2000: In appreciation. *Cambridge Journal of Economics*, 25, 439–442.

Obituary in *Daily Telegraph*, Sept. 4, 2000.

Obituary in *Guardian*, Sept. 1, 2000.

APPENDIX B

Data on Size Distributions

Below we provide a list of data sources, mostly of data in grouped form, along with information on the distributions that were fitted to these data. Our list is in no way exhaustive and merely pertains to the references presented in this book.

Sources for large data sets on individual incomes include the U.S. Panel Study of Income Dynamics (PSID), maintained by the University of Michigan (Hill, 1992) and the German Socio-Economic Panel (SOEP), maintained by the Deutsches Institut für Wirtschaftsforschung (DIW), Berlin (Burkhauser, Kreyenfeld, and Wagner, 1997). Further actuarial data sets may be found in textbooks on actuarial statistics such as Klugman, Panjer, and Willmot (1998).

Table B.1 Individual Data (Incomes)

Source	Description	No. of Observations	Distributions Fitted
Aggarwal and Singh (1984)	Kenyan annual earnings	200	No distribution fitted
Arnold (1983)	Lifetime earnings of professional golfers	50	Par, Par(II), Fisk, SM
Arnold (1983)	Texas county data (1969 total personal incomes)	157	Par, Par(II), Fisk, SM
Dyer (1981)	Annual wages of production-line workers under age 40 in large U.S. industrial firm	30	Par

Statistical Size Distributions in Economics and Actuarial Sciences, By Christian Kleiber and Samuel Kotz.
ISBN 0-471-15064-9

Table B.2 Grouped Data (Incomes)

Source	Description	No. of Groups	Distributions Fitted
Anand (1983)	Malaysia 1970	32	Par
Aoyama et al. (2000)	Japan 1998	18	Par
Bordley, McDonald, and Mantrala (1996)	United States 1970, 1975, 1980, 1985, 1990	9	15 distributions of gamma and beta type
Bowley (1926)	UK 1911–1912	11	Par
Bowman (1945)	United States 1935–1936	25	No distribution fitted
Brunazzo and Pollastri (1986)	Italy 1948	21	Generalized LN
Champernowne (1952)	United States 1918	18	4-parameter Champernowne, Davis
Champernowne (1952)	Norwegian townsmen 1930	11	3-parameter Champernowne
Champernowne (1952)	United States 1947	12	3-parameter Champernowne
Champernowne (1952)	Bohemia 1933	16	3-parameter Champernowne, Davis, Par, LN
Champernowne (1952)	UK 1938–1939	12	3-parameter Champernowne
Champernowne and Cowell (1998)	UK 1994–1995	16	No distribution fitted
Cowell and Mehta (1982)	Sweden 1977	21	No distritbution fitted
Creedy, Lye, and Martin (1997)	United States 1986	15	LN, generalized LN, GG, generalized Ga
Dagum (1980a)	United States 1969	10	LN, Ga, D
Dagum (1983)	United States 1978	21	LN, Ga, SM, D I-III
Dagum (1985)	Canada 1965, 1967, 1969, 1971, 1973, 1975, 1977, 1979, and 1981	16	D I, II, and III
Dagum and Lemmi (1989)	Italy 1977, 1980, and 1984	15	D I, II, and III
Davies and Shorrocks (1989)	Canada 1983, 1984	18, 12*	No distribution fitted
Davis (1941b)	United States 1918	73	Davis

Espinguet and Terraza (1983)	France 1978	11	D II
Fattorini and Lemmi (1979)	Italy 1967–1976	19	Champernowne, D I
Gastwirth and Smith (1972)	United States 1955	26	2- and 3-parameter LN, Fisk
Hayakawa (1951)	Japan 1932	34	Par
Iyengar (1960)	India 1955–1956	12	LN
Kakwani and Podder (1976)	Australia 1967–1978	11	No distribution fitted
Kalecki (1945)	UK 1938–1939	10	LN
Kloek and van Dijk (1978)	Netherlands 1973	16	LN, Ga, log-t, GG, Champernowne, log-Pearson IV
Kmietowicz and Webley (1975)	Kenya 1963–1964 (5 rural districts)	16	LN
Kmietowicz and Ding (1993)	China 1980, 1983, 1986	7, 9, 12	LN
Kordos (1990)	Poland 1973	10	LN, B2, B1, Ga
Kordos (1990)	Poland 1985	14	LN
Kordos (1990)	Poland 1987	10	LN
MacGregor (1936)	UK 1918–1919	27	Par
March (1898)	French, German, and American wages	≤ 70	Ga
McDonald and Mantrala (1993)	United States 1970, 1980, 1990	9, 11	Many gamma and beta types
Pham-Gia and Turkkan (1992)	Canada 1989	18	Ga, B1
Ransom and Cramer (1983)	United States 1960, 1969	10	Ga, LN, Par (with measurement error)
Stamp (1914)	UK 1801	9	Par
Steyn (1966)	South Africa 1951	15	LN
Suruga (1982)	Japan 1975	16	Par(II), LN, *Ga*, Beta, Fisk, SM
Vartia and Vartia (1980)	Finland 1967	15	B2
Yardi (1951)	United States 1922–1936	10	Par

Notes: *Two groupings for each year: Statistics Canada and optimal groupings.

Table B.3 Quantile Data (Incomes)

Source	Description	No. of Groups
Basmann et al. (1990)	United States 1977	100
Champernowne and Cowell (1998)	UK 1993–1994 (incomes before and after taxes)	10
Chotikapanich (1994)	Thailand 1981	10
Nygård and Sandström (1981)	Finland 1971	10
Nygård and Sandström (1981)	Sweden 1972	10
Shorrocks (1983)	19 countries (from all continents)	11

Table B.4 Bivariate Income Data

Source	Description	No. of Groups	Distributions Fitted
Anand (1983)	Malaysia 1970 (household income and size)	32×10	Only marginal distribution
Bakker and Creedy (1997)	New Zealand 1991 (earnings and age)	19×10	Only marginal distribution
Hart (1976a)	UK 1963, 1966, 1970*	≤ 14	Bivariate LN
Kmietowicz (1984)	Iraq 1971–1972 (household income and size)	9×8	Bivariate LN

*Constant sample of 800 men aged 30 in 1963.

Table B.5 Wealth Data

Source	Description	No. of Groups	Distributions Fitted
Bhattacharjee and Krishnaji (1985)	Indian landholdings 1961–1962	5	LN, log-Ga, Ga
Champernowne and Cowell (1998)	UK personal wealth 1994	13	No distribution fitted
Chesher (1979)	Ireland 1966	26	Par, LN
Dagum (1990b)	Italy 1977, 1980, 1984	15	D II
Sargan (1957)	British wealth 1911–1913, 1924–1930, 1935–1938, 1946–1947	6	LN
Steindl (1972)	Sweden 1955, 1968	12	Par
Steindl (1972)	Netherlands 1959, 1967	5	Par

Table B.6 Data on Actuarial Losses

Source	Description	No. of Observations	Distributions Fitted
Benckert and Jung (1974)	Fire insurance claims (Sweden 1958–1969)	17 groups	LN, Par
Benktander (1963)	Automobile insurance losses (France 1955–1958)	10 groups	Par
Cummins et al. (1990)	Aggregate fire losses*	23 individual observations	16 distributions of beta and gamma type
Cummins et al. (1990)	Fire claims*	12 groups	16 distributions of beta and gamma type
Ferrara (1971)	Industrial fire insurance losses (Italy 1963–1965)	10 groups	3-parameter LN
Hewitt and Lefkowitz (1979)	Automobile bodily injury losses	18 groups	LN, LN-Ga mixture
Hogg and Klugman (1983)	Hurricane losses	35 individual observations	LN, Wei, SM
Hogg and Klugman (1983)	Malpractice losses	23 groups	LN, Wei, Par(II), SM, B2
Klugman (1986)	Basic dental coverage	21 groups	Par(II), LN, SM

*Earlier given in Cummins and Freifelder (1978).

APPENDIX C

Size Distributions

For the convenience of the readers we collect here the basic properties (definitions and moments) of the main distributions studied in detail in the preceding chapters.

Pareto Type I Distribution

$$f(x) = \frac{\alpha x_0^\alpha}{x^{\alpha+1}}, \quad x \geq x_0 > 0. \tag{C.1}$$

Here $\alpha > 0$ is a shape and $x_0 > 0$ a scale parameter. The moments exist for $k < \alpha$ only and are given by

$$E(X^k) = \frac{\alpha x_0^k}{\alpha - k}. \tag{C.2}$$

Stoppa Distribution

$$f(x) = \theta \alpha x_0^\alpha x^{-\alpha-1} \left[1 - \left(\frac{x}{x_0} \right)^{-\alpha} \right]^{\theta-1}, \quad 0 < x_0 \leq x. \tag{C.3}$$

The kth moment exists for $k < \alpha$ and equals

$$E(X^k) = \theta x_0^k B\left(1 - \frac{k}{\alpha}, \theta \right). \tag{C.4}$$

Statistical Size Distributions in Economics and Actuarial Sciences, By Christian Kleiber and Samuel Kotz.
ISBN 0-471-15064-9

Lognormal Distribution

$$f(x) = \frac{1}{x\sqrt{2\pi}\sigma} \exp\left\{-\frac{1}{2\sigma^2}(\log x - \mu)^2\right\}, \quad x > 0. \tag{C.5}$$

Here $\exp \mu$ is a scale and $\sigma > 0$ a shape parameter. All the moments exist and are given by

$$E(X^k) = \exp\left(k\mu + \frac{1}{2}k^2\sigma^2\right). \tag{C.6}$$

Generalized Gamma Distribution

$$f(x) = \frac{a}{\beta^{ap}\Gamma(p)} x^{ap-1} e^{-(x/\beta)^a}, \quad x > 0. \tag{C.7}$$

Here $\beta > 0$ is a scale and $a, p > 0$ are shape parameters.

The moments exist for $-ap < k < \infty$ and are given by

$$E(X^k) = \frac{\beta^k \Gamma(p + k/a)}{\Gamma(p)}. \tag{C.8}$$

Gamma Distribution

$$f(x) = \frac{1}{\beta^p \Gamma(p)} x^{p-1} e^{-x/\beta}, \quad x > 0. \tag{C.9}$$

Here $\beta > 0$ is a scale and $p > 0$ a shape parameter.

The moments exist for $-p < k < \infty$ and are given by

$$E(X^k) = \frac{\beta^k \Gamma(p + k)}{\Gamma(p)}. \tag{C.10}$$

Inverse Gamma Distribution

$$f(x) = \frac{\beta^p}{\Gamma(p)} x^{-p-1} e^{-(\beta/x)}, \quad x > 0. \tag{C.11}$$

Here $\beta > 0$ is a scale and $p > 0$ a shape parameter.

The moments exist for $-\infty < k < p$ and are given by

$$E(X^k) = \frac{\Gamma(p)}{\Gamma(p-k)} \beta^k. \tag{C.12}$$

Loggamma Distribution

$$f(x) = \frac{\beta^\alpha}{\Gamma(\alpha)} x^{-\beta-1} \{\log(x)\}^{\alpha-1}, \quad 1 \le x, \tag{C.13}$$

where $\alpha_1 \beta > 0$, and both parameters are shape parameters.
The moments exist for $k < \beta$, in which case they are given by

$$E(X^k) = \left(\frac{\beta}{\beta - k} \right)^\alpha. \tag{C.14}$$

Weibull Distribution

$$f(x) = \frac{a}{\beta} \left(\frac{x}{\beta} \right)^{a-1} e^{-(x/\beta)^a}, \quad x > 0, \tag{C.15}$$

where a, $\beta > 0$. Here β is a scale and a a shape parameter.
Moments exist for $-a < k < \infty$ and are given by

$$E(X^k) = \Gamma\left(1 + \frac{k}{a} \right) \beta^k. \tag{C.16}$$

Log-Gompertz (Inverse Weibull) Distribution

$$f(x) = a\beta^a x^{-a-1} e^{-(x/\beta)^{-a}}, \quad x > 0. \tag{C.17}$$

Here $\beta > 0$ is a scale and $a > 0$ a shape parameter.
Moments exist for $-\infty < k < a$; they are

$$E(X^k) = \Gamma\left(1 - \frac{k}{a} \right) \beta^k. \tag{C.18}$$

Generalized Beta Distribution of the Second Kind

$$f(x) = \frac{ax^{ap-1}}{b^{ap} B(p, q)[1 + (x/b)^a]^{p+q}}, \quad x > 0, \tag{C.19}$$

where all four parameters a, b, p, q are positive. Here b is a scale and a, p, q are shape parameters.

The moments exist for $-ap < k < aq$ and are given by

$$E(X^k) = \frac{b^k B(p + k/a, q - k/a)}{B(p, q)} = \frac{b^k \Gamma(p + k/a)\Gamma(q - k/a)}{\Gamma(p)\Gamma(q)}. \tag{C.20}$$

Singh–Maddala Distribution

$$f(x) = \frac{aqx^{a-1}}{b^a[1 + (x/b)^a]^{1+q}}, \quad x > 0, \tag{C.21}$$

where all three parameters a, b, q are positive. Here b is a scale and a, q are shape parameters.

The moments exist for $-a < k < aq$ and are given by

$$E(X^k) = \frac{b^k \Gamma(1 + k/a)\Gamma(q - k/a)}{\Gamma(q)}. \tag{C.22}$$

Dagum Type I Distribution

$$f(x) = \frac{apx^{ap-1}}{b^{ap}[1 + (x/b)^a]^{p+1}}, \tag{C.23}$$

where all three parameters a, b, p are positive. Here b is a scale and a, p are shape parameters.

The moments exist for $-ap < k < a$ and are given by

$$E(X^k) = \frac{b^k \Gamma(p + k/a)\Gamma(1 - k/a)}{\Gamma(p)}. \tag{C.24}$$

Benini Distribution

$$f(x) = 2\beta x^{-1} \cdot \exp\left\{-\beta\left[\log\left(\frac{x}{x_0}\right)\right]^2\right\} \cdot \log\left(\frac{x}{x_0}\right), \quad x_0 \leq x. \tag{C.25}$$

All moments exist for this distribution.

Champernowne Distribution

$$f(x) = \frac{\alpha \sin\theta}{\theta x[(x/x_0)^{-\alpha} + 2\cos\theta + (x/x_0)^\alpha]}. \tag{C.26}$$

Here $x_0 > 0$ is a scale and $\alpha > 0$ and $0 < \theta < \pi$ are shape parameters.

The moments exist if $-\alpha < k < \alpha$ and are given by

$$E(X^k) = x_0^k \frac{\pi \sin(k\theta/\alpha)}{\theta \sin(k\pi/\alpha)}. \tag{C.27}$$

List of Symbols

Basic Notation

e, e_X	mean excess function (mean residual life)
F, F_X	cumulative distribution function (c.d.f.)
F^{-1}, F_X^{-1}	quantile function (generalized inverse of F)
$\bar{F}$, $\bar{F}_X$	survival function of F
f, f_X	probability density function (p.d.f.)
$F_{(k)}$	c.d.f. of kth moment distribution
$F_{i:n}$	c.d.f. of $X_{i:n}$
$F * H$	convolution of F and H
$F(\cdot;\theta)\bigwedge_\theta H(\cdot)$	mixture of F w.r.t. H
G	Gini coefficient
L, L_X	Lorenz curve
P	Pietra coefficient
r, r_X	hazard rate
sign x	sign function
tr A	trace of the matrix A
$X_{i:n}$	ith (smallest) order statistic
Z, Z_X	Zenga curve
$\stackrel{d}{=}$	equality in distribution
$\stackrel{d}{\to}$	convergence in distribution
$\stackrel{a.s.}{\to}$	almost sure convergence
$\lfloor t \rfloor$	largest integer not exceeding t
$\lvert A \rvert$	determinant of the matrix A
$(n)_p$	Pochhammer's symbol, $(n)_p = n(n+1)(n+2)\cdots(n+p-1)$
$s \wedge t$	$\min\{s, t\}$

Statistical Size Distributions in Economics and Actuarial Sciences, By Christian Kleiber and Samuel Kotz.
ISBN 0-471-15064-9 © 2003 John Wiley & Sons, Inc.

Special Functions

$\mathrm{erf}(x)$	error function, $\mathrm{erf}(x) = (2/\sqrt{\pi}) \int_0^x \exp(-t^2)\,dt$
$\Gamma(x)$	gamma function, $\Gamma(x) = \int_0^\infty t^{x-1}e^{-t}\,dt$
$\Gamma(\nu, x)$	incomplete gamma function, $\Gamma(\nu, x) = \int_x^\infty t^{\nu-1}e^{-t}\,dt$
$\gamma(\nu, x)$	incomplete gamma function, $\gamma(\nu, x) = \int_0^x t^{\nu-1}e^{-t}\,dt$
$\psi(x)$	psi function (digamma function), $\psi(x) = d\log\Gamma(x)/dx$
$B(x, y)$	beta function, $B(x, y) = \Gamma(x)\Gamma(y)/\Gamma(x+y)$
$B_z(x, y)$	incomplete beta function
${}_1F_1(a; b; x)$	confluent hypergeometric function (Kummer's function)
${}_2F_1(a_1, a_2; b; x)$	Gauss's hypergeometric series
${}_pF_q$	generalized hypergeometric function, ${}_pF_q(a_1, \ldots, a_p; b_1, \ldots, b_q; x) = \sum_{n=0}^\infty [((a_1)_n \cdots (a_p)_n)/((b_1)_n \cdots (b_q)_n)](x^n/n!)$, $\lvert x\rvert < 1$
$\zeta(p)$	Riemann zeta function, $\zeta(p) = \sum_{j=1}^\infty j^{-p}$, $p > 1$
$\Phi(x)$	c.d.f. of the standard normal distribution
$\varphi(x)$	p.d.f. of the standard normal distribution

Distributions

$\mathrm{B1}(p, q)$	beta distribution of the first kind with parameters p, q
$\mathrm{B2}(p, q)$	beta distribution of the second kind with parameters p, q
$\mathrm{D}(a, b, p)$	Dagum distribution with parameters a, b, p
$\mathrm{Exp}(\lambda, x_0)$	exponential distribution with parameters λ, x_0
$\mathrm{Fisk}(a, b)$	Fisk (log-logistic) distribution with parameters a, b
$\mathrm{Ga}(b, p)$	gamma distribution with parameters b, p
$\mathrm{GB1}(a, b, p, q)$	GB1 distribution with parameters a, b, p, q
$\mathrm{GB2}(a, b, p, q)$	GB2 distribution with parameters a, b, p, q
$\mathrm{genLN}(\mu, \sigma_r, r)$	generalized lognormal distribution with parameters μ, σ_r, r
$\mathrm{GG}(a, b, p)$	generalized gamma distribution with parameters a, b, p
$\mathrm{LN}(\mu, \sigma^2)$	two-parameter lognormal distribution with parameters μ, σ^2
$\mathrm{LN}(\mu, \sigma^2, \lambda)$	three-parameter lognormal distribution with parameters μ, σ^2, λ
$\mathrm{N}(\mu, \sigma^2)$	normal distribution with parameters μ, σ^2
$\mathrm{Par}(x_0, \alpha)$	(classical) Pareto distribution with parameters x_0, α
$\mathrm{Par(II)}(b, q)$	Pareto type II distribution (Lomax distribution) with parameters b, q
$\mathrm{SM}(a, b, q)$	Singh–Maddala distribution with parameters a, b, q
$\mathrm{U}[a, b]$	uniform distribution supported on $[a, b]$
$\mathrm{Wei}(a, b)$	Weibull distribution with parameters a, b

Stochastic Order Relations

$\geq_{\mathrm{FSD}}$	first-order stochastic dominance (usual stochastic order)
$\geq_{\mathrm{L}}$	Lorenz order
$\geq_{\mathrm{SSD}}$	second-order stochastic dominance
$\geq_{\mathrm{Z}}$	Zenga order

References

Aaberge, R. (2000). Characterizations of Lorenz curves and income distributions. *Social Choice and Welfare*, 17, 639–653.

Abberger, K., and Heiler, S. (2000). Simultaneous estimation of parameters for a generalized logistic distribution and application to time series models. *Allgemeines Statistisches Archiv*, 84, 41–49.

Adarkar, B. P., and Sen Gupta, S. N. (1936). The Pareto law and the distribution of incomes in India. *Economic Journal*, 46, 168–171.

Aebi, M., Embrechts, P., and Mikosch, T. (1992). A large claim index. *Bulletin of the Swiss Association of Actuaries*, 143–156.

Aggarwal, V., and Singh, R. (1984). On optimum stratification with proportional allocation for a class of Pareto distributions. *Communications in Statistics—Theory and Methods*, 13, 3107–3116.

Agrò, G. (1995). Maximum likelihood estimation for the exponential power function parameters. *Communications in Statistics—Simulation*, 24, 523–536.

Ahuja, J. C. (1969). On certain properties of the generalized Gompertz distribution. *Sankhyā*, B31, 541–544.

Aigner, D. J., and Goldberger, A. S. (1970). Estimation of Pareto's law from grouped observations. *Journal of the American Statistical Association*, 65, 712–723.

Aitchison, J., and Brown, J. A. C. (1954). On criteria for description of income distribution. *Metroeconomica*, 6, 88–107.

Aitchison, J., and Brown, J. A. C. (1957). *The Lognormal Distribution*. Cambridge: Cambridge University Press.

Al-Hussaini, E. K. (1991). A characterization of the Burr type XII distribution. *Applied Mathematics Letters*, 4, 59–61.

Ali, M. M., Woo, J., Lee, C., and Yoon, G.-E. (2001). Estimation of Lorenz curve and Gini index in a Pareto distribution. Working paper, Ball State University, IN.

Alzaid, A. A. (1990). Lorenz ranking of income distributions. *Statistical Papers*, 31, 209–224.

Statistical Size Distributions in Economics and Actuarial Sciences, By Christian Kleiber and Samuel Kotz.
ISBN 0-471-15064-9

Ammon, O. (1895). *Die Gesellschaftsordnung und ihre natürlichen Grundlagen*. Jena: Gustav Fischer.

Ammon, O. (1898). Some social applications of the doctrine of probability. *Journal of Political Economy*, 7, 204–237.

Amoroso, L. (1924–1925). Richerche intorno alla curva dei redditi. *Annali di Matematica Pura ed Applicata*, Series 4–21, 123–157.

Anand, S. (1983). *Inequality and Poverty in Malaysia*. New York and Oxford: Oxford University Press.

Andersson, H. (1971). An analysis of the development of the fire losses in the Northern countries after the second World War. *ASTIN Bulletin*, 6, 25–30.

Angle, J. (1986a). Coalitions in a stochastic process of wealth distribution. *ASA Proceedings of the Social Statistics Section*, 259–263.

Angle, J. (1986b). The surplus theory of social stratification and the size distribution of personal wealth. *Social Forces*, 65, 293–326.

Angle, J. (1990). A stochastic interacting particle system model of the size distribution of wealth and income. *ASA Proceedings of the Social Statistics Section*, 279–284.

Angle, J. (1992). The inequality process and the distribution of income to blacks and whites. *Journal of Mathematical Sociology*, 17, 77–98.

Angle, J. (1993a). Deriving the size distribution of personal wealth from "the rich get richer the poor get poorer." *Journal of Mathematical Sociology*, 18, 27–46.

Angle, J. (1993b). An apparent invariance of the size distribution of personal income conditioned on education. *ASA Proceedings of the Social Statistics Section*, 197–202.

Angle, J. (1994). Frequency spikes in income distributions. *ASA Proceedings of the Business and Economic Statistics Section*, 265–270.

Angle, J. (1996). How the gamma law of income distribution appears invariant under aggregation. *Journal of Mathematical Sociology*, 21, 325–358.

Angle. J. (1999). Pervasive competition: The dynamics of income and income distribution. *ASA Proceedings of the Social Statistics Section*, 331–336.

Angle. J. (2000). The binary interacting particle system (bips) underlying the maxentropic derivation of the gamma law of income distribution. *ASA Proceedings of the Social Statistics Section*, 270–275.

Anstis, G. R. (1978). *The Estimation of Lorenz Curves*. Thesis, Dept. of Economic Statistics, University of New England, Australia.

Aoyama, H., Souma, W., Nagahara, Y., Okazaki, M. P., Takayasu, H., and Takayasu, M. (2000). Pareto's law for income of individuals and debt of bankrupt companies. *Fractals*, 8, 293–300.

Arnold, B. C. (1971). Two characterizations of the exponential distribution. Mimeo, Iowa State University, Ames, Iowa.

Arnold, B. C. (1983). *Pareto Distributions*. Fairland, MD: International Co-operative Publishing House.

Arnold, B. C. (1985). Pareto distribution. In: S. Kotz, N. L. Johnson, and C. Read (eds.): *Encyclopedia of Statistical Sciences*, Vol. 6. New York: John Wiley, pp. 568–574.

Arnold, B. C. (1986). A class of hyperbolic Lorenz curves. *Sankhyā*, B48, 427–436.

Arnold, B. C. (1987). *Majorization and the Lorenz Order*. Lecture Notes in Statistics 43, Berlin and New York: Springer.

Arnold, B. C., and Balakrishnan, N. (1989). *Relations, Bounds and Approximations for Order Statistics*. Lecture Notes in Statistics 53, Berlin and New York: Springer.

Arnold, B. C., and Laguna, L. (1977). On generalized Pareto distributions with applications to income data. International Studies in Economics No. 10, Dept. of Economics, Iowa State University, Ames, Iowa.

Arnold, B. C., Robertson, C. A., Brockett, P. L., and Shu, B.-Y. (1987). Generating ordered families of Lorenz curves by strongly unimodal distributions. *Journal of Business and Economic Statistics*, 5, 305–308.

Arnold, B. C., Robertson, C. A., and Yeh, H. C. (1986). Some properties of a Pareto-type distribution. *Sankhyā*, A48, 404–408.

Asrabadi, B. R. (1990). Estimation in the Pareto distribution. *Metrika*, 37, 199–205.

Atkinson, A. B. (1970). On the measurement of inequality. *Journal of Economic Theory*, 2, 244–263.

Atkinson, A. B., and Bourguignon, F. (eds.) (2000). *Handbook of Income Distribution*. Amsterdam: Elsevier.

Atkinson, A. B., and Harrison, A. J. (1978). *Distribution of Personal Wealth in Britain*. Cambridge: Cambridge University Press.

Atoda, N., Suruga, T., and Tachibanaki, T. (1988). Statistical inference of functional form for income distribution. *Economic Studies Quarterly*, 39, 14–40.

Azlarov, T. A., Dzamirzaev, A. A., and Sultanova, M. M. (1972). Characterization properties of the exponential distribution, and their stability (in Russian). *Random Processes and Statistical Inference, No. II, Izdat. "Fan" Uzbek. SSR, Tashkent*, 10–19.

Azzalini, A., and Kotz, S. (2002). Log-skew-normal and log-skew-*t* distributions as models for family income data. Technical report, Dept. of Statistics, University of Padua, Italy.

Bakker, A., and Creedy, J. (1997). Age and the distribution of earnings. In: J. Creedy and V. L. Martin (eds.): pp. 111–127.

Bakker, A., and Creedy, J. (1998). Estimating the exponential family using grouped data: An application to the New Zealand distribution of income. *New Zealand Economic Papers*, 32, 111–127.

Bakker, A., and Creedy, J. (2000). Macroeconomic variables and income distribution. Conditional modelling with the generalised exponential. *Journal of Income Distribution*, 9, 183–197.

Balakrishnan, N., and Chen, W. W. S. (1999). *Handbook of Tables for Order Statistics from Lognormal Distributions with Applications*, Dordrecht, Boston, and London: Kluwer.

Bantilan, M. C. S., Bernal, N. B., de Castro, M. M., and Pattugalan, J. M. (1995). Income distribution in the Philippines, 1957–1988: An application of the Dagum model to the family income and expenditure survey (FIES) data. In: C. Dagum and A. Lemmi (eds.): pp. 11–43.

Barlow, R. E., and Proschan, F. (1981). *Statistical Theory of Reliability and Life Testing: Probability Models*, 2nd ed. Silver Spring, MD: To Begin With.

Barndorff-Nielsen, O.-E. (1977). Exponentially decreasing distributions for the logarithm of particle size. *Proceedings of the Royal Society London*, A353, 401–419.

Barndorff-Nielsen, O.-E. (1978). Hyperbolic distributions and distributions on hyperbolae. *Scandinavian Journal of Statistics*, 5, 151–157.

Barrett, G., and Donald, S. (2000). Statistical inference with generalized Gini indices of inequality and poverty. Discussion Paper, Dept. of Economics, University of New South Wales, Australia.

Barrett, G. F., and Pendakur, K. (1995). The asymptotic distribution of the generalized Gini indices of inequality. *Canadian Journal of Economics*, 28, 1042–1055.

Bartels, C. P. A. (1977). *Economic Aspects of Regional Welfare*. Leiden: Martinus Nijhoff.

Bartels, C. P. A., and Vries, O. M. (1977). Succinct analytical descriptions of income distribution using transformation functions. *Economie Appliquée*, 30, 369–390.

Basmann, R. L., Hayes, K. J., Slottje, D. J., and Johnson, J. D. (1990). A general functional form for approximating the Lorenz curve. *Journal of Econometrics*, 43, 77–90.

Baxter, M. A. (1980). Minimum variance unbiased estimation of the parameters of the Pareto distribution. *Metrika*, 27, 133–138.

Beach, C. M., and Davidson, R. (1983). Distribution-free statistical inference with Lorenz curves and income shares. *Review of Economic Studies*, 50, 723–735.

Beach, C. M., and Richmond, J. (1985). Joint confidence intervals for income shares and Lorenz curves. *International Economic Review*, 26, 439–450.

Becker, G. (1962). Investment in human capital: A theoretical analysis. *Journal of Political Economy*, 70, 9–49.

Becker, G. (1964). *Human Capital*. New York: Columbia University Press.

Beirlant, J., and Teugels, J. L. (1992). Modelling large claims in non-life insurance. *Insurance: Mathematics & Economics*, 11, 17–29.

Beirlant, J., Teugels, J., and Vynckier, P. (1996). *Practical Analysis of Extreme Values*. Leuven: Leuven University Press.

Bell, C., Klonner, S., and Moos, W. (1999). Inequality and Economic Performance in Rural India since 1955. Working paper, South-Asia Institute, Department of Development Economics, Universität Heidelberg, Germany.

Benckert, L.-G. (1962). The lognormal model for the distribution of one claim. *ASTIN Bulletin*, 2, 9–23.

Benckert, L.-G., and Jung, J. (1974). Statistical models of claim distributions in fire insurance. *ASTIN Bulletin*, 8, 1–25.

Benckert, L.-G., and Sternberg, I. (1957). An attempt to find an expression for the distribution of fire damage amount. *Transactions of the 15th International Congress of Actuaries*, 2, 288–294.

Benini, R. (1897). Di alcune curve descritte da fenomeni economici aventi relazione colla curva del reddito o con quella del patrimonio. *Giornale degli Economisti*, 14, 177–214. English translation in *Rivista di Politica Economica*, 87 (1997), 701–744.

Benini, R. (1901). *Principii di Demografia*. Florence: Barbera Editore.

Benini, R. (1905). I diagrammi a scala logaritmica (a proposito della graduazione per valore delle successioni ereditarie in Italia, Francia e Inghilterra). *Giornale degli Economisti*, Series II, 16, 222–231.

Benini, R. (1906). *Principii di Statistica Metodologica*. Torino: Unione Tipografico-Editrice Torinese.

Benktander, G. (1962). Notes sur la distribution conditionnee du montant d'un sinistre par rapport a l'hypothese qu'il y a eu un sinistre dans l'assurance automobile. *ASTIN Bulletin*, 2, 24–29.

Benktander, G. (1963). A note on the most "dangerous" and skewest class of distributions. *ASTIN Bulletin*, 2, 387–390.

Benktander, G. (1970). Schadenverteilungen nach Grösse in der Nicht-Lebensversicherung. *Bulletin of the Swiss Association of Actuaries*, 263–283.

Benktander, G., and Segerdahl, C. O. (1960). On the analytical representation of claim distributions with special reference to excess of loss reinsurance. *Transactions of the 16th International Congress of Actuaries*, 1, 626–648.

Berti, P., and Rigo, P. (1995). A note on Zenga concentration index. *Journal of the Italian Statistical Society*, 4, 397–404.

Bhattacharjee, M. C. (1988). Reliability ideas and applications on economics and social sciences. In: P. R. Krishnaiah and C. R. Rao (eds.): *Handbook of Statistics*, Vol. 7. Amsterdam: North-Holland, pp. 175–213.

Bhattacharjee, M. C., and Krishnaji, N. (1985). DFR and other heavy tail properties in modelling the distribution of land and some alternative measures of inequality. In: J. K. Ghosh (ed.): *Statistics: Applications and New Directions. Proceedings of the Indian Statistical Institute Golden Jubilee International Conference*. Calcutta: Indian Statistical Institute, pp. 100–115.

Bhattacharya, N. (1963). A property of the Pareto distribution. *Sankhyā*, B25, 195–196.

Bingham, N. H., Goldie, C. M., and Teugels, J. (1987). *Regular Variation*. Cambridge: Cambridge University Press.

Boissevain, C. H. (1939). Distribution of abilities depending upon two or more independent factors. *Metron*, 13, 49–58.

Bologna, S. (1985). Distribuzioni di alcune statistiche derivanti da una densita lognormale di primo ordine. *Statistica*, 45, 495–501.

Bomsdorf, E. (1977). The prize-competition distribution: A particular L-distribution as a supplement to the Pareto distribution. *Statistische Hefte*, 18, 254–264.

Bondesson, L. (1979). A general result on infinite divisibility. *Annals of Probability*, 7, 965–979.

Bordley, R. F., and McDonald, J. B. (1993). Estimating aggregate automotive income elasticities from the population income-share elasticity. *Journal of Business and Economic Statistics*, 11, 209–214.

Bordley, R. F., McDonald, J. B., and Mantrala, A. (1996). Something new, something old: Parametric models for the size distribution of income. *Journal of Income Distribution*, 6, 91–103.

Botargues, P., and Petrecolla, D. (1997). Funciones de Distribución del Ingreso en Gran Buenos Aires, Argentina, 1990–1996. Paper presented at the XXV Symposium of the Argentine Statistical Society, Mar del Plata, Argentina.

Botargues, P., and Petrecolla, D. (1999a). Funciones de distribución del ingreso y afluencia económica relativa para ocupados según nivel de educación en GBA, Argentina, 1992–1996. In: M. Cardenas Santa Maria and N. Lustig (eds.): *Pobreza y desigualdad en América Latina*, Santafé de Bogotá, D. C., Fedesarrollo, Lacea, Colciencias, Tercer Mundo. Also Documento de Trabajo Instituto Torcuato Di Tella, DTE 216.

Botargues, P., and Petrecolla, D. (1999b). Estimaciones parametricos y no parametricos de la distribucion del ingreso de los ocupados del Gran Buenos Aires, 1992–1997. *Economica* (National University of La Plata), XLV (no. 1), 13–34.

Bowley, A. L. (1926). *Elements of Statistics*, 5th ed. Westminster: P. S. King and Sons.

Bowman, K. O., and Shenton, L. R. (1988). *Properties of Estimators for the Gamma Distribution*. New York: Marcel Dekker.

Bowman, M. J. (1945). A graphical analysis of personal income distribution in the United States. *American Economic Review*, 35, 607–628.

Box, G. E. P., and Tiao, G. (1973). *Bayesian Inference in Statistical Analysis*. Reading, MA: Addison-Wesley.

Brachmann, K., Stich, A., and Trede, M. (1996). Evaluating parametric income distribution models. *Allgemeines Statistisches Archiv*, 80, 285–298.

Brambilla, F. (1960). *La Distribuzione dei Redditi*. Pavia: Fusi.

Brazauskas, V. (2002). Fisher information matrix for the Feller–Pareto distribution. *Statistics & Probability Letters*, 59, 159–167.

Brazauskas, V., and Serfling, R. (2001a). Robust estimation of tail parameters for two-parameter Pareto and exponential models via generalized quantile statistics. *Extremes*, 3, 231–249.

Brazauskas, V., and Serfling, R. (2001b). Small sample performance of robust estimators of tail parameters for Pareto and exponential models. *Journal of Statistical Computation and Simulation*, 70, 1–19.

Bresciani Turroni, C. (1905). Dell'influenza delle condizioni economiche sulla forma della curva dei redditi. *Giornale degli Economisti*, 31, 115–138. English translation in *Rivista di Politica Economica*, 87 (1997), 745–767.

Bresciani Turroni, C. (1910). Di un indice misuratore della disuguaglianza nella distribuzione della richezza. In: *Studi in onore di Biagio Brugi*, Palermo, pp. 797–812.

Bresciani Turroni, C. (1914). Osservazioni critiche sul "Metodo di Wolf" per lo studio della distribuzione dei redditi. *Giornale degli Economisti e Rivista di Statistica*, Series IV, 25, 382–394.

Bresciani Turroni, C. (1939a). On some methods of measuring the inequality of incomes. *Revue Al Qanoun Wal Iqtisad*, 371–403.

Bresciani Turroni, C. (1939b). Annual survey of statistical data: Pareto's law and the index of inequality of incomes. *Econometrica*, 7, 107–133.

Brunazzo, A., and Pollastri, A. (1986). Proposta di una nuova distribuzione: la lognormale generalizzata. In: *Scritti in Onore di Francesco Brambilla*, Vol. 1, Milano: Ed. Bocconi Comunicazioni, pp. 57–83.

Burkhauser, R. V., Kreyenfeld, M., and Wagner, G. G. (1997). The German-Socio-Economic Panel: A representative sample of reunited Germany and its parts. *Vierteljahrshefte zur Wirtschaftsforschung*, 66, 7–16.

Burnecki, K, Kukla, G., and Weron, R. (2000). Property insurance loss distributions. *Physica*, A287, 269–278.

Burr, I. W. (1942). Cumulative frequency functions. *Annals of Mathematical Statistics*, 13, 215–232.

Butler, R. J., and McDonald, J. B. (1989). Using incomplete moments to measure inequality. *Journal of Econometrics*, 42, 109–119.

Campano, F. (1987). A fresh look at Champernowne's five-parameter formula. *Economie Appliquée*, 40, 161–175.

Cammillieri, G. (1972). Di una distribuzione del prodotto di variabili stocastiche. Università di Palermo, Annali della Facoltà di Economia e Commercio, 25, 87–97.

Cantelli, F. (1921). Sulla deduzione delle leggi di frequenza da considerazioni di probabilità. *Metron*, 1, 83–91.

Cantelli, F. P. (1929). Sulla legge di distribuzione dei redditi. *Giornale degli Economisti e Rivista di Statistica*, Series IV, 44, 850–852.

Chakravarty, S. R. (1990). *Ethical Social Index Numbers*. Berlin and New York: Springer.

Champernowne, D. G. (1937). The theory of income distribution. *Econometrica*, 5, 379–381.

Champernowne, D. G. (1952). The graduation of income distributions. *Econometrica*, 20, 591–615.

Champernowne, D. G. (1953). A model of income distribution. *Economic Journal*, 53, 318–351.

Champernowne, D. G. (1956). Discussion of Hart and Prais (1956). *Journal of the Royal Statistical Society*, A119, 181–182.

Champernowne, D. G. (1973). *The Distribution of Income Between Persons*. Cambridge: Cambridge University Press.

Champernowne, D. G. (1974). A comparison of measures of inequality of income distribution. *Economic Journal*, 84, 787–816.

Champernowne, D. G., and Cowell, F. A. (1998). *Economic Inequality and Income Distribution*. Cambridge: Cambridge University Press.

Chandra, M., and Singpurwalla, N. D. (1981). Relationships between some notions which are common to reliability theory and economics. *Mathematics of Operations Research*, 6, 113–121.

Chen, Z. (1997). Exact confidence interval for the shape parameter of a log-logistic distribution. *Journal of Statistical Computation and Simulation*, 56, 193–211.

Cheng, R. C. H., and Amin, N. A. K. (1983). Estimating parameters in continuous univariate distributions with a shifted origin. *Journal of the Royal Statistical Society*, B44, 394–403.

Chesher, A. (1979). An analysis of the distribution of wealth in Ireland. *Economic and Social Review*, 11, 1–17.

Chipman, J. S. (1985). The theory and measurement of income distribution. In: R. L. Basmann and G. F. Rhodes (eds.): *Advances in Econometrics, Vol. 4: Economic Inequality. Survey Methods and Measurement*. London: JAI Press, pp. 135–165.

Chotikapanich, D. (1993). A comparison of alternative functional forms for the Lorenz curve. *Economics Letters*, 41, 129–138.

Chotikapanich, D. (1994). A note on the income distribution underlying the Rasche, Gaffney, Koo and Obst Lorenz curve. *Journal of Quantitative Economics*, 10, 235–246.

Clayton, D. G. (1978). A model for association in bivariate life tables and its application in epidemiological studies of family tendency in chronic disease incidence. *Biometrika*, 65, 141–151.

Cohen, A. C. (1951). Estimating parameters of logarithmic-normal distributions by maximum likelihood. *Journal of the American Statistical Association*, 46, 206–212.

Cohen, A. C. (1969). A generalization of the Weibull distribution. NASA Contractor Rept. No. 61293, NAS 8-11175, Marshall Space Flight Center.

Cohen, A. C. (1973). The reflected Weibull distribution. *Technometrics*, 15, 867–873.

Cohen, A. C., and Whitten, B. J. (1980). Estimation in the three-parameter lognormal distributions. *Journal of the American Statistical Association*, 75, 399–404.

Cohen, A. C., and Whitten, B. J. (1988). *Parameter Estimation in Reliability and Life Span Models*. New York: Marcel Dekker.

Colombi, R. (1990). A new model of income distribution: The Pareto–lognormal distribution. In: C. Dagum and M. Zenga (eds.): pp. 18–32.

Cook, R. D., and Johnson, M. E. (1981). A family of distributions for modelling non-elliptically symmetric multivariate data. *Journal of the Royal Statistical Society*, B43, 210–218.

Cowell, F. A. (1995). *Measuring Inequality*, 2nd ed. Hemel Hempstead: Prentice Hall/ Harvester Wheatsheaf.

Cowell, F. A. (2000). Measurement of inequality. In: A. B. Atkinson and F. Bourguignon (eds.): *Handbook of Income Distribution*, Amsterdam: Elsevier, pp. 87–166.

Cowell, F. A., Ferreira, F. H. G., and Litchfield, J. A. (1998). Income distribution in Brazil 1981–1990. Parametric and non-parametric approaches. *Journal of Income Distribution*, 8, 63–76.

Cowell, F. A., and Kuga, K. (1981). Additivity and the entropy concept: An axiomatic approach to inequality measurement. *Journal of Economic Theory*, 25, 131–143.

Cowell, F. A., and Mehta, F. (1982). The estimation and interpolation of inequality measures. *Review of Economic Studies*, 43, 273–290.

Cowell, F. A., and Victoria-Feser, M.-P. (1996). Robustness properties of inequality measures. *Econometrica*, 64, 77–101.

Cox, D. R. (1961). Tests of separate families of hypotheses. *Proceedings of the 4th Berkeley Symposium on Mathematical Statistics and Probability*, 1, 105–123.

Cramér, H. (1936). Über eine Eigenschaft der normalen Verteilungsfunktion. *Mathematische Zeitschrift*, 41, 405–414.

Cramer, J. S. (1978). A function for the size distribution of incomes: Comment. *Econometrica*, 46, 459–460.

Creedy, J. (1977). Pareto and the distribution of income. *Review of Income and Wealth*, 23, 405–411.

Creedy, J., Lye, J. N., and Martin, V. L. (1997). A model of income distribution. In: J. Creedy and V. L. Martin (eds.): pp. 29–45.

Creedy, J., and Martin, V. L. (eds.) (1997). *Nonlinear Economic Models. Cross-sectional, Time Series and Neural Network Applications*. Cheltenham, UK: Edward Elgar.

Cronin, D. C. (1979). A function for the size distribution of incomes: A further comment. *Econometrica*, 47, 773–774.

Crovella, M. E., Taqqu, M. S., and Bestavros, A. (1998). Heavy-tailed probability distributions in the World Wide Web. In: R. J. Adler, R. E. Feldman and M. S. Taqqu (eds.): *A Practical Guide to Heavy Tails*. Boston, Basel, and Berlin: Birkhäuser.

Crow, E. L., and Shimizu, K. S. (eds.) (1988). *Lognormal Distributions*. New York: Marcel Dekker.

Csörgő, M., and Zitikis, R. (1996). Strassen's LIL for the Lorenz curve. *Journal of Multivariate Analysis*, 59, 1–12.

Csörgő, M., and Zitikis, R. (1997). On the rate of strong consistency of Lorenz curves. *Statistics & Probability Letters*, 34, 113–121.

Csörgő, M., Gastwirth, J. L., and Zitikis, R. (1998). Asymptotic confidence bands for the Lorenz and Bonferroni curves based on the empirical Lorenz curve. *Journal of Statistical Planning and Inference*, 74, 65–91.

Cummins, J. D., and Freifelder, L. R. (1978). A comparative analysis of alternative maximum probable yearly aggregate loss estimators. *Journal of Risk and Insurance*, 35, 27–52.

Cummins, J. D., Dionne, G., McDonald, J. B., and Pritchett, B. M. (1990). Applications of the GB2 family of distributions in modeling insurance loss processes. *Insurance: Mathematics & Economics*, 9, 257–272.

D'Addario, R. (1932). Intorno alla curva dei redditi di Amoroso. *Rivista Italiano di Statistica, Economia e Finanza*, 4, 89–108.

D'Addario, R. (1934–1935). Curve di concentrazione, elasticità, flessibilità, densità media e densità marginale dei redditi. *Archivio Scientifico* (Bari), 9, 35–73.

D'Addario, R. (1936). Sulla curve dei redditi di Amoroso. *Annali dell'Istituto di Statistica dell'Università di Bari*, 10, 1–57.

D'Addario, R. (1938). Sulla rappresentazione analitica delle curve di frequenza. *Annali dell'Istituto di Statistica dell'Università di Bari*, 5–59.

D'Addario, R. (1939). Un metodo per la rappresentazione analitica delle distribuzione statistiche. *Annali dell'Istituto di Statistica dell'Università di Bari*, 16, 3–56.

D'Addario, R. (1949). Richerche sulla curva dei redditi. *Giornale degli Economisti e Annali di Economia*, 8, 91–114.

D'Addario, R. (1969). Sulle ripartizioni la cui media superiormente o inferiormente incompleta cresce linearmente col crescere della variabilità distributiva. *Giornale degli Economisti e Annali di Economia*, 28, 795–821.

D'Addario, R. (1974). Intorno ad una funzione di distribuzione. *Giornale degli Economisti e Annali di Economia*, 33, 205–214.

Dagum, C. (1975). A model of income distribution and the conditions of existence of moments of finite order. *Bulletin of the International Statistical Institute*, 46 (Proceedings of the 40th Session of the ISI, Warsaw, Contributed Papers), 199–205.

Dagum, C. (1977). A new model of personal income distribution: Specification and estimation. *Economie Appliquée*, 30, 413–437.

Dagum, C. (1980a). The generation and distribution of income, the Lorenz curve and the Gini ratio. *Economie Appliquée*, 33, 327–367.

Dagum, C. (1980b). Generating systems and properties of income distribution models. *Metron*, 38, 3–26.

Dagum, C. (1980c). Sistemas generadores de distribución de ingreso y la ley de Pareto. *El Trimestre Economico*, 47, 877–917. Reprinted in *Estadística*, 35 (1981), 143–183.

Dagum, C. (1983). Income distribution models. In: S. Kotz, N. L. Johnson, and C. Read (eds.): *Encyclopedia of Statistical Sciences*, Vol. 4. New York: John Wiley, pp. 27–34.

Dagum, C. (1985). Analysis of income distribution and inequality by education and sex in Canada. *Advances in Econometrics*, 4, 167–227.

Dagum, C. (1986). Book review of *Income Inequality and Poverty. Methods of Estimation and Policy Applications* by N. C. Kakwani. *Journal of Business and Economic Statistics*, 4, 391.

Dagum, C. (1990a). Generation and properties of income distribution functions. In: C. Dagum and M. Zenga (eds.): pp. 1–17.

Dagum, C. (1990b). A new model of net wealth distribution specified for negative, null, and positive wealth. A case study: Italy. In: C. Dagum and M. Zenga (eds.): pp. 42–56.

Dagum, C. (1996). A systemic approach to the generation of income distribution models. *Journal of Income Distribution*, 6, 105–126.

Dagum, C. (1999). Linking the functional and personal distributions of income. In: J. Silber (ed.): *Handbook on Income Inequality Measurement*, Boston, Dordrecht, and London: Kluwer, pp. 101–128.

Dagum, C., and Lemmi, A. (1989). A contribution to the analysis of income distribution and income inequality, and a case study: Italy. *Research on Economic Inequality*, 1, 123–157.

Dagum, C., and Lemmi, A. (eds.) (1995). *Research on Economic Inequality, Vol. 6: Income Distribution, Social Welfare, Inequality and Poverty.* Greenwich, CT: JAI Press.

Dagum, C. and Zenga, M. (eds.) (1990). *Income and Wealth Distribution, Inequality and Poverty: Proceedings of the Second International Conference on Income Distribution by Size: Generation, Distribution, Measurement and Applications*. New York, Berlin, London, and Tokyo: Springer.

Dancelli, L. (1986). Tendenza alla massima ed alla minima concentrazione nel modello di distribuzione del reddito personale di Dagum. In: *Scritti in Onore di Francesco Brambilla*, Vol. 1, Milano: Ed. Bocconi Comunicazioni, pp. 249–267.

Dardanoni, V., and Forcina, A. (1999). Inference for Lorenz curve orderings. *Econometrics Journal*, 2, 49–75.

Dargahi-Noubary, G. R. (1994). On distributions of large incomes. *Journal of Applied Statistical Science*, 1, 211–222.

David, H. A. (1968). Gini's mean difference rediscovered. *Biometrika*, 55, 573–574.

David, H. A. (1981). *Order Statistics*, 2nd ed. New York: John Wiley.

Davidson, R., and Duclos, J.-Y. (2000). Statistical inference for stochastic dominance and for the measurement of poverty and inequality. *Econometrica*, 68, 1435–1464.

Davies, J. B., and Shorrocks, A. F. (1989). Optimal grouping of income and wealth data. *Journal of Econometrics*, 42, 97–108.

Davis, H. T. (1941a). *The Analysis of Economic Time Series*. Bloomington, IN: Principia Press.

Davis, H. T. (1941b). *The Theory of Econometrics*. Bloomington, IN: Principia Press.

De Forest, E. (1882–1883). On an unsymmetrical probability curve. *The Analyst*, 9, 135–142, 161–168 (1882); 10, 1–7, 67–74 (1883).

Devroye, L. (1986). *Non-Uniform Random Variate Generation*. Berlin and New York: Springer.

Dimaki, C., and Xekalaki, E. (1990). Identifiability of income distributions in the context of damage and generating models. *Communications in Statistics—Theory and Methods*, 19, 2757–2766.

Dimaki, C., and Xekalaki, E. (1996). Towards a unification of certain characterizations by conditional expectations. *Annals of the Institute of Statistical Mathematics*, 48, 157–168.

Domański, C., and Jedrzejczak, A. (1998). Maximum likelihood estimation of the Dagum model parameters. *International Advances in Economic Research*, 4, 243–252.

Domma, F. (1994). Tendenza alla massima ed alla minima concentrazione nel modello Pareto generalizzato. *Quaderni di Statistica e Matematica Applicata alle Scienze Economico-Sociali*, 16, 111–127.

Domma, F. (1997). Mediana e range campionario per il modello di Dagum. *Quaderni di Statistica e Matematica Applicata alle Scienze Economico-Sociali*, 19, 195–204.

Donaldson, D., and Weymark, J. A. (1980). A single-parameter generalization of the Gini index of inequality. *Journal of Economic Theory*, 22, 67–86.

Donaldson, D., and Weymark, J. A. (1983). Ethically flexible Gini indices for income distributions in the continuum. *Journal of Economic Theory*, 29, 353–358.

Dorfman, R. (1979). A formula for the Gini coefficient. *Review of Economics and Statistics*, 61, 146–149.

Dubey, S. D. (1968). A compound Weibull distribution. *Naval Research Logistics Quarterly*, 15, 197–198.

Dubey, S. D. (1970). Compound gamma, beta and *F* distributions. *Metrika*, 16, 27–31.

DuMouchel, W. H., and Olshen, R. A. (1975). On the distribution of claims costs. In: P. M. Kahn (ed.): *Credibility: Theory and Applications* (Proceedings of the Berkeley Actuarial Research Conference on Credibility, Sept. 19–21, 1974), pp. 23–46.

Dunne, P., and Hughes, A. (1994). Age, size, growth and survival: UK companies in the 1980s. *Journal of Industrial Economics*, 42, 115–140.

Dutka, J. (1981). The incomplete beta function—a historical profile. *Archive for the History of Exact Sciences*, 24, 11–29.

Dyer, D. (1981). Structural probability bounds for the strong Pareto law. *Canadian Journal of Statistics*, 9, 71–77.

Edgeworth, F. Y. (1898). On the representation of statistics by mathematical formulae. *Journal of the Royal Statistical Society*, 1, 670–700.

El-Saidi, M. A., Singh, K. P., and Bartolucci, A. A. (1990). A note on a characterization of the generalized log-logistic distribution. *Environmetrics*, 1, 337–342.

Embrechts, P., Klüppelberg, C., and Mikosch, T. (1997). *Modelling Extremal Events*. Berlin and New York: Springer.

Engwall, L. (1968). Size distributions of firms—A stochastic model. *Swedish Journal of Economics*, 70, 138–159.

Erto, P. (1989). Genesi, proprietà ed identificazione del modello di sopravvivenza Weibull inverso. *Statistica Applicata*, 1, 117–128.

Espinguet, P., and Terraza, M. (1983). Essai d'extrapolation des distributions de salaires français. *Economie Appliquée*, 36, 535–561.

Esteban, J. M. (1986). Income-share elasticity and the size distribution of income. *International Economic Review*, 27, 439–444.

Evans, D. S. (1987a). Tests of some alternative theories of firm growth. *Journal of Political Economy*, 95, 657–674.

Evans, D. S. (1987b). The relationship between firm growth, size and age: estimates for 100 manufacturing industries. *Journal of Industrial Economics*, 35, 567–581.

Fakhry, M. E. (1996). Some characteristic properties of a certain family of probability distributions. *Statistics*, 28, 179–185.

Farebrother, R. W. (1999). A memoir on the life of Harold Thayer Davis (1892–1974). *The Manchester School*, 67, 603–610.

Fattorini, L., and Lemmi, A. (1979). Proposta di un modello alternativo per l'analisi della distribuzione personale del reddito. *Atti Giornate di Lavoro AIRO*, 28, 89–117.

Fechner, G. T. (1860). *Elemente der Psychophysik* (Vol. 1 & 2). Leipzig: Breitkopf & Härtel.

Fechner, G. T. (1897). *Kollektivmasslehre*. Leipzig: Engelmann.

Feller, W. (1971). *An Introduction to Probability Theory and Its Applications*, Vol. II, 2nd ed. New York: John Wiley.

Fellman, J. (1976). The effect of transformations on Lorenz curves. *Econometrica*, 44, 823–824.

Fellman, J. (1980). Transformations and Lorenz curves. Working Paper 48, Swedish School of Economics and Business Administration, Helsinki.

Ferrara, G. (1971). Distributions des sinistres incendie selon leur coût. *ASTIN Bulletin*, 6, 31–41.

Fisk, P. R. (1961a). The graduation of income distributions. *Econometrica*, 29, 171–185.

Fisk, P. R. (1961b). Estimation of location and scale parameters in a truncated grouped sech-square distribution. *Journal of the American Statistical Association*, 56, 692–702.

Fisz, M. (1958). Characterizations of some probability distributions. *Scandinavian Actuarial Journal*, 65–70.

Flehinger, B. J., and Lewis, P. A. (1959). Two-parameter lifetime distributions for reliability studies of renewal processes. *IBM Journal of Research*, 3, 58–73.

Fosam, E. B., and Sapatinas, T. (1995). Characterisations of some income distributions based on multiplicative damage models. *Australian Journal of Statistics*, 37, 89–93.

Foster, J. E., and Ok, E. A. (1999). Lorenz dominance and the variance of logarithms. *Econometrica*, 67, 901–907.

Frank, M. J. (1979). On the simultaneous associativity of $F(x, y)$ and $x + y - F(x, y)$. *Aequationes Mathematicae*, 19, 194–226.

Fréchet, M. (1927). Sur la loi de probabilité de l'écart maximum. *Annales de la Société Polonaise de Mathematique, Cracovie*, 6, 93–116.

Fréchet, M. (1939). Sur les formules de répartition des revenus. *Revue de l'Institut International de Statistique*, 7, 32–38.

Gail, M., and Gastwirth, J. L. (1978). A scale-free goodness-of-fit test based on the Lorenz curve. *Journal of the American Statistical Association*, 73, 787–793.

Galambos, J., and Kotz, S. (1978). *Characterizations of Probability Distributions*. New York and Berlin: Springer.

Galton, F. (1869). *Hereditary Genius: An Inquiry into Its Laws and Consequences*. London: Macmillan.

Galton, F. (1879). The geometric mean in vital and social statistics. *Proceedings of the Royal Society of London*, 29, 365–367.

Garg, M. L., Rao, B. R., and Redmond, C. K. (1970). Maximum-likelihood estimation of the parameters of the Gompertz survival function. *Applied Statistics*, 19, 152–159.

Gastwirth, J. L. (1971). A general definition of the Lorenz curve. *Econometrica*, 39, 1037–1039.

Gastwirth, J. L. (1972). The estimation of the Lorenz curve and Gini index. *Review of Economics and Statistics*, 54, 306–316.

Gastwirth, J. L. (1974). Large sample theory of some measures of income inequality. *Econometrica*, 42, 191–196.

Gastwirth, J. L., and Smith, J. T. (1972). A new goodness of fit test. *ASA Proceedings of the Business and Economic Statistics Section*, 320–322.

Ghitany, M. E. (1996). Some remarks on a characterization of the generalized log-logistic distribution. *Environmetrics*, 7, 277–281.

Gibrat, R. (1931). *Les Inégalités Économiques*. Paris: Librairie du Recueil Sirey.

Giesbrecht, F., and Kempthorne, O. (1976). Maximum likelihood estimation in the three-parameter lognormal distribution. *Journal of the Royal Statistical Society*, B38, 257–264.

Gini, C. (1909a). Il diverso accrescimento delle classi sociali e la concentrazione della ricchezza. *Giornale degli Economisti*, 38, 27–83.

Gini, C. (1909b). Concentration and dependency ratios (in Italian). English translation in *Rivista di Politica Economica*, 87 (1997), 769–789.

Gini, C. (1914). Sulla misura della concentrazione e della variabilità dei caratteri. *Atti del Reale Istituto Veneto di Scienze, Lettere ed Arti*, 73, 1203–1248.

Gini, C. (1932). Intorno alle curve di concentrazione. *Metron*, 9, 3–76.

Gini, C. (1936). On the Measure of Concentration with Special Reference to Income and Wealth. Cowles Commission Research Conference on Economics and Statistics, Colorado College Publication, General Series No. 208, 73–79.

Giorgi, G. M. (1990). A bibliographic portrait of the Gini ratio. *Metron*, 48, 183–221.

Glaser, R. E. (1980). Bathtub and related failure rate characterizations. *Journal of the American Statistical Association*, 75, 667–672.

Gnedenko, B. V. (1943). Sur la distribution limite du terme maximum d'une série aléatoire. *Annals of Mathematics*, 44, 423–453.

Goldie, C. (1977). Convergence theorems for empirical Lorenz curves and their inverses. *Advances in Applied Probability*, 9, 765–791.

Gompertz, B. (1825). On the nature of the function expressive of the law of human mortality. *Philosophical Transactions of the Royal Society of London*, A115, 513–580.

Greenwich, M. (1992). A unimodal hazard rate function and its failure distribution. *Statistical Papers*, 33, 187–202.

Griffiths, D. A. (1980). Interval estimation for the three-parameter lognormal distribution via the likelihood function. *Applied Statistics*, 29, 58–68.

Groeneveld, R. A. (1986). Skewness for the Weibull family. *Statistica Neerlandica*, 40, 135–140.

Guerrieri, G. (1969–1970). Sopra un nuovo metodo concernente la determinazione dei parametri della distribuzione lognormale e delle distribuzioni pearsoniane del III e del V tipo. *Annali dell'Istituto di Statistica dell'Università di Bari*, 34, 53–110.

Guerrieri, G., and Bonadies, P. (1987). Su alcuni modelli teorici per l'analisi della distribuzione personale dei redditi. In: M. Zenga (ed.): pp. 102–136.

Gupta, M. R. (1984). Functional form for estimating the Lorenz curve. *Econometrica*, 52, 1313–1314.

Hager, H. W., and Bain, L. J. (1970). Inferential procedures for the generalized gamma distribution. *Journal of the American Statistical Association*, 65, 334–342.

Hagstrœm, K.-G. (1925). La loi de Pareto et la réassurance. *Skandinavisk Aktuarietidskrift*, 8, 65–88.

Hall, B. E. (1987). The relationship between firm size and growth. *Journal of Industrial Economics*, 35, 583–606.

Hampel, F. R., Ronchetti, E. M., Rousseeuw, P., and Stahel, W. A. (1986). *Robust Statistics: The Approach Based on Influence Functions*. New York: John Wiley.

Harrison, A. J. (1974). Inequality of income and the Champernowne distribution. In: M. Parkin and A. R. Nobay (eds.): *Current Economic Problems*, Cambridge: Cambridge University Press, pp. 9–24.

Harrison, A. J. (1979). The upper tail of the earnings distribution: Pareto or lognormal? *Economics Letters*, 2, 191–195.

Harrison, A. J. (1981). Earnings by size: A tale of two distributions. *Review of Economic Studies*, 48, 621–631.

Hart, P. E. (1973). The comparative statics and dynamics of income distributions. University of Reading Discussion Papers in Economics, Series A, No. 49.

Hart, P. E. (1975). Moment distributions in economics: An exposition. *Journal of the Royal Statistical Society*, A138, 423–434.

Hart, P. E. (1976a). The comparative statics and dynamics of income distributions. *Journal of the Royal Statistical Society*, A139, 108–125.

Hart, P. E. (1976b). The dynamics of earnings 1963–1973. *Economic Journal*, 86, 551–565.

Hart, P. E., and Oulton, N. (1997). Zipf and the size distribution of firms. *Applied Economics Letters*, 4, 205–206.

Hart, P. E., and Prais, S. J. (1956). The analysis of business concentration: A statistical approach (with discussion). *Journal of the Royal Statistical Society*, A119, 150–175.

Hartley, M. J., and Revankar, N. S. (1974). On the estimation of the Pareto law from underreported data. *Journal of Econometrics*, 2, 327–341.

Hayakawa, M. (1951). The application of Pareto's law of income to Japanese data. *Econometrica* 19, 174–183.

Head, G. L. (1968). *Insurance to value: An analysis of coinsurance requirements in fire and selected allied lines of insurance*. Ph.D. thesis, University of Pennsylvania.

Heilmann, W.-R. (1985). Transformations of claim distributions. *Bulletin of the Swiss Association of Actuaries*, 57–69.

Henniger, C., and Schmitz, H.-P. (1989). Size distributions of incomes and expenditures: Testing the parametric approach. Discussion Paper A-219, SFB 303, Universität Bonn, Germany.

Hewitt, C. C., and Lefkowitz, B. (1979). Methods for fitting distributions to insurance loss data. *Proceedings of the Casualty Actuarial Society*, 66, 139–160.

Heyde, C. C. (1963). On a property of the lognormal distribution. *Journal of the Royal Statistical Society*, B25, 392–393.

Hill, B. M. (1963). The three-parameter lognormal distribution and Bayesian analysis of a point-source epidemic. *Journal of the American Statistical Association*, 58, 72–84.

Hill, M. S. (1992). *The Panel Study of Income Dynamics—A User's Guide*. Newbury Park, CA: Sage Publications.

Hinkley, D. V., and Revankar, N. S. (1977). Estimation of the Pareto law from underreported data. *Journal of Econometrics*, 5, 1–11.

Hogg, R. V., and Klugman, S. A. (1983). On the estimation of long-tailed skewed distributions with actuarial applications. *Journal of Econometrics*, 23, 91–102.

Hogg, R. V., and Klugman, S. A. (1984). *Loss Distributions*. New York: John Wiley.

Holcomb, E. W. (1973). Estimating parameters of stable distributions with application to nonlife insurance. Ph.D. thesis, University of Tennessee, Knoxville, TN.

Holm, J. (1993). Maximum entropy Lorenz curves. *Journal of Econometrics*, 59, 377–389.

Hossain, A. M., and Zimmer, W. J. (2000). Comparison of methods of estimation for a Pareto distribution of the first kind. *Communications in Statistics—Theory and Methods*, 29, 859–878.

Houthakker, H. S. (1992). Conic distributions of earnings and income. Discussion Paper No. 1598, Harvard Institute of Economic Research, Harvard University.

Huang, J. S. (1974). On a theorem of Ahsanullah and Rahman. *Journal of Applied Probability*, 11, 216–218.

Hwang, T.-Y., and Hu, C.-Y. (1999). On a characterization of the gamma distribution: The independence of the sample mean and the sample coefficient of variation. *Annals of the Institute of Statistical Mathematics*, 51, 749–753.

Hwang, T.-Y., and Hu, C.-Y. (2000). On some characterizations of population distributions. *Taiwanese Journal of Mathematics*, 4, 427–437.

Ijiri, Y., and Simon, H. A. (1977). *Skew Distributions and the Sizes of Business Firms*. Amsterdam: North-Holland.

Inoue, T. (1978). On income distributions, the welfare implications for the general equilibrium model, and the stochastic processes of income distribution formation. Ph.D. thesis, University of Minnesota.

Iritani, J., and Kuga, K. (1983). Duality between the Lorenz curves and the income distribution functions. *Economic Studies Quarterly*, 34, 9–21.

Iyengar, N. S. (1960). On the standard error of the Lorenz concentration ratio. *Sankhyā*, 22, 371–378.

Jain, S. (1975). *Size Distribution of Income: A Compilation of Data*. Washington, DC: World Bank.

Jakuszenkow, H. (1979). Estimation of the variance in the generalized Laplace distribution with quadratic loss function. *Demonstratio Mathematica*, 12, 581–591.

Johnson, N. L. (1949). Systems of frequency curves generated by methods of translation. *Biometrika*, 36, 147–176.

Johnson, N. L. (1954). Systems of frequency curves derived from the first law of Laplace. *Trabajos de Estadística*, 5, 283–291.

Johnson, N. L., and Kotz, S. (1972). Power transformations of gamma variables. *Biometrika*, 59, 226–229.

Johnson, N. L., Kotz, S., and Balakrishnan, N. (1994). *Continuous Univariate Distributions*, Vol. 1, 2nd ed. New York: John Wiley.

Johnson, N. L., Kotz, S., and Balakrishnan, N. (1995). *Continuous Univariate Distributions*, Vol. 2, 2nd ed. New York: John Wiley.

Johnson, N. L., and Tadikamalla, P. R. (1992). Translated families of distributions. In: N. Balakrishnan (ed.): *Handbook of the Logistic Distribution*, New York: Marcel Dekker, pp. 189–208.

Johnson, N. O. (1937). The Pareto law. *Review of Economics and Statistics*, 19, 20–26.

Jones, B., Waller, W. G., and Feldman, A. (1978). Root isolation using function values. *BIT*, 18, 311–319.

Kakwani, N. (1980a). On a class of poverty measures. *Econometrica*, 48, 437–446.

Kakwani, N. C. (1980b). *Income Inequality and Poverty. Methods of Estimation and Policy Applications*. New York: Oxford University Press.

Kakwani, N. (1984). Welfare ranking of income distributions. *Advances in Econometrics*, 3, 191–213.

Kakwani, N. C. (1990). Large sample distribution of several inequality measures: With application to Côte d'Ivoire. In: R. A. L. Carter, J. Dutta, and A. Ullah (eds.): *Contributions to Econometric Theory and Application. Essays in Honor of A. L. Nagar.* New York: Springer, pp. 50–81.

Kakwani, N. C., and Podder, N. (1973). On the estimation of Lorenz curves from grouped observations. *International Economic Review*, 14, 278–292.

Kakwani, N. C., and Podder, N. (1976). Efficient estimation of the Lorenz curve and associated inequality measures from grouped observations. *Econometrica*, 44, 137–148.

Kalbfleisch, J. D., and Prentice, R. L. (1980). *The Statistical Analysis of Failure Time Data*. New York: John Wiley.

Kalecki, M. (1945). On the Gibrat distribution. *Econometrica*, 13, 161–170.

Kang, S.-B., and Cho, Y.-S. (1997). Estimation of the parameters in a Pareto distribution by jackknife and bootstrap methods. *Journal of Information & Optimization Sciences*, 18, 289–300.

Kapteyn, J. C. (1903). *Skew Frequency Curves in Biology and Statistics*. Groningen: Noordhoff.

Kapur, J. N. (1989). *Maximum Entropy Models in Science and Engineering*. New Delhi: Wiley Eastern.

Kendall, M. G. (1956). Discussion of Hart and Prais (1956). *Journal of the Royal Statistical Society*, A119, 184–185.

Kern, D. M. (1983). Minimum variance unbiased estimation in the Pareto distribution. *Metrika*, 30, 15–19.

Khan, A. H., and Khan, I. A. (1987). Moments of order statistics from Burr distribution and its characterization. *Metron*, 45, 21–29.

Klefsjö, B. (1984). Reliability interpretations of some concepts from economics. *Naval Research Logistics Quarterly*, 31, 301–308.

Kleiber, C. (1994). Hyperbolische Pareto-Diagramme und Einkommensverteilungen. Diploma thesis, Dept. of Statistics, Universität Dortmund, Germany.

Kleiber, C. (1996). Dagum vs. Singh–Maddala income distributions. *Economics Letters*, 53, 265–268.

Kleiber, C. (1997). The existence of population inequality measures. *Economics Letters*, 57, 39–44. Corrigendum: *Economics Letters*, 71 (2001), 429.

Kleiber, C. (1999a). On the Lorenz order within parametric families of income distributions. *Sankhyā*, B61, 514–517.

Kleiber, C. (1999b). Tail ordering of income distributions. Paper presented at the Econometric Society European Meeting (ESEM 99), Santiago de Compostela, Spain.

Kleiber, C. (2000a). *Halbordnungen von Einkommensverteilungen*. Angewandte Statistik und Ökonometrie Vol. 47, Göttingen: Vandenhoeck & Ruprecht.

Kleiber, C. (2000b). A simple distribution without any moments. *The Mathematical Scientist*, 25, 59–60.

Kleiber, C. (2003a) The Burr distributions at sixty. Working paper, Universität Dortmund, Germany.

Kleiber, C. (2003b). On the structure of the stable distributions. Working paper, Universität Dortmund, Germany.

Kleiber, C., and Kotz, S. (2002). A characterization of income distributions in terms of generalized Gini coefficients. *Social Choice and Welfare*, 19, 789–794.

Kleiber, C., and Trede, M. (2003). Multivariate income distributions via copulas. Working paper, Universität Dortmund.

Kloek, T., and van Dijk, H. K. (1977). Further results on efficient estimation of income distribution parameters. *Economie Appliquée*, 30, 439–459.

Kloek, T., and van Dijk, H. K. (1978). Efficient estimation of income distribution parameters. *Journal of Econometrics*, 8, 61–74.

Klonner, S. (2000). The first-order stochastic dominance ordering of the Singh–Maddala distribution. *Economics Letters*, 69, 123–128.

Klüppelberg, C. (1988). Subexponential distributions and integrated tails. *Journal of Applied Probability*, 25, 132–141.

Klugman, S. A. (1986). Loss distributions. In: H. H. Panjer (ed.): *Actuarial Mathematics*, Proceedings of Symposia in Applied Mathematics, Vol. 35, Providence, RI: American Mathematical Society, pp. 31–55.

Klugman, S. A., Panjer, H. H., and Willmot, G. E. (1998). *Loss Models*. New York: John Wiley.

Klugman, S. A., and Parsa, R. (1999). Fitting bivariate loss distributions with copulas. *Insurance: Mathematics & Economics*, 24, 139–148.

Kmietowicz, T., and Webley, P. (1975). Statistical analysis of income distribution in the central province of Kenya. *Eastern Africa Economic Review*, 17, 1–25.

Kmietowicz, Z. W. (1984). The bivariate lognormal model for the distribution of household size and income. *The Manchester School*, 52, 196–210.

Kmietowicz, Z. W., and Ding, H. (1993). Statistical analysis of income distribution in the Jiangsu province of China. *The Statistician*, 42, 107–121.

Kordos, J. (1990). Research on income distribution by size in Poland. In: C. Dagum and M. Zenga (eds.): pp. 335–351.

Koshevoy, G., and Mosler, K. (1996). The Lorenz zonoid of a multivariate distribution. *Journal of the American Statistical Association*, 91, 873–882.

Koshevoy, G., and Mosler, K. (1997). Multivariate Gini indices. *Journal of Multivariate Analysis*, 60, 252–276.

Kotz, S., Kozubowski, T., and Podgórski, K. (2001). *The Laplace Distribution and Generalizations*. Boston, Basel, and Berlin: Birkhäuser.

Kotz, S., and Nadarajah, S. (2000). *Extreme Value Distributions. Theory and Applications*. London: Imperial College Press.

Kotz, S., and Shanbhag, D. N. (1980). Some new approaches to probability distributions. *Advances in Applied Probability*, 12, 903–921.

Kozubowski, T. (1998). Mixture representation of Linnik distribution revisited. *Statistics & Probability Letters*, 38, 157–160.

Krishnaji, N. (1970). Characterization of the Pareto distribution through a model of underreported incomes. *Econometrica*, 38, 251–255.

Krishnan, P., Ng, E., and Shihadeh, E. (1990). Some generalized forms of the Pareto curve to approximate income distributions. *ASA Proceedings of the Business and Economic Statistics Section*, 502–504.

Lancaster, H. O. (1966). Forerunners of the Pearson χ^2. *Australian Journal of Statistics*, 8, 117–126.

Landwehr, J. M., Matalas, N. C., and Wallis, J. R. (1979). Probability weighted moments compared with some traditional techniques in estimating Gumbel parameters and quantiles. *Water Resources Research*, 15, 1055–1064.

Latorre, G. (1987). Distribuzioni campionarie asintotiche di indici di concentrazione: Approccio parametrico. *Statistica*, 47, 573–587.

Latorre, G. (1988). Proprietà campionarie del modello di Dagum per la distribuzione dei redditi. *Statistica*, 48, 15–27.

Laurent, A. G. (1974). On a characterization of some distributions by truncation properties. *Journal of the American Statistical Association*, 69, 823–827.

Lawless, J. F. (1980). Inference in the generalized gamma and log gamma distributions. *Technometrics*, 22, 409–419.

Lehmann, E. L., and Casella, G. (1998). *Theory of Point Estimation*, 2nd ed. New York and Berlin: Springer.

Leipnik, R. B. (1990). A maximum relative entropy principle for distribution of personal income with derivations of several known income distributions. *Communications in Statistics—Theory and Methods*, 19, 1003–1036.

Leipnik, R. B. (1991). On lognormal random variables: I. The characteristic function. *Journal of the Australian Mathematical Society*, B32, 327–347.

Lemmi, A. (1987). Il modello kappa a tre parametri nell'analisi della distribuzione personale del reddito. In: M. Zenga (ed.): pp. 165–189.

Lenski, G. E. (1966). *Power and Privilege. A Theory of Social Stratification*. New York: McGraw-Hill.

Levy, M., and Solomon, S. (1997). New evidence for the power-law distribution of wealth. *Physica*, A242, 90–94.

Likeš, J. (1969). Minimum variance unbiased estimation of the parameters of power-function and Pareto's distribution. *Statistische Hefte*, 10, 104–110.

Lillard, L., Smith, J. P., and Welch, F. (1986). What do we really know about wages? The importance of nonreporting and census imputation. *Journal of Political Economy*, 94, 489–506.

Lin, G. D. (1988). Characterizations of distributions via relationships between two moments of order statistics. *Journal of Statistical Planning and Inference*, 19, 73–80.

Linnik, Yu. V. (1953). Linear forms and statistical criteria, I, II. *Ukrainskii Mat. Zhournal*, 5, 207–290 (in Russian). Also in *Selected Translations in Mathematical Statistics and Probability*, 3 (1963), 1–90.

Linnik, Yu. V. (1961). *Method of Least Squares and Principles of the Theory of Observations*. New York: Pergamon Press.

Lomax, K. S. (1954). Business failures: Another example of the analysis of failure data. *Journal of the American Statistical Association*, 49, 847–852.

Lorenz, M. O. (1905). Methods of measuring the concentration of wealth. *Journal of the American Statistical Association*, 9, 209–219.

Lucas, R. E. (1978). On the size distribution of business firms. *Bell Journal of Economics*, 9, 508–523.

Lukacs, E. (1965). A characterization of the gamma distribution. *Annals of Mathematical Statistics*, 26, 319–324.

Lunetta, G. (1963). Di una generalizzazione dello schema della curva normale. *Annali della Facolta di Economia e Commercio di Palermo*, 17, 237–244.

Lydall, H. F. (1959). The distribution of employment incomes. *Econometrica*, 27, 110–115.

Lydall, H. F. (1968). *The Structure of Earnings*. London: Oxford University Press.

Maasoumi, E., and Theil, H. (1979). The effect of the shape of the income distribution on two inequality measures. *Economics Letters*, 4, 289–291.

MacGregor, D. H. (1936). Pareto's law. *Economic Journal*, 46, 80–87.

Maddala, G. S., and Singh, S. K. (1977a). Estimation problems in size distributions of incomes. *Economie Appliquée*, 30, 461–480.

Maddala, G. S., and Singh, S. K. (1977b). A flexible functional form for Lorenz curves. *Economie Appliquée*, 30, 481–486.

Malik, H. J. (1967). Exact distribution of the quotient of independent generalized gamma variables. *Canadian Mathematical Bulletin*, 10, 463–465.

Majumder, A., and Chakravarty, S. R. (1990). Distribution of personal income: Development of a new model and its application to U.S. income data. *Journal of Applied Econometrics*, 5, 189–196.

Mandelbrot, B. (1959). Variables et processus stochastiques de Pareto-Lévy, et la répartition des revenus. *Comptes Rendus Acad. Sci. Paris*, 249, 613–615.

Mandelbrot, B. (1960). The Pareto–Lévy law and the distribution of income. *International Economic Review*, 1, 79–106.

Mandelbrot, B. (1961). Stable Paretian random functions and the multiplicative variation of income. *Econometrica*, 29, 517–543.

Mandelbrot, B. (1964). Random walks, fire damage amount and other Paretian risk phenomena. *Operations Research*, 12, 582–585.

Mandelbrot, B. (1997). A case against the lognormal distribution. In: *Fractals and Scaling in Finance* (Selected Works of B. B. Mandelbrot, Volume E), New York and Berlin: Springer, Chapter 9, pp. 252–269.

Mansfield, E. (1962). Entry, Gibrat's law, innovations, and the growth of firms. *American Economic Review*, 52, 1023–1051.

March, L. (1898). Quelques exemples de distribution des salaires. *Journal de la Société statistique de Paris*, 193–206 and 241–248.

Marron, J. S., and Schmitz, H.-P. (1992). Simultaneous density estimation of several income distributions. *Econometric Theory*, 8, 476–488.

Marsaglia, G. (1989). The $X + Y$, X/Y characterization of the gamma distribution. In: L. J. Bleier, M. D. Perlman, S. J. Press, and A. R. Sampson (eds.): *Contributions to Probability and Statistics*, New York: Springer, pp. 91–98.

Mazzoni, P. (1957). Sulla curva di Amoroso per la distribuzione dei redditi. In: *Scritti Matematici in Onore di Filippo Sibirani*, Bologna: Cesare Buffi, pp. 181–193.

McAlister, D. (1879). The law of the geometric mean. *Proceedings of the Royal Society of London*, 29, 367–375.

McDonald, J. B. (1981). Some issues associated with the measurement of income inequality. *Statistical Distributions in Scientific Work*, 6, 161–179.

McDonald, J. B. (1984). Some generalized functions for the size distribution of income. *Econometrica*, 52, 647–663.

McDonald, J. B., and Butler, R. J. (1987). Some generalized mixture distributions with an application to unemployment duration. *Review of Economics and Statistics*, 69, 232–240.

McDonald, J. B., and Jensen, B. (1979). An analysis of some properties of alternative measures of income inequality based on the gamma distribution function. *Journal of the American Statistical Association*, 74, 856–860.

McDonald, J. B., and Mantrala, A. (1993). Apples, oranges and distribution trees. Discussion paper, Brigham Young University, Provo, Utah.

McDonald, J. B., and Mantrala, A. (1995). The distribution of income: Revisited. *Journal of Applied Econometrics*, 10, 201–204.

McDonald, J. B., and Ransom, M. (1979a). Functional forms, estimation techniques, and the distribution of income. *Econometrica*, 47, 1513–1526.

McDonald, J. B., and Ransom, M. (1979b). Alternative parameter estimators based on grouped data. *Communications in Statistics—Theory and Methods*, 8, 899–917.

McDonald, J. B., and Richards, D. O. (1987). Hazard rates and generalized beta distributions. *IEEE Transactions on Reliability*, 36, 463–466.

McDonald, J. B., and Xu, Y. J. (1995). A generalization of the beta distribution with applications. *Journal of Econometrics*, 66, 133–152. Erratum: *Journal of Econometrics*, 69, 427–428.

Meidell, B. (1912). Zur Theorie des Maximums. *Transactions of the 7th International Congress of Actuaries*, Vol. 1, 85–99.

Merkies, A. H. Q. M., and Steyn, I. J. (1993). Income distribution, Pareto laws, and regular variation. *Economics Letters*, 43, 177–182.

Metcalf, C. E. (1969). The size distribution of personal income during the business cycle. *American Economic Review*, 59, 657–668.

Mielke, P. W. (1973). Another family of distributions for describing and analyzing precipitation data. *Journal of Applied Meteorology*, 12, 275–280.

Mielke, P. W., and Johnson, E. S. (1973). Three-parameter kappa distribution maximum likelihood estimates and likelihood ratio tests. *Monthly Weather Review*, 101, 701–707.

Mielke, P. W., and Johnson, E. S. (1974). Some generalized beta distributions of the second kind having desirable application features in hydrology and meteorology. *Water Resources Research*, 10, 223–226.

Mills, J. A., and Zandvakili, S. (1997). Statistical inference via bootstrapping for measures of inequality. *Journal of Applied Econometrics*, 12, 133–150.

Mincer, J. (1958). Investment in human capital and personal income distribution. *Journal of Political Economy*, 66, 281–302.

Montroll, E. W., and Shlesinger, M. F. (1982). On $1/f$ noise and other distributions with long tails. *Proceedings of the National Academy of Sciences, U.S.A.*, 79, 3380–3383.

Montroll, E. W., and Shlesinger, M. F. (1983). Maximum entropy formalism, fractals, scaling phenomena, and $1/f$ noise: A tale of tails. *Journal of Statistical Physics*, 32, 209–230.

Moothathu, T. S. K. (1981). On Lorenz curves of lognormal and Pareto distributions. *Journal of the Indian Statistical Association*, 19, 103–108.

Moothathu, T. S. K. (1985). Sampling distributions of Lorenz curve and Gini index of the Pareto distribution. *Sankhyā*, B47, 247–258.

Moothathu, T. S. K. (1986). The best estimators of quantiles and the three means of the Pareto distribution. *Calcutta Statistical Association Bulletin*, 35, 111–121.

Moothathu, T. S. K. (1988). On the best estimators of coefficients of variation, skewness and kurtosis of Pareto distribution and their variances. *Calcutta Statistical Association Bulletin*, 37, 29–39.

Moothathu, T. S. K. (1989). On unbiased estimation of Gini index and Yntema–Pietra index of lognormal distribution and their variances. *Communications in Statistics—Theory and Methods*, 18, 661–672.

Moothathu, T. S. K. (1990a). Lorenz curve and Gini index. *Calcutta Statistical Association Bulletin*, 40, 307–324.

Moothathu, T. S. K. (1990b). A characterizing property of Weibull, exponential and Pareto distributions. *Journal of the Indian Statistical Association*, 28, 69–74.

Moothathu, T. S. K. (1990c). The best estimator and a strongly consistent asymptotically normal unbiased estimator of Lorenz curve, Gini index and Theil entropy index of Pareto distribution. *Sankhyā*, B52, 115–127.

Moothathu, T. S. K. (1991). On a sufficient condition for two non-intersecting Lorenz curves. *Sankhyā*, B53, 268–274. Editorial note: *Sankhyā*, B60 (1998), 380.

Moothathu, T. S. K. (1992). On unbiased estimation of Gini, Piesch and Mehran inequality indices of Log-Laplace distribution. *Calcutta Statistical Association Bulletin*, 42, 163–175.

Moothathu, T. S. K. (1993). A characterization of the α-mixture Pareto distribution through a property of the Lorenz curve. *Sankhyā*, B55, 130–134.

Moothathu, T. S. K., and Christudas, D. (1992). On unbiased estimation of Gini, Piesch and Mehran inequality indices of log-Laplace distribution. *Calcutta Statistical Association Bulletin*, 42, 163–175.

Mortara, G. (1917). *Elementi di Statistica*. Rome: Athenaeum.

Mortara, G. (1949). Representaçao analitica das distribuçoes dos contribuintes a das respectivas rendas liquidas determinadas para a aplicaçao do imposto de renda em funçao do valor da renda liquida. *Revista Brasileira de Economia*, 7–34.

Mosler, K. (1994). Majorization in economic disparity measures. *Linear Algebra and Its Applications*, 199, 91–114.

Moyes, P. (1987). A new concept of Lorenz domination. *Economics Letters*, 23, 203–207.

Muliere, P., and Scarsini, M. (1989). A note on stochastic dominance and inequality measures. *Journal of Economic Theory*, 49, 314–323.

Nair, N. U., and Hitha, N. (1990). Characterizations of Pareto and related distributions. *Journal of the Indian Statistical Association*, 28, 75–79.

Näslund, B. (1977). *An Analysis of Economic Size Distributions*. Berlin and New York: Springer.

Nalbach-Leniewska, A. (1979). Measures of dependence of the multivariate lognormal distribution. *Mathematische Operationsforschung—Serie Statistik*, 10, 381–387.

Nelsen, R. B. (1998). *An Introduction to Copulas*. Lecture Notes in Statistics 139, New York and Berlin: Springer.

Nygård, F., and Sandström, A. (1981). *Measuring Income Inequality*. Stockholm: Almqvist & Wiksell International.

Ogwang, T., and Rao, U. L. G. (1996). A new functional form for approximating the Lorenz curve. *Economics Letters*, 52, 21–29.

Ogwang, T., and Rao, U. L. G. (2000). Hybrid models for the Lorenz curve. *Economics Letters*, 69, 39–44.

Okun, A. M. (1975). *Equality and Efficiency*. Washington, DC.

Okuyama, K., Takayasu, M., and Takayasu, H. (1999). Zipf's law in income distribution of companies. *Physica*, A269, 125–131.

Ord, J. K. (1975). Statistical models for personal income distribution. *Statistical Distributions in Scientific Work*, 2, 151–158.

Ord, J. K. (1985). Pearson system of distributions. In: S. Kotz, N. L. Johnson, and C. Read (eds.): *Encyclopedia of Statistical Sciences*, Vol. 6. New York: John Wiley, pp. 655–659.

Ord, J. K., Patil, G. P., and Taillie, C. (1981). The choice of a distribution to describe personal incomes. *Statistical Distributions in Scientific Work*, 6, 193–201.

Ord, J. K., Patil, G. P., and Taillie, C. (1983). Truncated distributions and measures of income inequality. *Sankhyā*, B45, 413–430.

Ortega, P. Martín, Fernández, Ladoux, M., and García, A. (1991). A new functional form for estimating Lorenz curves. *Review of Income and Wealth*, 37, 447–452.

Pakes, A. G. (1981). On income distributions and their Lorenz curves. Technical report, Dept. of Mathematics, University of Western Australia, Nedlands, WA.

Pakes, A. G. (1983). Remarks on a model of competitive bidding for employment. *Journal of Applied Probability*, 20, 349–357.

Pakes, A. G., and Khattree, R. (1992). Length-biasing characterizations of laws and the moment problem. *Australian Journal of Statistics*, 34, 307–326.

Pareto, V. (1895). La legge della domanda. *Giornale degli Economisti*, 10, 59–68. English translation in *Rivista di Politica Economica*, 87 (1997), 691–700.

Pareto, V. (1896). La courbe de la répartition de la richesse. Reprinted 1965 in G. Busoni (ed.): *Œeuvres complètes de Vilfredo Pareto, Tome 3: Écrits sur la courbe de la répartition de la richesse*, Geneva: Librairie Droz. English translation in *Rivista di Politica Economica*, 87 (1997), 647–700.

Pareto, V. (1897a). *Cours d'économie politique*. Lausanne: Ed. Rouge.

Pareto, V. (1897b). Aggiunta allo studio della curva delle entrate. *Giornale degli Economisti*, Series 2, 14, 15–26. English translation in *Rivista di Politica Economica*, 87 (1997), 645–700.

Parker, R. N., and Fenwick, R. (1983). The Pareto curve and its utility for open-ended income distributions in survey research. *Social Forces*, 61, 872–885.

Parker, S. C. (1997). The distribution of self-employment income in the United Kingdom, 1976–1991. *Economic Journal*, 107, 455–466.

Parker, S. C. (1999a). The generalised beta as a model for the distribution of earnings. *Economics Letters*, 62, 197–200.

Parker, S. C. (1999b). The beta as a model for the distribution of earnings. *Bulletin of Economic Research*, 53, 243–251.

Patel, J. K. (1973). Complete sufficient statistics and MVU estimators. *Communications in Statistics*, 2, 327–336.

Paulson, A. S., Holcomb, E. W., and Leitch, R. A. (1975). The estimation of parameters of the stable laws. *Biometrika*, 62, 163–170.

Pearson, K. (1895). Contributions to the mathematical theory of evolution. II. Skew variation in homogeneous material. *Transactions of the Royal Society London*, A186, 343–415.

Pederzoli, G., and Rathie, P. N. (1980). Distribution of product and quotient of Pareto variates. *Metrika*, 27, 165–169.

Perks, W. F. (1932). On some experiments in the graduation of mortality statistics. *Journal of the Institute of Actuaries*, 58, 12–57.

Persky, J. (1992). Pareto's law. *Journal of Economic Perspectives*, 6, 181–192.

Peterson, L., and von Foerster, H. (1971). Cybernetics of income taxation: The optimization of economic participation. *Journal of Cybernetics*, 1, 5–22.

Pham, T. G., and Turkkan, N. (1994). The Lorenz curve and the scaled total-time-on-test-transform curves: A unified approach. *IEEE Transactions on Reliability*, 43, 76–84.

Pham-Gia, T., and Turkkan, N. (1992). Determination of the beta distribution from its Lorenz curve. *Mathematical and Computer Modelling*, 16, 73–84.

Pham-Gia, T., and Turkkan, N. (1997). Change in income distribution in the presence of reporting errors. *Mathematical and Computer Modelling*, 25, 33–42.

Piesch, W. (1967). Konzentrationsmaße von aggregierten Verteilungen. In: A. E. Ott (ed.): *Theoretische und empirische Beiträge zur Wirtschaftsforschung*. Tübingen, pp. 269–280.

Piesch, W. (1971). Lorenzkurve und inverse Verteilungsfunktion. *Jahrbücher für Nationalökonomie und Statistik*, 185, 209–234.

Piesch, W. (1975). *Statistische Konzentrationsmaße*. Tübingen: J. C. B. Mohr (Paul Siebeck).

Pietra, G. (1915). Delle relazioni fra indici di variabilità, note I e II. *Atti del Reale Istituto Veneto di Scienze, Lettere ed Arti*, 74, 775–804.

Piketty, T. (2000). Book review of *Economic Inequality and Income Distribution* by D. G. Champernowne and F. A. Cowell. *Economica*, 67, 461–462.

Podder, N. (1972). Distribution of household income in Australia. *Economic Record*, 48, 181–200.

Polisicchio, M. (1990). Sulla interpretazione dei parametri di modelli analitici per la distribuzione del reddito personale. *Statistica*, 50, 383–397.

Polisicchio, M. (1995). Sulla ordinamenti parziali basati sulla curva di Lorenz e sulla misura puntuale $Z(p)$. *Quaderni di Statistica e Matematica Applicata alle Scienze Economico-Sociali*, 15, 63–86.

Polisicchio, M., and Zenga, M. (1996). Kurtosis diagram for continuous variables. *Metron*, 55, 21–41.

Pollastri, A. (1987a). Characteristics of Zenga's concentration measure ξ_2. In: M. Zenga (ed.): pp. 214–229.

Pollastri, A. (1987b). Le cruve dil concentrazione L_p e Z_p nella distribuzione lognormale generalizzata. *Giornale degli Economisti e Annali di Economia* (New Series), 46, 639–663.

Pollastri, A. (1990). A comparison of the traditional estimators of the parameter α of the Pareto distribution. In C. Dagum and M. Zenga (eds.): pp. 183–193.

Pollastri, A. (1997). Lo studio della curtosi della distribuzione lognormale generalizzata. *Quaderni di Statistica e Matematica Applicata alle Scienze Economio-Sociali*, 19, 79–88.

Prentice, R. L. (1974). A log-gamma model and its maximum likelihood estimation. *Biometrika*, 61, 539–544.

Prentice, R. L. (1975). Discrimination among some parametric models. *Biometrika*, 62, 607–614.

Prentice, R. L., and El Shaarawi, A. (1973). A model for mortality rates and a test of fit for the Gompertz force of mortality. *Applied Statistics*, 22, 301–314.

Prudnikov, A. P., Brychkov, Yu. A., and Marichev, O. I. (1986). *Integrals and Series. Vol. 1: Elementary Functions*. New York: Gordon and Breach.

Quandt, R. E. (1966a). On the size distribution of firms. *American Economic Review*, 56, 416–432.

Quandt, R. E. (1966b). Old and new methods of estimation and the Pareto distribution. *Metrika*, 10, 55–82.

Rahman, M., and Gokhale, D. V. (1996). On estimation of parameters of the exponential power family of distributions. *Communications in Statistics—Simulation and Computation*, 25, 291–299.

Ramachandran, G. (1969). The Poisson process and fire loss distribution. *Bulletin of the International Statistical Institute*, 53 (Proceedings of the 37th Session of the ISI), 220–222.

Ramachandran, G. (1974). Extreme value theory and large fire losses. *ASTIN Bulletin*, 7, 293–310.

Ramlau-Hansen, H. (1988). A solvency study in non-life insurance. Part 1. Analyses of fire, windstorm, and glass claims. *Scandinavian Actuarial Journal*, 3–34.

Ransom, M. R., and Cramer, J. S. (1983). Income distributions with disturbances. *European Economic Review*, 22, 363–372.

Rao, C. R., and Zhao, L. C. (1995). Strassen's law of the iterated logarithm for the Lorenz curves. *Journal of Multivariate Analysis*, 54, 239–252.

Rao, U. L. G., and Tam, A. Y.-P. (1987). An empirical study of selection and estimation of alternative models for the Lorenz curve. *Journal of Applied Statistics*, 14, 275–280.

Rasche, R. H., Gaffney, J., Koo, A. Y. C., and Obst, N. (1980). Functional forms for estimating the Lorenz curve. *Econometrica*, 48, 1061–1062.

Ratz, H. C., and van Scherrenberg, M. (1981). The distribution of incomes for professional engineers. *IEEE Transactions on Systems, Man, and Cybernetics*, 11, 322–325.

Reed, W. J. (2001a). The Pareto, Zipf and other power laws. *Economics Letters*, 74, 15–19.

Reed, W. J. (2001b). The double Pareto-lognormal distribution—a new parametric model for size distribution. Working paper, University of Victoria, Canada.

Reed, W. J. (2003). The Pareto law of incomes—an explanation and an extension. *Physica* A319, 469–486.

Revankar, N. S., Hartley, M. J., and Pagano, M. (1974). A characterization of the Pareto distribution. *Annals of Statistics*, 2, 599–601.

Rhodes, E. C. (1944). The Pareto distribution of incomes. *Economica* (New Series), 11, 1–11.

Rodriguez, R. N. (1977). A guide to the Burr type XII distributions. *Biometrika*, 64, 129–134.

Rodriguez, R. N. (1983). Burr distributions. In: S. Kotz and N. L. Johnson (eds.): *Encyclopedia of Statistical Sciences*, Vol. 1, New York: John Wiley, pp. 335–340.

Rosin, P., and Rammler, B. (1933). The laws governing the fineness of powdered coal. *Journal of the Institute of Fuels*, 7, 29–36.

Rousu, D. N. (1973). Weibull skewness and kurtosis as a function of the shape parameter. *Technometrics*, 15, 927–930.

Rutherford, R. S. G. (1955). Income distribution: A new model. *Econometrica*, 23, 277–294.

Ryu, H. K., and Slottje, D. J. (1996). Two flexible functional forms for approximating the Lorenz curve. *Journal of Econometrics*, 72, 251–274.

Rytgaard, M. (1990). Estimation in the Pareto distribution. *ASTIN Bulletin*, 20, 201–216.

Saksena, S. K., and Johnson, A. M. (1984). Best unbiased estimators for the parameters of a two-parameter Pareto distributions. *Metrika*, 31, 77–83.

Salem, A. B., and Mount, T. D. (1974). A convenient descriptive model of income distribution: The gamma density. *Econometrica*, 42, 1115–1127.

Sarabia, J.-M. (1997). A hierarchy of Lorenz curves based on the generalized Tukey's lambda distribution. *Econometric Reviews*, 16, 305–320.

Sarabia, J.-M., Castillo, E., and Slottje, D. J. (1999). An ordered family of Lorenz curves. *Journal of Econometrics*, 91, 43–60.

Sarabia, J.-M., Castillo, E., and Slottje, D. J. (2001). An exponential family of Lorenz curves. *Southern Economic Journal*, 67, 748–756.

Sarabia, J.-M., Castillo, E., and Slottje, D. J. (2002). Lorenz ordering between McDonald's generalized functions of the income size distribution. *Economics Letters*, 75, 265–270.

Sargan, J. D. (1957). The distribution of wealth. *Econometrica*, 25, 568–590.

Saving, T. R. (1965). The four-parameter lognormal, diseconomies of scale and the size distribution of manufacturing establishments. *International Economic Review*, 6, 105–114.

Schader, M., and Schmid, F. (1986). Optimal grouping of data from some skewed distributions. *Computational Statistics Quarterly*, 3, 151–159.

Schader, M., and Schmid, F. (1994). Fitting parametric Lorenz curves to grouped income distributions—a critical note. *Empirical Economics*, 19, 361–370.

Scheid, S. (2001). Die verallgemeinerte Lognormalverteilung. Diploma thesis, Dept. of Statistics, Universität Dortmund, Germany.

Schmittlein, D. C. (1983). Some sampling properties of a model for income distribution. *Journal of Business and Economic Statistics*, 1, 147–153.

Seal, H. (1980). Survival probabilities based on Pareto claim distributions. *ASTIN Bulletin*, 11, 61–71.

Sen, A. K. (1997). *On Economic Inequality. Expanded Edition*. Oxford: Clarendon Press.

Sendler, W. (1979). On statistical inference in concentration measurement. *Metrika*, 26, 109–122.

Shackett, J. R., and Slottje, D. J. (1986). Labor supply decisions, human capital attributes, and inequality in the size distribution of earnings in the U.S., 1952–1981. *Journal of Human Resources*, 22, 82–100.

Shah, B. K., and Dave, P. H. (1963). A note on log-logistic distribution. *Journal of the M.S. University of Baroda*, 12, 15–20.

Shah, A., and Gokhale, D. V. (1993). On maximum product of spacings (MPS) estimation for Burr XII distributions. *Communications in Statistics—Simulation*, 22, 615–641.

Shanbhag, D. N. (1970). The characterization for exponential and geometric distributions. *Journal of the American Statistical Association*, 65, 1256–1259.

Shaked, M., and Shanthikumar, G. (1994). *Stochastic Orders and Their Applications*. San Diego, CA: Academic Press.

Sharma, D. (1984). on estimating the variance of a generalized Laplace distribution. *Metrika*, 31, 85–88.

Shimizu, K. S., and Crow, E. L. (1988). History, genesis, and properties. In: E. L. Crow and K. S. Shimizu (1988): pp. 1–25.

Shirras, G. F. (1935). The Pareto law and the distribution of income. *Economic Journal*, 45, 663–681.

Shorrocks, A. F. (1975). On stochastic models of size distributions. *Review of Economic Studies*, 42, 631–642.

Shorrocks, A. F. (1983). Ranking income distributions. *Economica*, 50, 3–17.

Shoukri, M. M., Mian, I. U. H., and Tracy, D. S. (1988). Sampling properties of estimators of the log-logistic distribution with application to Canadian precipitation data. *Canadian Journal of Statistics*, 16, 223–236.

Shpilberg, D. C. (1977). The probability distribution of fire loss amount. *Journal of Risk and Insurance*, 44, 103–115.

Shpilberg, D. (1982). *Statistical Decomposition of Industrial Fire Loss*. Huebner Foundation Monograph No. 11, University of Pennsylvania, Philadelphia.

Simon, H. A. (1955). On a class of skew distribution functions. *Biometrika*, 52, 425–440.

Singh, M., and Singh, R. P. (1992). Comparing coefficients of variation in income distributions. *Journal of the Indian Society of Agricultural Statistics*, 44, 199–215.

Singh, S. K., and Maddala, G. S. (1975). A stochastic process for income distribution and tests for income distribution functions. *ASA Proceedings of the Business and Economic Statistics Section*, 551–553.

Singh, S. K., and Maddala, G. S. (1976). A function for the size distribution of incomes. *Econometrica*, 44, 963–970.

Singh, V. P., and Guo, H. (1995). Parameter estimation for 2-parameter Pareto distribution by Pome. *Water Resources Management*, 9, 81–93.

Slottje, D. J. (1987). Relative price changes and inequality in the size distribution of various components of income. *Journal of Business and Economic Statistics*, 5, 19–26.

Smith, R. L. (1985). Maximum likelihood estimation in a class of nonregular cases. *Biometrika*, 72, 67–90.

Stacy, E. W. (1962). A generalization of the gamma distribution. *Annals of Mathematical Statistics*, 33, 1187–1192.

Stamp, J. C. (1914). A new illustration of Pareto's law. *Journal of the Royal Statistical Society*, 77, 200–204.

Stanley, M. H. R., Buldryev, S. V., Havlin, S., Mantegna, R. N., Salinger, M. A., and Stanley, H. E. (1995). Zipf plots and the size distribution of firms. *Economics Letters*, 49, 453–457.

Steindl, J. (1965). *Random Processes and the Growth of Firms*. London: Griffin.

Steindl, J. (1972). The distribution of wealth after a model of Wold and Whittle. *Review of Economic Studies*, 39, 263–280.

Steyn, H. S. (1959). A model for the distribution of incomes. *South African Journal of Economics*, 27, 149–156.

Steyn, H. S. (1966). On the departure from the logarithmic normal distribution for income. *South African Journal of Economics*, 34, 225–232.

Stigler, S. M. (1974). Linear functions of order statistics with smooth weight functions. *Annals of Statistics*, 2, 676–693.

Stigler, S. M. (1978). Mathematical statistics in the early States. *Annals of Statistics*, 6, 239–265.

Stoppa, G. (1985). Un metodo non-parametrico per la stima delle distribuzioni statistiche. *Atti IX Convegno A.M.A.S.E.S.*, Bologna: Ed. Pitagora, 363–377.

Stoppa, G. (1989). Una estensione al teorema di Revankar, Hartley e Pagano. *Atti XIII Convegno A.M.A.S.E.S.*, Verona, 13–15 Settembre 1989, 843–848.

Stoppa, G. (1990a). A new generating system of income distribution models. *Quaderni di Statistica e Matematica Applicata alle Scienze Economico-Sociali*, 12, 47–55.

Stoppa, G. (1990b). A new model for income size distributions. In: C. Dagum and M. Zenga (eds.): *Income and Wealth Distribution, Inequality and Poverty: Proceedings of the 2nd International Conference on Income Distribution by Size: Generation, Distribution, Measurement and Applications*. New York, Berlin, London, and Tokyo: Springer, pp. 33–41.

Stoppa, G. (1990c). Proprietà campionarie di un nuovo modello Pareto generalizzato. *Atti XXXV Riunione Scientifica della Società Italiana di Statistica*, Padova: Cedam, 137–144.

Stoppa, G. (1993). Una tavola per modelli di probabilità. *Metron*, 51, 99–117.

Stoppa, G. (1994). Le distribuzioni campionarie della curva di concentrazione $Z(p)$ e degli indici λ e ξ nel modello di Pareto. *Statistica*, 54, 51–59.

Stoppa, G. (1995). Explicit estimators for income distributions. In: C. Dagum and A. Lemmi (eds.): *Research on Economic Inequality, Vol. 6: Income Distribution, Social Welfare, Inequality and Poverty*. Greenwich, CT: JAI Press, pp. 393–405.

Subbotin, M. T. (1923). On the law of frequency of error. *Mathematicheskii Sbornik*, 31, 296–300.

Suruga, T. (1982). Functional forms of income distributions: The case of yearly income groups in the "Annual Report on the Family Income and Expenditure Survey" (in Japanese). *Economic Studies Quarterly*, 33, 79–85.

Sutton, J. (1997). Gibrat's legacy. *Journal of Economic Literature*, 35, 40–59.

Sweet, A. L. (1990). On the hazard rate of the lognormal distribution. *IEEE Transactions on Reliability*, 39, 325–328.

Tachibanaki, T., Suruga, T., and Atoda, N. (1997). Estimations of income distribution parameters for individual observations by maximum likelihood method. *Journal of the Japan Statistical Society*, 27, 191–203.

Tadikamalla, P. R. (1977). An approximation to the moments and the percentiles of gamma order statistics. *Sankhyā*, B39, 372–381.

Tadikamalla, P. R. (1980). A look at the Burr and related distributions. *International Statistical Review*, 48, 337–344.

Tadikamalla, P. R., and Johnson, N. L. (1982). Systems of frequency curves generated by transformation of logistic variables. *Biometrika*, 69, 461–465.

Taguchi, T. (1968). Concentration curve methods and structures of skew populations. *Annals of the Institute of Statistical Mathematics*, 20, 107–141.

Taguchi, T. (1972a). On the two-dimensional concentration surface and extensions of concentration coefficient and Pareto distribution to the two-dimensional case—I. *Annals of the Institute of Statistical Mathematics*, 24, 355–382.

Taguchi, T. (1972b). On the two-dimensional concentration surface and extensions of concentration coefficient and Pareto distribution to the two-dimensional case—II. *Annals of the Institute of Statistical Mathematics*, 24, 599–619.

Taguchi, T. (1980). On an interpretation and estimation of shape parameters of generalized gamma distributions. *Metron*, 43, 27–40.

Taguchi, T., Sakurai, H., and Nakajima, S. (1993). A concentration analysis of income distribution model and consumption pattern. Introduction of logarithmic gamma distribution and statistical analysis of Engel elasticity. *Statistica*, 53, 31–57.

Taillie, C. (1981). Lorenz ordering within the generalized gamma family of income distributions. *Statistical Distributions in Scientific Work*, 6, 181–192.

Takahasi, K. (1965). Note on the multivariate Burr's distribution. *Annals of the Institute of Statistical Mathematics*, 17, 257–260.

Takayasu, H. (ed.) (2002). *Empirical Science of Financial Fluctuations. The Advent of Econophysics*. Tokyo, Berlin, Heidelberg, and New York: Springer.

Takayasu, H., and Okuyama, K. (1998). Country dependence on company size distributions and numerical model based on competition and cooperation. *Fractals*, 6, 67–79.

Temme, N. M. (1996). *Special Functions. An Introduction to the Classical Functions of Mathematical Physics*. New York: John Wiley.

Theil, H. (1967). *Economics and Information Theory*. Amsterdam: North-Holland.

Thistle, P. D. (1989a). Duality between generalized Lorenz curves and distribution functions. *Economic Studies Quarterly*, 40, 183–187.

Thistle, P. D. (1989b). Ranking distributions with generalized Lorenz curves. *Southern Economic Journal*, 56, 1–12.

Thistle, P. D. (1990). Large sample properties of two inequality indices. *Econometrica*, 58, 725–728.

Thompson, W. A., Jr. (1976). Fisherman's luck. *Biometrics*, 32, 265–271.

Thurow, L. (1970). Analyzing the American income distribution. *American Economic Review* (Papers and Proceedings), 48, 261–269.

Tiao, G. C., and Lund, D. R. (1970). The use of OLUMV estimators in inference robustness studies of the location parameter of a class of symmetric distributions. *Journal of the American Statistical Association*, 65, 370–386.

Tinbergen, J. (1956). On the theory of income distribution. *Weltwirtschaftliches Archiv*, 77, 155–175.

van der Wijk, J. (1939). *Inkomens- en Vermogensverdeling*. Publications of the Nederlandsch Economisch Instituut, Vol. 26, Haarlem: De Erven F. Bohn N.V.

van Dijk, H. K., and Kloek, T. (1978). Empirical evidence on Pareto–Lévy and log stable income distributions. Report 7812/E, Econometric Institute, Erasmus University, Rotterdam.

van Dijk, H. K., and Kloek, T. (1980). Inferential procedures in stable distributions for class frequency data on incomes. *Econometrica*, 48, 1139–1148.

van Praag, B., Hagenaars, A., and van Eck, W. (1983). The influence of classification and observability errors on the measurement of income inequality. *Econometrica*, 51, 1093–1108.

van Uven, M. J. (1917). Logarithmic frequency distributions. *Koninklijke Akademie van Wetenschappen te Amsterdam, Proceedings*, 19(4), 670–694.

van Zwet, W. R. (1964). *Convex Transformations of Random Variables*. Mathematical Centre Tracts 7, Amsterdam.

Vartia, P. L. I., and Vartia, Y. O. (1980). Description of the income distribution by the scaled *F* distribution model. In: N. A. Klevmarken and J. A. Lybeck (eds.): *The Statics and Dynamics of Income*. Clevedon, UK: Tieto, pp. 23–36.

Venter, G. (1983). Transformed beta and gamma distributions and aggregate losses. *Proceedings of the Casualty Actuarial Society*, 70, 156–193.

Vianelli, S. (1963). La misura della variabilita condizionata in uno schema generale delle curve normali di frequenza. *Statistica*, 23, 447–474.

Vianelli, S. (1982a). Sulle curve lognormali di ordine *r* quali famiglie di distribuzioni di errori di proporzione. *Statistica*, 42, 155–176.

Vianelli, S. (1982b). Una nota sulle distribuzioni degli errori di proporzione con particolare riguardo alla distribuzione corrispondente alla prima legge degli errori additivi di Laplace. *Statistica*, 42, 371–380.

Vianelli, S. (1983). The family of normal and lognormal distributions of order *r*. *Metron*, 41, 3–10.

Victoria-Feser, M.-P. (1993). Robust methods for personal income distribution models. Ph.D. thesis, University of Geneva, Switzerland.

Victoria-Feser, M.-P. (1995). Robust methods for personal income distribution models with applications to Dagum's model. In: C. Dagum and A. Lemmi (eds.): pp. 225–239.

Victoria-Feser, M.-P. (2000). Robust methods for the analysis of income distribution, inequality and poverty. *International Statistical Review*, 68, 277–293.

Victoria-Feser, M.-P. (2001). Robust income distribution estimation with missing data. Discussion Paper DARP 57, STICERD, London School of Economics.

Victoria-Feser, M.-P., and Ronchetti, E. (1994). Robust methods for personal income distribution models. *Canadian Journal of Statistics*, 22, 247–258.

Victoria-Feser, M.-P., and Ronchetti, E. (1997). Robust estimation for grouped data. *Journal of the American Statistical Association*, 92, 333–340.

Villaseñor, J. A., and Arnold, B. C. (1989). Elliptical Lorenz curves. *Journal of Econometrics*, 40, 327–338.

Vinci, F. (1921). Nuovi contributi allo studio della distribuzione dei redditi. *Giornale degli Economisti e Rivista di Statistica*, 61, 365–369.

Voinov, V. G., and Nikulin, M. S. (1993). *Unbiased Estimators and Their Applications. Vol. 1: Univariate Case*. Dordrecht, Boston, and London: Kluwer.

Voit, J. (2001). The growth dynamics of German business firms. Preprint, Dept. of Physics, Universität Bayreuth, Germany.

Walter, W. (1981). Über ein Modell zur Pareto-Verteilung. *Applicable Analysis*, 11, 233–239.

Watkins, A. J. (1999). An algorithm for maximum likelihood estimation in the three parameter Burr XII distribution. *Computational Statistics & Data Analysis*, 32, 19–27.

Watson, G. S., and Wells, W. T. (1961). On the possibility of improving the mean useful life of items by eliminating those with short lives. *Technometrics*, 3, 281–298.

Weibull, W. (1939a). A statistical theory of the strength of material. Report No. 151, Ingeniörs Vetenskaps Akademiens Hadligar, Stockholm.

Weibull, W. (1939b). The phenomenon of rupture in solids. Report No. 153, Ingeniörs Vetenskaps Akademiens Hadligar, Stockholm.

Wilfling, B. (1996a). A sufficient condition for Lorenz ordering. *Sankhyā*, B58, 62–69.

Wilfling, B. (1996b). Lorenz ordering of generalized beta-II income distributions. *Journal of Econometrics*, 71, 381–388.

Wilfling, B. (1996c). Lorenz ordering of power-function order statistics. *Statistics and Probability Letters*, 30, 313–319.

Wilfling, B., and Krämer, W. (1993). Lorenz ordering of Singh–Maddala income distributions. *Economics Letters*, 43, 53–57.

Wingo, D. R. (1979). Estimation in a Pareto distribution: Theory & computation. *IEEE Transactions on Reliability*, 28, 35–37.

Wingo, D. R. (1983). Maximum likelihood methods for fitting the Burr type XII distribution to life test data. *Biometrical Journal*, 25, 77–84.

Wingo, D. R. (1984). Fitting three-parameter lognormal models by numerical global optimization—an improved algorithm. *Computational Statistics & Data Analysis*, 2, 13–25.

Wingo, D. R. (1987a). Computing globally optimal maximum-likelihood estimates of generalized gamma distribution parameters—some new numerical approaches and analytical results. *Proceedings of the 19th Symposium on the Interface of Computing Science and Statistics*, pp. 454–457.

Wingo, D. R. (1987b). Computing maximum-likelihood parameter estimates of the generalized gamma distribution by numerical root isolation. *IEEE Transactions on Reliability*, 36, 586–590.

Winkler, W. (1950). The corrected Pareto law and its economic meaning. *Bulletin of the International Statistical Institute*, 32, 441–449.

Wold, H. O. A., and Whittle, P. (1957). A model explaining the Pareto distribution of wealth. *Econometrica*, 25, 591–595.

Wolff, E. N. (ed.) (1987). *International Comparisons of Household Wealth*. Oxford: Clarendon Press.

Woo, J., and Kang, S.-B. (1990). Estimation for functions of two parameters in the Pareto distributions. *Youngnam Statistical Letters*, 1, 67–76.

Xu, K. (2000). Inference for the generalized Gini indices using the iterated-bootstrap method. *Journal of Business and Economic Statistics*, 18, 223–227.

Yardi, M. R. (1951). The validity of Pareto's law. *Bulletin of the International Statistical Institute*, 33, 133–146.

Yitzhaki, S. (1983). On an extension of the Gini inequality index. *International Economic Review*, 24, 617–628.

Zandonatti, A. (2001). Distribuzioni de Pareto Generalizzate. Tesi di Laurea, Dept. of Economics, University of Trento, Italy.

Zelterman, D. (1987). Parameter estimation in the generalized logistic distribution. *Computational Statistics & Data Analysis*, 5, 177–184.

Zenga, M. (1984). Proposta per un indice di concentrazione basato sui rapporti fra quantili di popolazione e quantili reddito. *Giornale degli Economisti e Annali di Economia*, 48, 301–326.

Zenga, M. (1985). Un secondo indice di concentrazione basato sui rapporti fra quantili di reddito e quantili di popolazione. *Rivista di Statistica Applicata*, 3, 143–154.

Zenga, M. (ed.) (1987). *La Distribuzione Personale del Reddito: Problemi di Formazione, di Ripartizione e di Misurazione*. Milan: Publicazioni dell'Unversità Cattolica di Milano.

Zenga, M. (1990). Concentration curves and concentration indices derived from them. In: C. Dagum and M. Zenga (eds.): pp. 94–110.

Zenga, M. (1996). La curtosi. *Statistica*, 56, 87–102.

Zheng, B. (2002). Testing Lorenz curves with non-simple random samples. *Econometrica*, 70, 1235–1244.

Ziebach, T. (2000). *Die Modellierung der personellen Einkommensverteilung mit verallgemeinerten Pareto-Kurven*. Lohmar: Josef Eul.

Zipf, G. K. (1949). *Human Behavior and the Principle of Least Effort*. Cambridge, MA: Addison-Wesley.

Zitikis, R., and Gastwirth, J. L. (2002). The asymptotic distribution of the *S*-Gini index. *Australian & New Zealand Journal of Statistics*, 44, 439–446.

Zolotarev, V. M. (1986). *One-dimensional Stable Distributions*. Providence, RI: American Mathematical Society.

Author Index

Subject Index

WILEY SERIES IN PROBABILITY AND STATISTICS
ESTABLISHED BY WALTER A. SHEWHART AND SAMUEL S. WILKS

The ***Wiley Series in Probability and Statistics*** is well established and authoritative. It covers many topics of current research interest in both pure and applied statistics and probability theory. Written by leading statisticians and institutions, the titles span both state-of-the-art developments in the field and classical methods.

Reflecting the wide range of current research in statistics, the series encompasses applied, methodological and theoretical statistics, ranging from applications and new techniques made possible by advances in computerized practice to rigorous treatment of theoretical approaches.

This series provides essential and invaluable reading for all statisticians, whether in academia, industry, government, or research.

ABRAHAM and LEDOLTER · Statistical Methods for Forecasting
AGRESTI · Analysis of Ordinal Categorical Data
AGRESTI · An Introduction to Categorical Data Analysis
AGRESTI · Categorical Data Analysis, *Second Edition*
ANDĚL · Mathematics of Chance
ANDERSON · An Introduction to Multivariate Statistical Analysis, *Second Edition*
*ANDERSON · The Statistical Analysis of Time Series
ANDERSON, AUQUIER, HAUCK, OAKES, VANDAELE, and WEISBERG · Statistical Methods for Comparative Studies
ANDERSON and LOYNES · The Teaching of Practical Statistics
ARMITAGE and DAVID (editors) · Advances in Biometry
ARNOLD, BALAKRISHNAN, and NAGARAJA · Records
*ARTHANARI and DODGE · Mathematical Programming in Statistics
*BAILEY · The Elements of Stochastic Processes with Applications to the Natural Sciences
BALAKRISHNAN and KOUTRAS · Runs and Scans with Applications
BARNETT · Comparative Statistical Inference, *Third Edition*
BARNETT and LEWIS · Outliers in Statistical Data, *Third Edition*
BARTOSZYNSKI and NIEWIADOMSKA-BUGAJ · Probability and Statistical Inference
BASILEVSKY · Statistical Factor Analysis and Related Methods: Theory and Applications
BASU and RIGDON · Statistical Methods for the Reliability of Repairable Systems
BATES and WATTS · Nonlinear Regression Analysis and Its Applications
BECHHOFER, SANTNER, and GOLDSMAN · Design and Analysis of Experiments for Statistical Selection, Screening, and Multiple Comparisons
BELSLEY · Conditioning Diagnostics: Collinearity and Weak Data in Regression

*Now available in a lower priced paperback edition in the Wiley Classics Library.

BELSLEY, KUH, and WELSCH · Regression Diagnostics: Identifying Influential Data and Sources of Collinearity
BENDAT and PIERSOL · Random Data: Analysis and Measurement Procedures, *Third Edition*
BERRY, CHALONER, and GEWEKE · Bayesian Analysis in Statistics and Econometrics: Essays in Honor of Arnold Zellner
BERNARDO and SMITH · Bayesian Theory
BHAT and MILLER · Elements of Applied Stochastic Processes, *Third Edition*
BHATTACHARYA and JOHNSON · Statistical Concepts and Methods
BHATTACHARYA and WAYMIRE · Stochastic Processes with Applications
BILLINGSLEY · Convergence of Probability Measures, *Second Edition*
BILLINGSLEY · Probability and Measure, *Third Edition*
BIRKES and DODGE · Alternative Methods of Regression
BLISCHKE AND MURTHY (editors) · Case Studies in Reliability and Maintenance
BLISCHKE AND MURTHY · Reliability: Modeling, Prediction, and Optimization
BLOOMFIELD · Fourier Analysis of Time Series: An Introduction, *Second Edition*
BOLLEN · Structural Equations with Latent Variables
BOROVKOV · Ergodicity and Stability of Stochastic Processes
BOULEAU · Numerical Methods for Stochastic Processes
BOX · Bayesian Inference in Statistical Analysis
BOX · R. A. Fisher, the Life of a Scientist
BOX and DRAPER · Empirical Model-Building and Response Surfaces
*BOX and DRAPER · Evolutionary Operation: A Statistical Method for Process Improvement
BOX, HUNTER, and HUNTER · Statistics for Experimenters: An Introduction to Design, Data Analysis, and Model Building
BOX and LUCEÑO · Statistical Control by Monitoring and Feedback Adjustment
BRANDIMARTE · Numerical Methods in Finance: A MATLAB-Based Introduction
BROWN and HOLLANDER · Statistics: A Biomedical Introduction
BRUNNER, DOMHOF, and LANGER · Nonparametric Analysis of Longitudinal Data in Factorial Experiments
BUCKLEW · Large Deviation Techniques in Decision, Simulation, and Estimation
CAIROLI and DALANG · Sequential Stochastic Optimization
CHAN · Time Series: Applications to Finance
CHATTERJEE and HADI · Sensitivity Analysis in Linear Regression
CHATTERJEE and PRICE · Regression Analysis by Example, *Third Edition*
CHERNICK · Bootstrap Methods: A Practitioner's Guide
CHERNICK and FRIIS · Introductory Biostatistics for the Health Sciences
CHILÈS and DELFINER · Geostatistics: Modeling Spatial Uncertainty
CHOW and LIU · Design and Analysis of Clinical Trials: Concepts and Methodologies
CLARKE and DISNEY · Probability and Random Processes: A First Course with Applications, *Second Edition*
*COCHRAN and COX · Experimental Designs, *Second Edition*
CONGDON · Bayesian Statistical Modelling
CONOVER · Practical Nonparametric Statistics, *Second Edition*
COOK · Regression Graphics
COOK and WEISBERG · Applied Regression Including Computing and Graphics

*Now available in a lower priced paperback edition in the Wiley Classics Library.

COOK and WEISBERG · An Introduction to Regression Graphics
CORNELL · Experiments with Mixtures, Designs, Models, and the Analysis of Mixture Data, *Third Edition*
COVER and THOMAS · Elements of Information Theory
COX · A Handbook of Introductory Statistical Methods
*COX · Planning of Experiments
CRESSIE · Statistics for Spatial Data, *Revised Edition*
CSÖRGÖ and HORVÁTH · Limit Theorems in Change Point Analysis
DANIEL · Applications of Statistics to Industrial Experimentation
DANIEL · Biostatistics: A Foundation for Analysis in the Health Sciences, *Sixth Edition*
*DANIEL · Fitting Equations to Data: Computer Analysis of Multifactor Data, *Second Edition*
DASU and JOHNSON · Exploratory Data Mining and Data Cleaning
DAVID · Order Statistics, *Second Edition*
*DEGROOT, FIENBERG, and KADANE · Statistics and the Law
DEL CASTILLO · Statistical Process Adjustment for Quality Control
DETTE and STUDDEN · The Theory of Canonical Moments with Applications in Statistics, Probability, and Analysis
DEY and MUKERJEE · Fractional Factorial Plans
DILLON and GOLDSTEIN · Multivariate Analysis: Methods and Applications
DODGE · Alternative Methods of Regression
*DODGE and ROMIG · Sampling Inspection Tables, *Second Edition*
*DOOB · Stochastic Processes
DOWDY and WEARDEN · Statistics for Research, *Second Edition*
DRAPER and SMITH · Applied Regression Analysis, *Third Edition*
DRYDEN and MARDIA · Statistical Shape Analysis
DUDEWICZ and MISHRA · Modern Mathematical Statistics
DUNN and CLARK · Applied Statistics: Analysis of Variance and Regression, *Second Edition*
DUNN and CLARK · Basic Statistics: A Primer for the Biomedical Sciences, *Third Edition*
DUPUIS and ELLIS · A Weak Convergence Approach to the Theory of Large Deviations
*ELANDT-JOHNSON and JOHNSON · Survival Models and Data Analysis
ENDERS · Applied Econometric Time Series
ETHIER and KURTZ · Markov Processes: Characterization and Convergence
EVANS, HASTINGS, and PEACOCK · Statistical Distributions, *Third Edition*
FELLER · An Introduction to Probability Theory and Its Applications, Volume I, *Third Edition,* Revised; Volume II, *Second Edition*
FISHER and VAN BELLE · Biostatistics: A Methodology for the Health Sciences
*FLEISS · The Design and Analysis of Clinical Experiments
FLEISS · Statistical Methods for Rates and Proportions, *Second Edition*
FLEMING and HARRINGTON · Counting Processes and Survival Analysis
FULLER · Introduction to Statistical Time Series, *Second Edition*
FULLER · Measurement Error Models
GALLANT · Nonlinear Statistical Models
GHOSH, MUKHOPADHYAY, and SEN · Sequential Estimation
GIFI · Nonlinear Multivariate Analysis
GLASSERMAN and YAO · Monotone Structure in Discrete-Event Systems

*Now available in a lower priced paperback edition in the Wiley Classics Library.

GNANADESIKAN · Methods for Statistical Data Analysis of Multivariate Observations, *Second Edition*
GOLDSTEIN and LEWIS · Assessment: Problems, Development, and Statistical Issues
GREENWOOD and NIKULIN · A Guide to Chi-Squared Testing
GROSS and HARRIS · Fundamentals of Queueing Theory, *Third Edition*
*HAHN and SHAPIRO · Statistical Models in Engineering
HAHN and MEEKER · Statistical Intervals: A Guide for Practitioners
HALD · A History of Probability and Statistics and their Applications Before 1750
HALD · A History of Mathematical Statistics from 1750 to 1930
HAMPEL · Robust Statistics: The Approach Based on Influence Functions
HANNAN and DEISTLER · The Statistical Theory of Linear Systems
HEIBERGER · Computation for the Analysis of Designed Experiments
HEDAYAT and SINHA · Design and Inference in Finite Population Sampling
HELLER · MACSYMA for Statisticians
HINKELMAN and KEMPTHORNE: · Design and Analysis of Experiments, Volume 1: Introduction to Experimental Design
HOAGLIN, MOSTELLER, and TUKEY · Exploratory Approach to Analysis of Variance
HOAGLIN, MOSTELLER, and TUKEY · Exploring Data Tables, Trends and Shapes
*HOAGLIN, MOSTELLER, and TUKEY · Understanding Robust and Exploratory Data Analysis
HOCHBERG and TAMHANE · Multiple Comparison Procedures
HOCKING · Methods and Applications of Linear Models: Regression and the Analysis of Variance, *Second Edition*
HOEL · Introduction to Mathematical Statistics, *Fifth Edition*
HOGG and KLUGMAN · Loss Distributions
HOLLANDER and WOLFE · Nonparametric Statistical Methods, *Second Edition*
HOSMER and LEMESHOW · Applied Logistic Regression, *Second Edition*
HOSMER and LEMESHOW · Applied Survival Analysis: Regression Modeling of Time to Event Data
HØYLAND and RAUSAND · System Reliability Theory: Models and Statistical Methods
HUBER · Robust Statistics
HUBERTY · Applied Discriminant Analysis
HUNT and KENNEDY · Financial Derivatives in Theory and Practice
HUSKOVA, BERAN, and DUPAC · Collected Works of Jaroslav Hajek—with Commentary
IMAN and CONOVER · A Modern Approach to Statistics
JACKSON · A User's Guide to Principle Components
JOHN · Statistical Methods in Engineering and Quality Assurance
JOHNSON · Multivariate Statistical Simulation
JOHNSON and BALAKRISHNAN · Advances in the Theory and Practice of Statistics: A Volume in Honor of Samuel Kotz
JUDGE, GRIFFITHS, HILL, LÜTKEPOHL, and LEE · The Theory and Practice of Econometrics, *Second Edition*
JOHNSON and KOTZ · Distributions in Statistics
JOHNSON and KOTZ (editors) · Leading Personalities in Statistical Sciences: From the Seventeenth Century to the Present
JOHNSON, KOTZ, and BALAKRISHNAN · Continuous Univariate Distributions, Volume 1, *Second Edition*

*Now available in a lower priced paperback edition in the Wiley Classics Library.

JOHNSON, KOTZ, and BALAKRISHNAN · Continuous Univariate Distributions, Volume 2, *Second Edition*
JOHNSON, KOTZ, and BALAKRISHNAN · Discrete Multivariate Distributions
JOHNSON, KOTZ, and KEMP · Univariate Discrete Distributions, *Second Edition*
JUREČKOVÁ and SEN · Robust Statistical Procedures: Asymptotics and Interrelations
JUREK and MASON · Operator-Limit Distributions in Probability Theory
KADANE · Bayesian Methods and Ethics in a Clinical Trial Design
KADANE AND SCHUM · A Probabilistic Analysis of the Sacco and Vanzetti Evidence
KALBFLEISCH and PRENTICE · The Statistical Analysis of Failure Time Data, *Second Edition*
KASS and VOS · Geometrical Foundations of Asymptotic Inference
KAUFMAN and ROUSSEEUW · Finding Groups in Data: An Introduction to Cluster Analysis
KEDEM and FOKIANOS · Regression Models for Time Series Analysis
KENDALL, BARDEN, CARNE, and LE · Shape and Shape Theory
KHURI · Advanced Calculus with Applications in Statistics, *Second Edition*
KHURI, MATHEW, and SINHA · Statistical Tests for Mixed Linear Models
KLEIBER and KOTZ · Statistical Size Distributions in Economics and Actuarial Sciences
KLUGMAN, PANJER, and WILLMOT · Loss Models: From Data to Decisions
KLUGMAN, PANJER, and WILLMOT · Solutions Manual to Accompany Loss Models: From Data to Decisions
KOTZ, BALAKRISHNAN, and JOHNSON · Continuous Multivariate Distributions, Volume 1, *Second Edition*
KOTZ and JOHNSON (editors) · Encyclopedia of Statistical Sciences: Volumes 1 to 9 with Index
KOTZ and JOHNSON (editors) · Encyclopedia of Statistical Sciences: Supplement Volume
KOTZ, READ, and BANKS (editors) · Encyclopedia of Statistical Sciences: Update Volume 1
KOTZ, READ, and BANKS (editors) · Encyclopedia of Statistical Sciences: Update Volume 2
KOVALENKO, KUZNETZOV, and PEGG · Mathematical Theory of Reliability of Time-Dependent Systems with Practical Applications
LACHIN · Biostatistical Methods: The Assessment of Relative Risks
LAD · Operational Subjective Statistical Methods: A Mathematical, Philosophical, and Historical Introduction
LAMPERTI · Probability: A Survey of the Mathematical Theory, *Second Edition*
LANGE, RYAN, BILLARD, BRILLINGER, CONQUEST, and GREENHOUSE · Case Studies in Biometry
LARSON · Introduction to Probability Theory and Statistical Inference, *Third Edition*
LAWLESS · Statistical Models and Methods for Lifetime Data, *Second Edition*
LAWSON · Statistical Methods in Spatial Epidemiology
LE · Applied Categorical Data Analysis
LE · Applied Survival Analysis
LEE and WANG · Statistical Methods for Survival Data Analysis, *Third Edition*
LePAGE and BILLARD · Exploring the Limits of Bootstrap
LEYLAND and GOLDSTEIN (editors) · Multilevel Modelling of Health Statistics
LIAO · Statistical Group Comparison

*Now available in a lower priced paperback edition in the Wiley Classics Library.

LINDVALL · Lectures on the Coupling Method
LINHART and ZUCCHINI · Model Selection
LITTLE and RUBIN · Statistical Analysis with Missing Data, *Second Edition*
LLOYD · The Statistical Analysis of Categorical Data
MAGNUS and NEUDECKER · Matrix Differential Calculus with Applications in Statistics and Econometrics, *Revised Edition*
MALLER and ZHOU · Survival Analysis with Long Term Survivors
MALLOWS · Design, Data, and Analysis by Some Friends of Cuthbert Daniel
MANN, SCHAFER, and SINGPURWALLA · Methods for Statistical Analysis of Reliability and Life Data
MANTON, WOODBURY, and TOLLEY · Statistical Applications Using Fuzzy Sets
MARDIA and JUPP · Directional Statistics
MASON, GUNST, and HESS · Statistical Design and Analysis of Experiments with Applications to Engineering and Science, *Second Edition*
McCULLOCH and SEARLE · Generalized, Linear, and Mixed Models
McFADDEN · Management of Data in Clinical Trials
McLACHLAN · Discriminant Analysis and Statistical Pattern Recognition
McLACHLAN and KRISHNAN · The EM Algorithm and Extensions
McLACHLAN and PEEL · Finite Mixture Models
McNEIL · Epidemiological Research Methods
MEEKER and ESCOBAR · Statistical Methods for Reliability Data
MEERSCHAERT and SCHEFFLER · Limit Distributions for Sums of Independent Random Vectors: Heavy Tails in Theory and Practice
*MILLER · Survival Analysis, *Second Edition*
MONTGOMERY, PECK, and VINING · Introduction to Linear Regression Analysis, *Third Edition*
MORGENTHALER and TUKEY · Configural Polysampling: A Route to Practical Robustness
MUIRHEAD · Aspects of Multivariate Statistical Theory
MURRAY · X-STAT 2.0 Statistical Experimentation, Design Data Analysis, and Nonlinear Optimization
MYERS and MONTGOMERY · Response Surface Methodology: Process and Product Optimization Using Designed Experiments, *Second Edition*
MYERS, MONTGOMERY, and VINING · Generalized Linear Models. With Applications in Engineering and the Sciences
NELSON · Accelerated Testing, Statistical Models, Test Plans, and Data Analyses
NELSON · Applied Life Data Analysis
NEWMAN · Biostatistical Methods in Epidemiology
OCHI · Applied Probability and Stochastic Processes in Engineering and Physical Sciences
OKABE, BOOTS, SUGIHARA, and CHIU · Spatial Tesselations: Concepts and Applications of Voronoi Diagrams, *Second Edition*
OLIVER and SMITH · Influence Diagrams, Belief Nets and Decision Analysis
PANKRATZ · Forecasting with Dynamic Regression Models
PANKRATZ · Forecasting with Univariate Box-Jenkins Models: Concepts and Cases
*PARZEN · Modern Probability Theory and Its Applications
PEÑA, TIAO, and TSAY · A Course in Time Series Analysis
PIANTADOSI · Clinical Trials: A Methodologic Perspective

*Now available in a lower priced paperback edition in the Wiley Classics Library.

PORT · Theoretical Probability for Applications
POURAHMADI · Foundations of Time Series Analysis and Prediction Theory
PRESS · Bayesian Statistics: Principles, Models, and Applications
PRESS · Subjective and Objective Bayesian Statistics, *Second Edition*
PRESS and TANUR · The Subjectivity of Scientists and the Bayesian Approach
PUKELSHEIM · Optimal Experimental Design
PURI, VILAPLANA, and WERTZ · New Perspectives in Theoretical and Applied Statistics
PUTERMAN · Markov Decision Processes: Discrete Stochastic Dynamic Programming
*RAO · Linear Statistical Inference and Its Applications, *Second Edition*
RENCHER · Linear Models in Statistics
RENCHER · Methods of Multivariate Analysis, *Second Edition*
RENCHER · Multivariate Statistical Inference with Applications
RIPLEY · Spatial Statistics
RIPLEY · Stochastic Simulation
ROBINSON · Practical Strategies for Experimenting
ROHATGI and SALEH · An Introduction to Probability and Statistics, *Second Edition*
ROLSKI, SCHMIDLI, SCHMIDT, and TEUGELS · Stochastic Processes for Insurance and Finance
ROSENBERGER and LACHIN · Randomization in Clinical Trials: Theory and Practice
ROSS · Introduction to Probability and Statistics for Engineers and Scientists
ROUSSEEUW and LEROY · Robust Regression and Outlier Detection
RUBIN · Multiple Imputation for Nonresponse in Surveys
RUBINSTEIN · Simulation and the Monte Carlo Method
RUBINSTEIN and MELAMED · Modern Simulation and Modeling
RYAN · Modern Regression Methods
RYAN · Statistical Methods for Quality Improvement, *Second Edition*
SALTELLI, CHAN, and SCOTT (editors) · Sensitivity Analysis
*SCHEFFE · The Analysis of Variance
SCHIMEK · Smoothing and Regression: Approaches, Computation, and Application
SCHOTT · Matrix Analysis for Statistics
SCHUSS · Theory and Applications of Stochastic Differential Equations
SCOTT · Multivariate Density Estimation: Theory, Practice, and Visualization
*SEARLE · Linear Models
SEARLE · Linear Models for Unbalanced Data
SEARLE · Matrix Algebra Useful for Statistics
SEARLE, CASELLA, and McCULLOCH · Variance Components
SEARLE and WILLETT · Matrix Algebra for Applied Economics
SEBER and LEE · Linear Regression Analysis, *Second Edition*
SEBER · Multivariate Observations
SEBER and WILD · Nonlinear Regression
SENNOTT · Stochastic Dynamic Programming and the Control of Queueing Systems
*SERFLING · Approximation Theorems of Mathematical Statistics
SHAFER and VOVK · Probability and Finance: It's Only a Game!
SMALL and MCLEISH · Hilbert Space Methods in Probability and Statistical Inference
SRIVASTAVA · Methods of Multivariate Statistics
STAPLETON · Linear Statistical Models
STAUDTE and SHEATHER · Robust Estimation and Testing

*Now available in a lower priced paperback edition in the Wiley Classics Library.

STOYAN, KENDALL, and MECKE · Stochastic Geometry and Its Applications, *Second Edition*

STOYAN and STOYAN · Fractals, Random Shapes and Point Fields: Methods of Geometrical Statistics

STYAN · The Collected Papers of T. W. Anderson: 1943–1985

SUTTON, ABRAMS, JONES, SHELDON, and SONG · Methods for Meta-Analysis in Medical Research

TANAKA · Time Series Analysis: Nonstationary and Noninvertible Distribution Theory

THOMPSON · Empirical Model Building

THOMPSON · Sampling, *Second Edition*

THOMPSON · Simulation: A Modeler's Approach

THOMPSON and SEBER · Adaptive Sampling

THOMPSON, WILLIAMS, and FINDLAY · Models for Investors in Real World Markets

TIAO, BISGAARD, HILL, PEÑA, and STIGLER (editors) · Box on Quality and Discovery: with Design, Control, and Robustness

TIERNEY · LISP-STAT: An Object-Oriented Environment for Statistical Computing and Dynamic Graphics

TSAY · Analysis of Financial Time Series

UPTON and FINGLETON · Spatial Data Analysis by Example, Volume II: Categorical and Directional Data

VAN BELLE · Statistical Rules of Thumb

VIDAKOVIC · Statistical Modeling by Wavelets

WEISBERG · Applied Linear Regression, *Second Edition*

WELSH · Aspects of Statistical Inference

WESTFALL and YOUNG · Resampling-Based Multiple Testing: Examples and Methods for *p*-Value Adjustment

WHITTAKER · Graphical Models in Applied Multivariate Statistics

WINKER · Optimization Heuristics in Economics: Applications of Threshold Accepting

WONNACOTT and WONNACOTT · Econometrics, *Second Edition*

WOODING · Planning Pharmaceutical Clinical Trials: Basic Statistical Principles

WOOLSON and CLARKE · Statistical Methods for the Analysis of Biomedical Data, *Second Edition*

WU and HAMADA · Experiments: Planning, Analysis, and Parameter Design Optimization

YANG · The Construction Theory of Denumerable Markov Processes

*ZELLNER · An Introduction to Bayesian Inference in Econometrics

ZHOU, OBUCHOWSKI, and McCLISH · Statistical Methods in Diagnostic Medicine

*Now available in a lower priced paperback edition in the Wiley Classics Library.